Comprehensive Laboratory Manual of Life Science

About the Author

Dr. Mamta Baunthiyal completed her Master's degree in Biochemistry from Kurukshetra University, India and PhD in Biotechnology from Banasthali University, Raj. She has 23 years of experience of teaching undergraduate and post-graduate classes in Biotechnology. Her research is focused on environmental biotechnology, nano biotechnology, herbal biotechnology and she has about 45 research articles published in reputed international journals. Besides she has written popular articles for magazines like Energy Future, Everyman's Science etc. Presently she is working as Head & Associate Professor in the Dept. of Biotechnology, GB Pant Institute of Engineering and Technology, Pauri, Uttarakhand, India.

Dr. Indu Ravi is presently working as Assistant Regional Director, IGNOU Regional Centre, Jaipur. She has worked as Associate professor in the Dept. of Bioscience and Biotechnology, Banasthali University, Banasthali, Raj. and has 14 years of teaching experience of UG and PG classes in the field of biotechnology. She acquired her post graduation in Biotechnology from Indian Veterinary Research Institute (IVRI), Izatnagar, Post Graduate Diploma in Distance Education from IGNOU and Ph.D. in Biotechnology from Banasthali University. Her area of specialization is Plant Molecular Biology. She has written two books and published a number of research papers in National and International journals.

Dr. Jyoti Saxena is a Professor and Head in Biochemical Engineering Department at Bipin Tripathi Kumaon Institute of Technology, Dwarahat (Uttarakhnad). She has completed her M.Sc and Ph.D from Kumaun University, Nainital and post doc studies from UNL, U.S.A. She has 24 years of experience of teaching U.G. and P.G. classes in Bioscience, Biotechnology and Biochemical Engineering fields. Dr Saxena has 4 books and around 100 research papers in reputed National and International journals to her credit.

Comprehensive Laboratory Manual of Life Science

Mamta Baunthiyal
Indu Ravi
Jyoti Saxena

Published by:

SCIENTIFIC PUBLISHERS (INDIA)
5A, New Pali Road, P.O. Box 91
Jodhpur 342 001, India
E-mail: info@scientificpub.com
Website: www.scientificpub.com

ISBN: 978-93-87893-93-1
eISBN: 978-93-87991-57-6

Reprinted and revised from the book Laboratory Manual of Microbiology, Biochemistry and Molecular Biology by Jyoti Saxena, Mamta Baunthiyal & Indu Ravi (Pages: 380 Page and Year: 2015)

Printed in India

PREFACE

After the enormous success of first edition of Laboratory Manual of Microbiology, Biochemistry and Molecular Biology, we are happy to bring the second book with some additional sections. However, our goal remains the same. Our focus in writing this manual has been to bring together the protocols in simple and understandable manner to provide the students the laboratory practices as an enjoyable learning experience. We have tried to clear the basic concept behind the particular section or experiment.

This is based on the feedback of students and other users of the laboratory manual and has been written from the point of view of making this book a comprehensive laboratory manual having the experiments from all important sections prescribed in syllabi of most of the universities for practical training of UG and PG students. First book from authors had protocols from Microbiology, Biochemistry and Molecular Biology which covered a major percentage of practicals, however it was felt that Immunology and Environmental Biotechnology also have important share in many courses. Therefore, these two sections were added Immunology is the basis of all disease diagnosis in today's scenario. Seven immunology related experiments have been included to form the basis of understanding the immune response. Although ready to use kits are available these days for the diagnosis of diseases and pathogens causing disease the experiments given in the book will definitely help in developing in understanding about the practical aspects of the subject.

Environmental science is an interdisciplinary field. Because environmental disharmonies occur as a result of the interaction between humans and the natural world, we must include both when seeking solutions to environmental problems. Therefore, we found it a necessity to include a separate section on Environment Biology. This Environmental science section includes the analysis of samples of air, water, hazardous waste and sediment/ soil. The laboratory experiments emphasize sampling, extraction and instrumental analysis. Environmental Biotechnology section also covers all the practicals commonly followed by different universities.

The first part is the Introduction; the second part is about Instruments: Principle and Precautions. The third part is about the Experiments which includes many protocols related to Microbiology, Biochemistry, Molecular Biology, Immunology, and Environmental Science. In the end, we have included the Appendix and Index also.

This book will definitely be more useful to the graduate and postgraduate science students.

Mamta Baunthiyal
Indu Ravi
Jyoti Saxena

CONTENTS

Preface v

Abbreviations xvii

1 INTRODUCTION **1–21**

1. Laboratory Etiquette and Safety 1
2. Good Laboratory Practices 2
3. Maintenance of Distillation Apparatus 4
4. Washing of Glassware 5
5. Washing of Plastic Apparatus 6
6. Dry Sterilization 6
7. Handling of Pipette 7
8. Molecular, Empirical and Formula Weight 10
9. Planning a solution of a particular molarity 11
10. Making solutions from hydrated compounds 11
11. Measurement of Liquids with Normality/Molarity 12
12. Numbers Found on Chemical Bottles 13
13. Accuracy and Calibration 14
14. Buffers 15

2 INSTRUMENTS : PRINCIPLE AND PRECAUTIONS **22–45**

1. Weighing Balance 22
2. Spectrophotometer 25
3. Laminar Flow Hood 27
4. Autoclave 29
5. Centrifuge 30
6. pH Meter 36
7. Incubator 38
8. Hot Air Oven 40
9. Compound Microscope 41
10. ELISA Reader 43

3. EXPERIMENTS **46–381**

A. MICROBIOLOGY . . . 46–129

Introduction **46**

1.1 Micrometery and Measurement of Micro-organisms **47**

1.1.1. Calibration of Microscope by using Ocular and Stage Micrometer 47

1.1.2. Measurement of Size of Micro-organism/Spore 49

1.2 Preparation of Culture Media **50**

1.2.1 Preparation of Liquid Culture Medium (Broth) 51

1.2.2 Preparation of Solid Medium 52

1.2.3 Preparation of Agar Plates and Agar Slants 53

1.2.4 Demonstration of Selective and Differential Media 55

1.2.4.1 Eosine Methylene Blue (EMB) medium 55

1.2.4.2 Mannitol Salt Agar 56

1.2.4.3 Pikovskaya Agar 56

1.3 Aseptic Methods **57**

1.3.1 Physical Methods of Sterilization 57

1.3.2 Chemical Methods of Sterilization 58

1.3.3 Gaseous Sterilization 59

1.4 Cultivation Techniques **59**

1.4.1 To isolate the Microorganisms by Streak Plate Method 59

1.4.2 To Isolate the Microorganisms by Pour Plate Technique 61

1.4.3 To Isolate the Microorganisms by Spread Plate Technique 63

1.4.4 To Isolate the Microorganisms by Serial Dilution Technique (or viable plate count method) 64

1.5 Preservation and Maintenance of Culture **67**

1.5.1 Maintenance by Sub-culturing 67

1.5.2 To Prepare Agar Slants with Mineral Oil (Paraffin Method) 68

1.5.3 Storage in Soil 68

1.5.4 To Store the Cultures in Liquid Nitrogen 69

1.5.5 To Store the Culture at -70°C 70

1.5.6 To Preserve the Cultures by Lyophilization (Freeze drying) 70

1.6 Staining Techniques **72**

1.6.1 Preparation of Bacterial Smear and Fixation 73

1.6.2 Simple Staining of Bacteria 74

1.6.3 Negative Staining of Bacteria 75
1.6.4 Gram Staining of Bacteria 75
1.6.5 Acid-fast Staining 77
1.6.6 Endospore Staining 78
1.6.6.1 Endospore staining by malachite green 78
1.6.6.2 Endospore staining by Dorner's method 79
1.6.7 Capsule Staining to Detect the Capsule or Slime Layer in Bacteria 80
1.6.8 Flagella Staining 81
1.6.8.1 Lee's method 81
1.6.8.2 Leifson's method 82
1.6.9 To Test the Viability of Bacteria by Staining 83
1.6.10 Staining of Fungi 83
1.6.11 Staining of Arbuscular Mycorrhiza (AM) 84
1.7 Isolation and Enumeration of Microbes 85
1.7.1 Isolation and Enumeration of Fungi from Soil 86
1.7.1.1 Isolation of fungi by dilution plate method 86
1.7.1.2 Isolation of fungi by Warcup method 87
1.7.2 Isolation of Seed Microflora 88
1.7.2.1 Agar plate method (ISTA, 1966) 88
1.7.2.2 Blotter method 89
1.7.3 Isolation of Aeromycoflora 89
1.7.4 Isolation of Arbuscular Mycorrhizal (AM) Spores from the Soil 90
1.7.4.1 Wet-sieving method 90
1.7.4.2 Floatation method 91
1.7.5 Isolation and Enumeration of Microorganisms from Rhizosphere 91
1.7.6 Isolation of Microorganisms from Rhizoplane 93
1.7.7 Isolation of Protozoa from Soil 93
1.7.8 Isolation of Cyanobacteria 94
1.7.9. Isolation of Aquatic Fungi by Baiting Method 95
1.7.10 Isolation of Yeast 96
1.7.11 Isolation of Rhizobia from the Root Nodules 96
1.7.12 Isolation and Cultivation of Anaerobic Bacteria 97
1.7.12.1 Candle jar method 97
1.7.12.2 Gaspak anaerobic jar method 98
1.7.13 Analysis and Enumeration of Bacteria in Milk 99

1.7.14 Presumptive Test for Coliforms to Check the Quality of Milk 100

1.7.15 Microbiological Examination of Cheese 100

1.7.16 Microbiological Examination of Butter 101

1.8 Counting of Cells/Spores 102

1.8.1 Measurement of Cells/Spores by Counting Chamber (Hemocytometer procedure) 102

1.8.2 Counting of Cells by Serial Dilution Technique (Viable cell count) 104

1.8.3 Counting of Bacterial Population Spectrophotometrically 105

1.9 Microbial Growth 106

1.9.1 Determination of Bacterial Growth Curve by Spectrophotometric Method 106

1.9.2 Measurement of Fungal Growth by Colony Diameter Method 107

1.9.3 Measurement of Fungal Growth by Dry Weight of Mycelium 108

1.9.4 Estimation of Biomass 109

1.9.4.1 Dry cell weight estimation 109

1.9.4.2 Packed cell volume (PCV) determination 110

1.9.5 Effect of Temperature on Growth of Fungi/Bacteria 111

1.9.6 Effect of pH on Growth 113

1.9.7 Determination of Antibiotic Sensitivity by Disc Method 114

1.10 Fermentation Technology 114

1.10.1 Estimation of Lactic Acid in Curd 115

1.10.2 Production and Estimation of Citric Acid by Aspergillus niger 116

1.10.3 Estimation of Growth and Product Yield During Bioconversion of Tannic Acid to Gallic Acid by Aspergillus niger MTCC 282 118

1.10.4 Determination of Alcohol Production in the Fermented Broth 120

1.10.5 Alpha-amylase Biosynthesis and Measurement of its Activity 122

1.11 Mushroom Cultivation 125

1.11.1 Production of Spawn for White Button Mushroom (Agaricus bisporus) 125

1.11.2 Cultivation of White Button Mushroom 127

B. BIOCHEMISTRY . . . [130–255]

Introduction **130**

2.1 Carbohydrates **130**

2.1.1 Qualitative test for Carbohydrates 131

2.1.1.1 Molisch's Test 131

2.1.1.2 Iodine Test 132

2.1.1.3 Barfoed's Test 133

2.1.1.4. Seliwanoff's Test 133

2.1.1.5 Fehling's Test 134

2.1.1.6 Benedict's Test 134

2.1.1.7 Bial's Test 135

2.1.1.8 Osazone Test 135

2.1.1.9 Test for non-reducing sugars 136

2.1.2 Quantitative Test for Carbohydrates 137

2.1.2.1 Determination of total carbohydrate by Anthrone reagent 137

2.1.2.2 Determination of reducing sugar by Nelson-Somogyi method 138

2.1.2.3 Determination of Cellulose 140

2.2 Amino Acids and Proteins **141**

2.2.1 Colour Reaction of Proteins/Amino Acids 144

2.2.2 Titration Curve of Amino Acids 150

2.2.3 Protein Assay Methods 152

2.2.3.1 λ max for protein and amino acids 153

2.2.3.2 Determination of molar absorbance coefficient (ε) of L-tyrosine 154

2.2.3.3 Lowry method of protein assay 155

2.2.3.4 Bradford protein assay 158

2.2.3.5 Bicinchoninic Acid (BCA) Protein Assay 159

2.2.3.6 Biuret Protein Assay 161

2.2.4 The Isolation of Casein from Milk 162

2.2.5 Ammonium Sulphate Fractionation of Proteins 164

2.3 Enzymes **167**

2.3.1 Extraction and Purification of Enzyme: General Strategies 167

2.3.2 Expression of enzyme activity 172

2.3.2.1 To study the effect of pH and temperature on activity of acid phosphatase (AP) 174

2.3.2.2 Purification of acid phosphatase from raw wheat germ 181

2.3.2.3 To determine the Km and Vmax value of acid phosphatase 184

2.3.3 Extraction and assay of enzymes 187

2.3.3.1 Malate dehydrogenase (EC 1.1.1.37) 187

2.3.3.2 Peroxidase (E.C. 1.11.1.7) 189

2.3.3.3 Polyphenol Oxidase (EC 1.14.18.1) 191

2.3.3.4 Determination of salivary amylase activity 193

2.3.3.5 Papain (Papainase, EC 3.4.22.2) 194

2.3.4 Immobilization of an Enzyme 196

2.4 Lipids **198**

2.4.1 Qualitative Tests of Lipids 199

2.4.1.1 Solubility Test for Lipids 200

2.4.1.2 Grease Spot Test 200

2.4.1.3 Acrolein Test for Glycerol 201

2.4.1.4 Liebermann-Burchard Test for Cholesterol 201

2.4.2 Determination of Free Fatty Acids 202

2.4.3 Determination of Iodine Number 203

2.4.4 Determination of Saponification value 205

2.4.5 Estimation of Serum Cholesterol (Total) 207

2.5 Nucleic Acids **209**

2.5.1 Determination of λ max of DNA 209

2.5.2 Determination of DNA by Diphenylamine Method 210

2.5.3 Determination of RNA by Orcinol Method 211

2.6 Cell Organelles **212**

2.6.1 Isolation of Mitochondria 212

2.6.1.1 To isolate mitochondria from plant leaves 212

2.6.1.2 To Isolate Mitochondria from Rat Liver 214

2.6.2 Isolation of Chloroplasts from Plant Leaves 214

2.7 Plant Pigments and Phenolics **217**

2.7.1 Chlorophylls 217

2.7.2 Carotenes 218

2.7.3 Lycopene 219

2.7.4 Phenols 221

2.8 Vitamins **223**

2.8.1 Determination of Thiamine 223

2.8.2 Estimation of Ascorbic Acid in Lemon Juice 224

2.8.3 Measurement of Riboflavin (Vitamin B_2) in Human Urine 225

2.8.4 Estimation of Vitamin A 227

2.9 Biochemical Separation Techniques **229**

2.9.1 Paper Chromatography 229

2.9.1.1 Separation of Amino Acids by one Dimensional Paper Chromatography 234

2.9.2 Separation of Plant Pigments by Thin Layer Chromatography (TLC) 236

2.9.3 Separation of Blue-dextran and Potassium Dichromate by Gel–Filtration Chromatography 239

2.9.4 To Separate Proteins by SDS – PAGE 241

2.9.5 Silver Staining of Proteins 248

2.9.6 Separation of Proteins by Isoelectric Focusing 250

C. MOLECULAR BIOLOGY . . . [256–329]

Introduction **256**

3.1 Isolation of Nucleic Acid **259**

3.1.1 Isolation of DNA 259

3.1.1.1 Rapid Method For Isolation of Plasmid DNA 260

3.1.1.2 Isolation E. coli genomic DNA 265

3.1.1.3 Isolation of bacteriophage □ DNA 266

3.1.1.4 Extraction of plant DNA using CTAB 269

3.1.1.5 Isolation of mammalian DNA 271

3.1.1.6 Isolation of chloroplast DNA 275

3.1.1.7 Isolation of mitochondrial DNA 277

3.1.1.8 Agarose gel electrophoresis of DNA 278

3.1.2 RNA Isolation 282

3.1.2.1 Isolation of total RNA 282

3.1.2.2 Isolation of messenger RNA 284

3.1.2.3 Agarose gel electrophoresis of RNA 285

3.1.2.4 Quantification of DNA/RNA 286

3.1.2.5 Determination of DNA quality 286

3.2 Denaturation and Renaturation of DNA **287**

3.2.1 Determination of Melting Temperature (Tm value)/ Melting Curve of DNA 287

3.2.2. Determination of GC/AT (Base composition) Content of DNA 289

3.3 Restriction Digestion **290**

3.3.1 Restriction Digestion of DNA 290

3.3.2 Preparation of Agarose Gels and Running of Digested and Undigested DNA (Gel Electrophoresis) 291

3.4 Polymerase Chain Reaction (PCR) **292**

3.4.1 DNA Amplification by PCR 292

3.4.2 RAPD (Random Amplified Polymorphic DNA) Analysis of Different Plant Genotypes 298

3.4.3 PCR Amplification of 16s Ribosomal RNA 300

3.5 Blotting Techniques and Hybridization **302**

3.5.1 Southern Hybridization 306

3.5.2 Northern Blotting and Hybridization 313

3.5.3 Western Blotting and Hybridization 313

3.6 Ligation and Transformation of Bacterial DNA (*E. Coli*) **315**

3.6.1 Ligation 316

3.6.2 Preparation of Competent Cells 317

3.6.3 Transformation of *E. coli* 318

3.6.4 Transformation of Plant Cells using Agrobacterium tumifaciens 318

3.6.5 Isolation of Plant Protoplasts 322

3.7 Selection of Recombinant Cells **324**

3.7.1 Selection by Insertional Inactivation of Antibiotic Resistance Gene 324

3.7.2 Lac Z' gene Selection 325

3.7.3 GUS Assay 327

D. IMMUNOLOGY . . . [330–353]

Introduction **330**

4.1 To collect the Serum from blood 334

4.2 To prepare the blood film and identify leucocytes 335

4.3 To determine ABO blood types by slide agglutination reaction 338

4.4 To isolate and purify sheep IgG from whole sheep serum and determine its concentration 340

4.5 To perform double Immunodiffusion by Ouchterlony method. 343

4.6 To perform Immunoelectrophoresis 346

4.7 To perform Enzyme linked immunosorbent assay (ELISA) 347

4.7.1 To detect the antibodies using direct ELISA 350

4.7.2 To detect the antibodies using indirect ELISA 351

E. ENVIRONMENTAL BIOTECHNOLOGY . . . [354–381]

Introduction **354**

5.1 To determine the organic carbon content in soil 355

5.2 Determination of Total Dissolved Solids in Waste Water Sample 357

5.3 To determine the Dissolved Oxygen (DO) in given water sample by Winkler's method 359

5.4 To determine the Biological Oxygen Demand (BOD_5 at 20°C) of given sample of waste water 362

5.5 To determine Chemical Oxygen Demand (COD) in waste water sample 366

5.6 Analysis of total hardness of waste water sample 369

5.7 Detection of coliforms for determination of the purity of potable water 372

5.8 Estimation of the pesticide residues in soil sample 374

5.9 Determination of LD_{50} of a given pesticide 377

5.10 Determination of Fluoride Content in Water Sample 378

REFERENCES **382–383**

APPENDIX **384–453**

Appendix IV A : Microbiology 384

Appendix IV B : Biochemistry 410

Appendix IV C : Molecular Biology 422

Appendix IV D : Immunology 445

Appendix IV E : Environmental Biotechnology 447

INDEX **454–458**

Index for Microbiology 454

Index for Biochemistry 456

Index for Molecular Biology 457

Index for Immunology 457

Index for Environmental Biology 458

ABBREVIATIONS

A	–	Absorbance
AB	–	Antibody
AIDS	–	Acquired immunodeficiency Syndrome
Ag	–	Antigen
APS	–	Ammonium persulphate
BOD	–	Biological Oxygen demand
BPB	–	Bromophenol blue
BSA	–	Bovine Serum Albumin
BTH	–	Benzo (1,2,3) thiadiozole -7-carbothionic acid S-methyl ester
bp	–	Base pairs
CAT	–	Chloramphenicol acetyl transferase
COD	–	Chemical Oxygen demand
CTAB	–	Cetyltrimethylammonium bromide
Da	–	Daltons
DAB	–	3,3'-Diamino benzidine
DEPC	–	Diethyl pyrocarbonate
DO	–	Dissolved oxygen
DMAB	–	*p*-Dimethyl amino benzaldehyde
DMF	–	N, N-dimethyl formamide
DMSO	–	Dimethyl sulfoxide
DTT	–	Dithiothreitol
DW	–	Distilled water
EDTA	–	Ethylenediamine tetra acetic acid
EIA	–	Enzyme immunoassay
ELISA	–	Enzyme linked immunosorbent assay
EMB	–	Eosine- methylene blue
FW	–	Formula weight
g	–	gram
GAA	–	Glacial acetic acid
GFP	–	Green fluorescent protein
GUS	–	β- Glucuronidase
H	–	Hours
HEPA	–	High Efficiency Particulate Air
HEPES	–	N-2- hydroxyethylpiperazine-N'-2-ethanesulphonic acid

HIV – Human immunodeficiency virus
H_2O_2 – Hydrogen peroxide
HCl – Hydrochloric acid
HPLC – High performance liquid chromatography
IgG – Immunoglobulin G
IR – Infra red
IPTG – Isopropyl-thiagalactoside
Kbp – kilobase pairs
KDa – kilodaltons
Km – Michaelis-Menton constant
L - litre
LB media - Luria Bertani media
LD_{50} – Lethal Dose 50
LMW – Low molecular weight marker
M – Molar
mg – milligram
min – minutes
ml – milliliter
Mm – Mili molar
Mr – Relative molecular weight
MW – Molecular weight
μg – microgram
μkat – microkatal
μl – microlitre
μm – micromoles
N – Normal
Nm – nano meters
OD – Optical Density
PBS – phosphate buffer saline
PCR – Polymerase chain reaction
PDA – Potato Dextrose Agar
Pfu – Plaque forming units
Pi – inorganic phosphate
Ppm – parts per million
Psi – Pounds per square inch
PVDF – Polyvinylidene difluoride
RBC – Red blood cell
RCF – Relative centrifugal force

Rf	–	Relative front
RPM	–	Revolutions per minute
S	–	Svedberg unit
SDS	–	PAGE-Sodium dodecyl sulphate polyacrylamide gel electrophoresis
Sec	–	seconds
TBS	–	Tris buffered saline
TE	–	Tris EDTA
TEMED	–	N,N,N', N'-Tetra methyl ethylene diamine (hydroxyl methyl) amino methane
Tm	–	melting temperature
TLC	–	Thin layer chromatography
TOC	–	Total Organic Carbon
Tris	–	Tris (hydroxymethyl) aminomethane
TS	–	Total solids
UV	–	Ultra violet
V	–	velocity
Vmax	–	maximum velocity
v/v	–	volume by volume
WBC	–	White blood cell
w/v	–	weight by volume
X-gal	–	5-bromo-4-chloro-3-indolyl-β-D-galactopyranoside

Chapter - 1

I. INTRODUCTION

1. Laboratory Etiquette and Safety

Sincerity, honesty and inquisitiveness are the virtues which must be followed by the students while working in a laboratory. Proper utilization and planning of time is expected to complete the assigned work within a specific period. Laboratory provisions should be handled with utmost care. At the end of the lab period, working space should be left clean and in order.

There are some guidelines and instructions which every student is obliged to respect for proper utilization of the facilities available in the lab and to safeguard oneself and the fellow students. The aim of practical education should be to learn safe handling of chemicals in laboratory and protection from any risk involved.

First of all it is essential to formulate the specific operating instructions for any laboratory course. The operating instructions shall contain the specific aspects of the particular lab course.

- A list of all chemicals used in laboratory is mandatory.
- There should be operating manuals for all kinds of apparatus and a log book should be filled by the students after the use of the particular apparatus.
- Detailed experimental and operating instructions with a proper protocol should be provided to the students.
- More important is that instructions for laboratory course must be read by all students.
- Laboratories are accident prone, therefore, students must have knowledge of fire extinguishers and first aid kit.
- No food or drink in laboratory should be stored or consumed due to risk of contamination and hands should be washed with antiseptic soap before leaving the laboratory.
- Proper attire: long sleeved lab coat, eye protection and footware is must.
- Many chemicals and biological materials should not be handled without using hand gloves; as most of the chemicals used in

molecular biology experiments are carcinogenic, thus appropriate type of gloves should be selected. Mostly disposable gloves are used in lab, but one must remember that these gloves are resistant but not impervious. They should also be changed often.

- Proper disposal of broken glass, leftover reagents and chemical, biological and radioactive waste is important.
- Designated units or trash cans should be used for proper disposal of broken glassware, etc.
- Hazardous chemicals and reagents containing them should not be drained. Biological waste, microbial culture, etc., should be autoclaved before discarding them.
- Chemicals and samples if kept for sometime, should be labeled.
- If all the precautions are taken properly there is no reason why working in a laboratory would not be an enjoyable experience.

Specific precautions for microbiology laboratory

- Working table should be cleaned with some disinfectant, *e.g.*, 90% ethanol or 1:100 phenol.
- Drinking, eating and smoking is not allowed in laboratory.
- Microbial cultures should never be opened in laboratory except in the laminar flow.
- If a live culture broth is spilled or a disk broken down, the area should be covered with disinfectant for 15 minutes and then it should be cleaned.
- Special care should be taken for burners or spirit lamps.
- All cultures must be labeled.
- Pathogenic cultures should not be opened without taking adequate precautions.
- Laboratory should be fumigated time to time by using $KMnO_4$ and formalin.

2. Good Laboratory Practices

Number of laboratory operations and precautions related to analysis of water quality parameters must be routinely performed in a laboratory to obtain reliable information. These practices are termed good lab practices.

Chemicals and reagents

The following rules should always be followed while handling reagents and chemicals.

- As far as possible, use the smallest packing of chemical that would supply the desired quantity.
- Replace the cap or stopper of the bottle immediately after taking out your requirement.
- Stoppers of reagent bottles should never be placed on the desktop.
- Do not insert spatulas or pipettes in reagent bottles. Take out a slightly excess amount in another container to use the required quantity.
- Never return any excess chemical or reagent to bottle.

The following general cautions should always be observed in the laboratory:

- Become completely familiar with the properties of the substances used before beginning the procedure.
- The absence of a warning does not necessarily means that the material is safe, since information may not always be complete or available.
- For specific guidelines on appropriate gloves, consult your local safety office.
- Handle concentrated acids and bases with great care. Wear goggles and appropriate gloves. A face shield should be worn when handling large quantities. Do not mix strong acids with organic solvents as they may react. Especially, sulfuric acid and nitric acid, may react highly exothermically and cause fires and explosions. Do not mix strong bases with halogenated solvent as they may form reactive carbenes which can lead to explosions.
- Never pipette solutions using mouth suction. This method is not sterile and can be dangerous. Always use a pipette aid or bulb.
- Keep halogenated and nonhalogenated solvents separately, *e.g.*, mixing chloroform and acetone can cause unexpected reactions in the presence of bases. Halogenated solvents are organic solvents, such as chloroform, dichloromethane, trichlorotrifluoroethane and dichloroethane. Some nonhalogenated solvents are pentane, heptane, ethanol, methanol, benzene, toluene, N,N-dimethy1formamide (DMF), dimethyl sulfoxide (DMSO) and acetonitrile.
- Flash lamps, due to their light intensity can be harmful to the eyes. They also may explode on occasion. Wear appropriate eye protection and follow the manufacturer's guidelines.
- Photographic fixatives and developers also contain chemicals that can be harmful. Handle them with care and follow manufacturers' directions.

- Power supplies and electrophoresis equipment pose serious fire hazards and electrical shock hazards if not used properly.
- Microwave ovens and autoclaves in the lab require certain precautions. Accidents have occurred involving their use. If the screw top is not completely removed and there is not enough space for the steam to vent, the bottles can explode and cause severe injury when the containers are removed from the microwave or autoclave. Always completely remove bottle caps before microwaving or autoclaving. An alternative method for routine agarose gels that do not require sterile agar is to weigh out the agar and place the solution in a flask.
- Use extreme caution when handling cutting devices, such as microtome blades, scalpels, razor blades, or needles. Microtome blades are extremely sharp and care should be taken when sectioning. If you are unfamiliar with their use, have someone to demonstrate proper procedures. For proper disposal, use the "sharps" disposal container in your lab. Discard used unshielded needles, with the syringe still attached. This prevents injuries (and possible infections) while manipulating used needles, since many accidents occur while trying to replace the needle shield. Injuries may also be caused by broken Pasteur pipettes, cover slips or slides.

3. Maintenance of Distillation Apparatus

Distilled water is must for performing every experiment. Therefore, the distillation assembly is always in use and gets out of order very frequently. The following precautions can be taken by the students to increase the life of the distillation apparatus:

- The various joints of the distillation assembly should be properly greased.
- Cold water need to be circulated through the condenser and it should be stopped after one hour of stopping power supply to heating mantle.
- Glass beads or boiling chips need to be added in the boiling water during distillation to avoid bumping and effective heat transfer.
- Store the distilled water in the clean glass/polyethylene/other suitable plastic container which do not affects the quality of the water. No air space needs to be left in the container before sealing the bottle.
- All stills are prone to scaling and accumulation of precipitated salts, which if not attended, results in carry over of salts in

distilled apparatus and will ultimately lead to overheating and damage the heating elements. The scales should be periodically removed by dissolving in dilute HCl.

4. Washing of Glassware

- Do not leave reagents and dirt on glassware.
- Do not leave detergent on glassware.
- Do not leave tap water on glassware.
- Do not damage glassware.

(i) General Glassware

- Put soaked glassware in water in a vat. Do not keep glassware out of water and do not leave air inside, otherwise, dried dirt will be difficult to wash off.
- Soak the glassware in diluted neutral detergent. Do not desiccate salt solution or protein solution, otherwise it will be very difficult to get it off. The concentration of solution differs according to deter-gent, but soak in solution of 2–5 percent concentration for 2 to 24 hours. If it is very dirty, use solution of higher concentration. Soaking time can be shortened to 5–15 minutes by raising the temperature of detergent solution to 40–50°C. Ultrasonic washing will take 5–45 minutes at room temperature. Always use neutral detergent because alkali can damage glass.
- Wash the glassware by a brush with detergent 3 times. Use a brush and a sponge suitable for the glassware. Always wear a pair of gloves to protect hands and to prevent dirt from one's hand getting on the glassware. Hands become slippery with gloves on, so be cautious not to let the glassware slip and drop.
- Wash it by a brush with water more than 3 times. Wash detergent off and brush the glassware with running tap water.
- Rinse it with running water more than 3 times without the brush with deionised water 2 times. Pour deionised water thorou-ghly using a washing bottle. Rinse the glassware with distilled water one more time if necessary.
- Place its bottom upward on a basket and put in a drier.

(ii) Pipette

- Soak the pipettes in the water in a vat or soak them end downward in the water in a bucket. Be cautious because pipettes easily get clogged particularly at the end and the inside of the pipettes can not be washed with a brush.

- Soak them end upward in diluted detergent in a cylinder with a basket. Place a bucket in a pipette cylinder and soak pipette with the top upward in it. The concentration of solution varies according to detergents. Soak them in 5 percent concentrated solution for more than 16 hours. Glass can be damaged by alkali, so always use neutral detergent. When soaking pipettes, lift the basket, put pipettes gently in it and replace the basket slowly. When putting pipettes directly in a cylinder, make sure that it is done carefully and the cylinder lid is shut tightly to prevent detergent liquid from escaping or splashing over the top.
- Transfer the basket containing pipettes to the pipette washer by catching water with the lid and wash them for more than 6 hours.
- Rinse them with ion exchange water more than 3 times. Wear a pair of gloves when handling the washed pipettes. Put pipettes with the top upward in a 2 litre plastic measuring cylinder and pour ion exchange water thoroughly over the top to save ion exchange water.
- Place them with their end upward on the basket and put in a drier.

5. Washing of Plastic Apparatus

Most of the plastic apparatus can be washed in the same way as glassware. But, do not scrub and use a cleanser as they are easily damaged. For washing of electrophoretic chamber, rinse well with tap water and put distilled water after use. Water rinsing is enough to avoid touching the platinum wires exposed inside the electrophoretic tank.

6. Dry Sterilization

A dry sterilizer is an oven tolerant to over 200°C. It enables to kill bacteria and fungi, and also to denature organic matter and enzymes, such as DNase and RNase. It is available only for apparatus using materials tolerant to 200°C.

A medium glass bottle with resinous cap cannot be sterilized by this method. A Teflon-lined stirrer bar (a Teflon coated magnet) is resistant to high temperature and possible for dry sterilization.

The dry sterilizer is available for glassware, metallic and porcelain, and Teflon-lined apparatus.

(1) Cover or wrap clean and dry apparatus with aluminum foil. Classify pipettes according to size and put in sterilization cans.

- Leave the lid tight to keep the condition of sterilized apparatus intact up until use.
- Generally lids and parts are wrapped separately and sterilized in order to prevent them from cracking and sticking due to expansion when heated. For example, glass fitting lids and homogenizers should be wrapped separately and sterilized.
- Keep glass petri dishes and lids together in pairs to keep off dust.
- Put measuring pipettes with tip upward.
- Pasteur pipettes can be kept in flat cans.

(2) Small pieces of apparatus should be put in a big can.

(3) Label an indicator tape for dry sterilization to record date and existence of silicon, etc. Indicator tapes for dry sterilization are not as well known as those for autoclaves, but they are very useful because dry sterilization shows little changes before and after sterilization.

(4) Sterilize them in an oven at 160°C for 2 hours. Time and temperature depend on experiment and apparatus. Do not put inflammable objects, such as paper in an oven as it can cause fire.

(5) Turn off the oven and let it cool down. Wait until it cools down to a certain degree. Sterilization is only halfway done while at high temperature.

(6) Sterilized apparatus and unsterilized ones should be kept separately.

Typical Sterilization Program

General sterilization: 160 ~ 170°C, 2 ~ 4 hours

or

180°C ~ 200°C, 0.5 ~ 1 hour

RNase free sterilization: 180°C, more than 8 hours

or

250°C, more than 30 min.

7. Handling of Pipette

(1) Take the pipette of proper volume and hold it with fingers.

- Do not touch other pipettes as fingers may be contaminated.
- There are pipettes with standard graduation and full scale

graduation. It is recommended that a laboratory uses only one kind of pipettes.

- Touch only the rear end. The part with graduations should not be touched fundamentally as it is possible that a pipette contaminates reagents in bottles.
- Pipettes are long and unsteady. Be careful that their tips will not touch something.

(2) Open the bottle of the reagent and put the pipette into the bottle. In case of aseptic handling, there are some methods like heating over flame.

(3) Suck the reagent a bit over required amount. Put only the tip of the pipette into the reagent. It prevents from contamination and calculation errors caused by the reagents attaching to the exterior of the pipette and measuring errors caused by water pressure.

(4) Holding the rear end by the tip of the index finger, bring graduations to read in front of your eyes. At the same time, put the tip of the pipette along the inner wall of the bottle.

- Be careful not to let the reagent flow over when big pipettes like 10 ml is used. If the tip goes out of the reagent, it allows the reagent to go into your mouth. Lower the pipette as you suck the reagent.
- Suck gently, otherwise air bubbles might, enter the pipette or your mouth.
- In case the reagent goes into your mouth, wash out your mouth with a lot of water as soon as possible.
- Make sure to hold the end of the tip by the index finger (not other fingers). This method is best in terms of promptness and preciseness of handling.

(5) Adjust the amount of the reagent by reducing power of the index finger.

- In order to release the excessive amount of the reagent rapidly, the index finger can be removed. However, a slight move of the finger is enough near the target graduation.
- Adjust the meniscus accurately.

(6) Make sure that the reagent will not leak out of the pipette by holding the rear end. Put the tip of the pipette along the inner wall of the container.

- Putting the tip along the inner wall of the container makes the pipette steady. Besides, it enables smoother injection.

(7) Loose the strength of index finger and inject the reagent.

- Put the tip along the inner wall of the container, so that drop of the reagent will not remain at the tip of the pipette. These drops may cause errors.
- For aliquot, drop the reagent into each container by watching the level of the reagent horizontally in front of your face. In this method, you must measure exact amount each time.

(8) Drops remaining at the tip can be taken out by following methods:

- Cover the rear end of the pipette with a finger and warm a middle part of the pipette by hand or over flame. Expanded air pushes them out.
- Fit a sucker to the rear end of the pipette by vacant hand and blow them out.
- It is not necessary to fit the sucker tightly. Air-blow in a moment is enough.
- Blow them out with breath.

(9) Immerse the pipettes in water soon after using. It is difficult to remove crystal of bases and film of protein once the pipette gets dried.

Accurate Measurement by Pipette

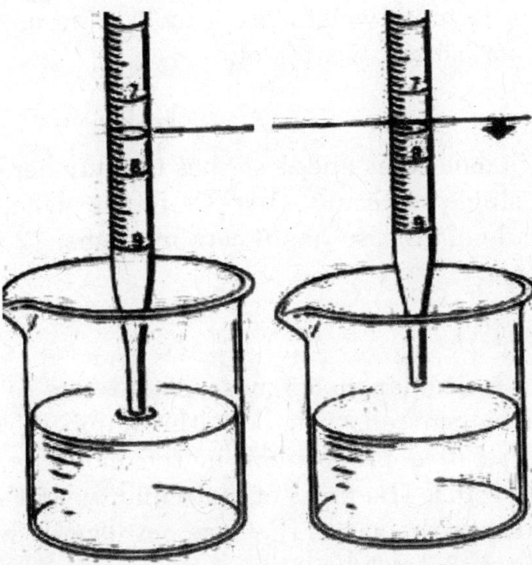

Fig. I.1 : Proper way of measurement by pipette.

The level goes up by pressure, if the pipette is in the solution. When graduations are read in this condition, the measured amount will be less than required. Moreover, if the tip of the pipette is unsteady in the air, a drop at the tip will interfere the accurate measurement.

Standard Graduation and Full Scale

A pipette that measures solution by blowing out all content is called a standard graduation. On the other hand, full scale type pipette with the zero (0) graduation upper than the tip must remain the solution from the graduation up to the tip. The full scale pipette is more popular than the standard graduation recently.

8. Molecular, Empirical and Formula Weight

Empirical Formula

The lowest whole number ratio of atoms in a compound, *e.g.,* the empirical formula for hydrogen peroxide (molecular formula H_2O_2) is HO. This term is also used when describing ionic compounds, such as NaCl, where the smallest units of the substance are ions, not molecules.

Formula Weight (FW)

Also called the empirical molar mass, this refers to the mass of the empirical formula, which tells only the relative proportion of each atom and thus is useful for talking about ionic compounds. The formula weight of NaCl is 58.44 (because Na^+ has a mass of 29.99 and Cl^- has a mass of 35.45). The units of formula weight are g $mole^{-1}$ and as they are always the same they are not usually written out.

Molecular Formula

This term applies to molecules and describes the number of atoms of each element within a single molecule. $C_6H_{12}O_6$ is the molecular formula of glucose, one molecule of glucose has 6 carbon atoms, 12 hydrogen atoms and 6 oxygen atoms.

Molecular Weight (MW)

It is also called the molecular molar mass. This refers to the mass of the molecular formula, which tells you both the number of atoms of each element and their relative proportions in the molecule. The molecular weight of glucose is 180.2. The units of molecular weight are g $mole^{-1}$ and because they are always the same, they are not usually written out. This term is now sharing space with "relative molecular weight" (Mr).

Relative Molecular Weight (Mr)

This is the molecular weight of a molecule expressed as a ratio of the mass of the molecule to the mass of ½ of a ^{12}C atom. Since 1/12 of a ^{12}C atom has a mass of 1 D, the magnitude of the Mr is the same as the magnitude of the MW. Since Mr is a ratio, it is dimensionless, which simplifies comparisons. In practice, because the magnitude of the number is the same as the molecular weight, there can be confusion. If you are reporting the relative molecular weight, you should write Mr 10,000 (no units because Mr is dimensionless) and refer to it as the 10,000 Mr protein.

9. Planning a solution of a particular molarity

One simple way to plan the making of a solution from a dry chemical is as follows:

i. Figure out the mass you would need if making a full litre.

ii. Figure out what fraction of a litre you are making.

iii. Use that same fraction of the mass of the chemical.

$$M \times MW \times V = g$$

M = Final molarity of compound (Mole $litre^{-1}$)

MW = Molecular weight of compound (Listed on the bottle as FW) [g $mole^{-1}$]

V = Desired final volume (liter)

G = Grams of compound to add to solvent (g)

Q.S. = Add sufficient solvent to bring the volume to final volume.

When you are actually making up the solution, put only about 80 percent of the total volume of solvent into the beaker before adding the chemical. Once the chemicals have been added and dissolved, only then you should transfer the contents to a graduated cylinder and Q.S. to the final total volume.

Q.S. = quantity sufficient. To Q.S. a solution properly, it must be measured in an appropriately calibrated vessel, *i.e.*, a graduated cylinder or volumetric flask and not a beaker.

10. Making solutions from hydrated compounds

Some chemicals come with water molecules attached. The first thing to remember is that the MW (listed as FW on the bottle) of such a compound includes the mass of the water, whenever you would have used the MW of the unhydrated compound in the equations above, use instead, the MW of the hydrated compound. The second thing to remember when using a hydrated compound is that the attached water molecules contribute water to the solution, potentially diluting the final concentrations (if the solvent is water). Therefore, you must account for the calculation of water from the hydrated compound when determining the volume of solvent (water) to add.

$M \times V \times H_2O \times 18.015$ = ml H_2O contributed by the hydrated compound

M = Desired molarity of compound (Mole/ litre^{-1})

V = Desired final volume of solution (litre)

H_2O= Number of H_2O molecules in the hydrated compound

18.015 Millilitres per mole of H_2O (ml/mole)

11. Measurement of Liquids with Normality/Molarity

$$\text{Volume of impure liquied} = \frac{\text{Molarity of pure liquid} \times \text{Volume of pure liquid} \times \text{Mass}}{\%\ \text{purity} \times \text{density}}$$

$$e.g.,\ 0.1\text{N HCL} = \frac{010\text{N } 1\text{L} \times 36.46}{0.35 \times 1.2 \text{ g}}$$

$$= 8.8 \text{ ml --to 1 litre}$$

35% pure

The easiest way to measure chemicals, however, is by mass, since mass is what lab balances report. So, to make a solution, typically you have to determine the mass of chemical that you need, based on knowing the desired final concentration (usually molarily), the molecular weight of the chemical and the final volume of solution.

When making any solution, you should start by putting about 80 percent of the total volume of solvent into your mixing vessel and then add the chemicals one by one. Once all the chemicals have been added, transfer your almost finished solution to a graduated cylinder, then Q.S. to the final total volume. You do it that way because the chemical you are adding will contribute some volume to the solution and don't overshoot the desired final total volume.

Molality (m)

A solution containing 1 mole of the solute dissoloved in 1 kg of the solvent is said to be a molal solution. Thus,

$$\text{Molality} = \frac{\text{Weight of a solution in g/kg of solvent}}{\text{Molecular weight of solute}}$$

Mass percentage or % (w/w)

This represents the component which is present in 100 parts by weight of the solution.

If a solution contains 10g of NaCl in 40g of water then mass % of NaCl will be

$$= \frac{10 \times 100}{(10 + 40)} = 20\%$$

Percentages by volume or % (v/v)

This represents the volume of the component in 100 parts by volume of the solution.

If a solution contains 10ml of glycerol in 90ml of DW, the % volume of glycerol will be 10%

$$= \frac{10 \times 100}{(10 + 90)} = 10\%$$

Parts Per Million (ppm)

This represents gram of a solute per millions grams of solution or gram of a solute per million ml of the solution.

$$\text{ppm} = \frac{\text{g or ml of solute or substance}}{\text{g or ml of solution}} \times 10^6$$

12. Numbers Found on Chemical Bottles

Important numbers are found on the labels of chemical bottles and in chemical company catalog listings. Many of the numbers included on chemical bottles have obvious meanings, and others do not, but each provides information that may be extremely useful.

- Product Number: This is the company catalog number.
- Lot Number: It is a good idea to record the lot number of a reagent in your lab notebook, especially when using reagents that may vary significantly from batch to batch, such as antibodies or enzymes. Having a lot number easily accessible can be very useful for trouble shooting your experiments. In case a reagent/enzyme has been contaminated or is of low activity, the company will need to know the lot number.

Storage Temperature

This tells you the temperature to store a reagent to avoid things like loss of activity or decay. Also watch labels for notes indicating whether a chemical should be stored desiccated or in the dark. Pay attention to the storage temperature and expiration dates of chemicals, because chemicals really can go bad.

- –70°C means keep it in the ultra cold freezer (usually –70°C to –80°C).
- –20°C means keep it in the deep freezer.
- –20°C – 0°C means keep it in the freezer compartment of the refrigerator.
- 0°C means keep it in the freezer compartment of the refrigerator.
- 2°C – 8°C means keep it in the refrigerator.
- RT means keep it at room temperature.

Formula Weight (FW)

This tells you the mass in grams mole^{-1} of the chemical as listed in the formula on the label; in this context, FW is being used interchan-geably with molecular weight. If it is a hydrated compound, the FW includes the mass of the water.

Purity

Sometimes trace amounts of by-products of the method used to isolate a chemical compound cannot be entirely eliminated. Purity is a measure of the purity of a chemical, *i.e.*, it tells you how much is not contaminated. Purity is typically represented as a percent and the percentage of what you bought is actually what you want.

Risk and Safety Numbers

These numbers refer to lists found in the catalogs of companies that sell chemicals. They are like mini-Material Safely Data Sheets (MSDS) but the MSDS is the only reliable source for complete safety information. MSDSs are available free online at a number of sites on the internet.

13. Accuracy and Calibration

Accuracy is a measure of truth, or the nearer of your measurement to correct value. To improve the accuracy of your measurements, calibrate your equipment by measuring standards of known values. Once the device gives the correct measure of a standard of known value, then you have to trust that it will give you a measure of equal accuracy, while evaluating something unknown. To evaluate and report accuracy, you report a statistic (usually the standard deviation and the probability value) that indicates the probability of the measurement being within a certain distance of the true value. The single most important thing you can do to improve the accuracy of your measurements is to calibrate your equipment.

Calibration is the process of testing and adjusting a measuring device to determine the relationship between it's output and the true value of the variable being measured. Calibration is accomplished by measuring independently supplied standards, the actual values of the standards are known prior. For example, when calibrating a pH meter, you adjust the device itself, according to the standard, so that the readout is accurate. In some cases, when calibrating a spectrophotometer, you determine the relationship of the devices output to true values by graphing output as a function of the value of the standard.

It is important while calibrating a device to use a range of standards that will bracket the values you expect to measure. For example, if you are measuring the pH of biological reagents, the values you are measuring will be around 7, thus, you should use a pH 4 standard and pH 10 standard to calibrate the pH meter. Similarly, to calibrate a spectrophotometer you should measure standards that bracket the expected concentrations of the unknowns.

For some instruments, an important part of calibrating is telling this instrument what value to call "Zero". This process is called **tarring** or **zeroing**. Spectrophotometers, for example, must be zeroed before every reading; balances must be tarred before each weighing. In both cases, you are telling the machine how much of the signal is not relevant. For a balance, you are telling it to subtract the mass of the weigh boat; for the spectrophotometer, you are telling it to subtract the absorption by the medium and the cuvette. It is by tarring or zeroing that you make sure that only the value of the unknown is reported, not the value of the measuring container plus the unknown.

14. Buffers

A buffer is a solution that can maintain a nearly constant pH if it is diluted, or if strong acids or bases are added. A buffer solution consists of a mixture of a weak acid and its conjugate base (or a weak base and its conjugate acid).

Consider a solution containing both acetic acid (CH_3COOH), and acetate ions (CH_3COO^-).

Any strong base that is added to the solution is neutralized by acetic acid: $CH_3COOH\,(aq) + OH^-\,(aq) \rightleftharpoons CH_3COO^-\,(aq) + H_2O\,(aq)$

Any strong acid that is added to the solution is neutralized by acetate: $CH_3COO^-(aq) + H^+\,(aq) \rightleftharpoons CH_3COOH\,(aq)$

The amount of strong acid or base that a buffer can neutralize is called the buffer capacity.

- Since the buffer concentration is usually high, the [H$^+$] concentration can be calculated from the equilibrium expression assuming that the concentrations of the conjugate acid-base pair do not change appreciably.
- Most effective buffering capacity is around the pK, *i.e.*, the pH does not change drastically with the addition of OH$^-$.
- Generally, the effective range of a buffer is 1 pH unit above and 1 below its pK.
- Biological buffers are based on phosphate and bicarbonate buffers.
- Proteins can also act as buffers. Haemoglobin is also a strong buffer.

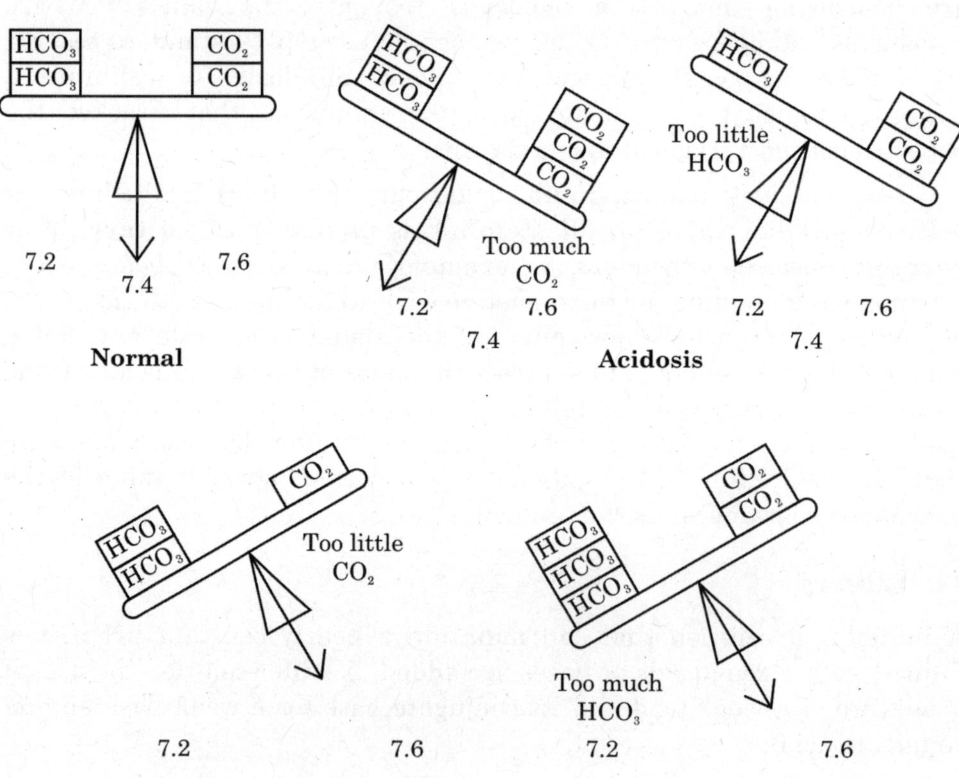

HEPES pKa = 7.6

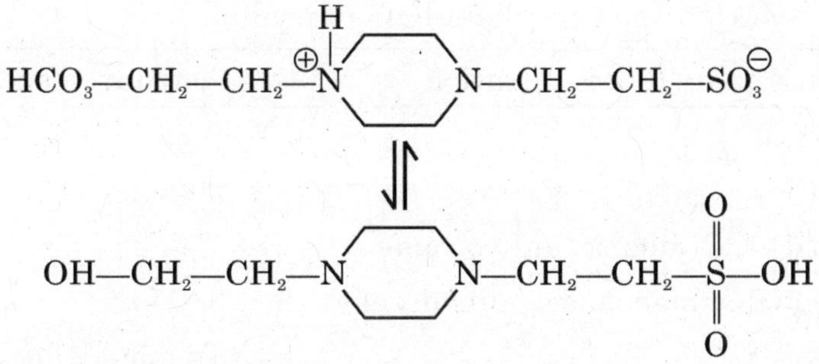

$$HCO_3—CH_2—CH_2—\overset{\oplus}{N}H \langle \rangle N—CH_2—CH_2—SO_3^{\ominus}$$

$$\rightleftharpoons$$

$$OH—CH_2—CH_2—N \langle \rangle N—CH_2—CH_2—S(=O)_2—OH$$

For 50mM HEPES buffer (pH 7.0) mix 25 ml of 200mM HEPES (52.06g) litre of HEPES Na salt, 11.8 ml of 100 ml of 100 mM NaOH and distilled H_2O to 100 ml. Sterilize the solution by passage through a 0.22 µm filter.

0.1 M Phosphate Buffer at 20°C

To prepare 1 M stock, add 136 g of monobasic (KH_2PO_4) or 174.2 g of dibasic (K_2HPO_4) potassium phosphate to 1 litre of distilled H_2O. Combine the following volumes of the 1 M stock to achieve the desired pH. Dilute the combined 1 M stock solution to 1000 ml with distilled H_2O.

pH	Volume of 1M K_2HPO_4	Volume of 1M KH_2PO_4
5.8	8.5	91.5
6.0	13.2	86.8
6.2	19.2	80.8
6.4	27.8	72.2
6.6	38.1	61.9
6.8	49.7	50.3
7.0	61.5	38.5
7.2	71.7	28.3
7.4	80.2	19.8
7.6	86.6	13.4
7.8	90.8	9.2
8.0	94.0	6.0

Phosphate Buffered Saline (PBS) (pH 7.2 – 7.4)

Earlier, this commonly used reagent was sodium phosphate buffered saline (150 mM NaCl plus 10 mM sodium phosphate). PBS has been modified to suit different applications. One common preparation of PBS (including potassium) and another (omitting potassium) are provided. This is Ca^{++} / Mg^{++} free phosphate-buffered saline.

Component and final concentration	Amount to add per 1 litre
137 mM NaCl	8 g
2.7 mM KCl	200 mg
10 mM Na_2PO_4O (dibasic, anhydrous)	1.44 g
2 mM KH_2PO_4O (monobasic, anhydrous)	240 mg
H_2O	To make 1 litre

Dissolve the components in approximately 0.9 litre of H_2O. Adjust the pH to 7.2–7.4 with HCl and then adjust the final volume to 1 litre with H_2O. Dispense in convenient volumes and sterilize by autoclaving for 20 min. at 15 psi on liquid cycle. This can be stored at room temperature.

When used for tissue culture, the pH is 7.2. For other uses (*e.g.*, *in situ* hybridization), the pH is 7.4. PBS for tissue culture can be purchased already made up as a sterile 10× or 1× solution. If Ca^{++} and Mg^{++} are added, the amounts are 1.33 g of $CaCl_2.2H_2O$ and 0.10 g of $MgCl_2.6H_2O$ per litre.

50 mM Tricine Buffer (pH 7)

For 50 mM tricine buffer (pH 7 at 20°C), mix 25 ml of 200 mM solution of Tris tricine (35.83 g/liter) and 2.7 ml of 100 mM NaOH to 100 ml with distilled H_2O.

1M Tris - HCl Buffer

Mixing the volumes of concentrated HCl (11.6 N) designated in the table below with 121 g of Tris base (MW = 121 g/mole) results in 1 litre of 1 M Tris-Cl buffer of the desired pH at 25°C. To prepare the 1 M buffer, dissolve 121 g of Tris base in approximately 0.9 litre of H_2O. Add the appropriate volume of concentrated HCl and adjust the final volume to 1 litre with H_2O.

Volume of concentrated HCl (ml)	pH
8.6	9.0
14	8.8
21	8.6
28.5	8.4
38	8.2
46	8.0

Volume of concentrated HCl (ml)	pH
56	7.8
66	7.6
71.3	7.4
76	7.2

Tris solutions should be kept in a closed container, because they will absorb carbon dioxide from the air, should be colourless. Tris–HCl buffers can also be made by mixing Tris base and Tris–HCl in appropriate quantities, but these must be desiccated before being weighed.

50 mM Tris Buffer

Tris buffers (0.05 M) of the desired pH can be made by mixing 50 ml of 0.1 M Tris base with the indicated volume of 0.1 N HCl and then adjusting the volume of the mixture to 100 ml with H_2O.

Desired pH (25°C)	Volume of 0.1 N HCl
7.10	45.7
7.20	44.7
7.30	43.4
7.40	42.0
7.50	40.3
7.60	38.5
7.70	36.6
7.80	34.5
7.90	32.0
8.00	29.2
8.10	26.2
8.20	22.9
8.30	19.9
8.40	17.2
8.50	14.7
8.60	12.4
8.70	10.3
8.80	8.5
8.90	7.0

Tris-buffered Saline (TBS)

TBS is a commonly used buffer and is used as an alternative biological substrate to PBS for a variety of protein analysis and immunocytochemical approaches. Although the standard concentration is 10 mM Tris, 150 mM NaCl, different protocols may utilize different concentrations of Tris in their version of TBS. Tris-based buffers use concentrated HCl to achieve the desired pH.

Tris- buffered Saline (TBS) (pH 7.15) from Stock Solutions (10 mM Tris, 150 mM NaCl)

	Final (1X)	Mix for 5X
Tris-HCl pH 7.15	10 mM	50ml of 1 M
NaCl	150 mM	150 ml of 5 M
Distilled H_2O to 1 litre		

Add sodium azide (1%) to prevent microbial contamination. Dilute one volume of stock and four volumes of distilled H_2O for use.

1× TBS (50 mM Tris-HCl [pH 8.1]) from powder

Dissolve 48.4 g of Tris base and 70.1 g of NaCl in sterile distilled H_2O. Adjust the pH to 8.1 with concentrated HCl. Adjust the volume to 8 litres with distilled H_2O. Store at room temperature.

Tris-buffered Saline (TBS) (10 mM Tris, 100 mM NaCl)

	Volume	Final concentration
1 M Tris-HCl (pH 7.9)	10 ml	10 mM
5 M NaCl	20 ml	100 mM

Adjust the volume to 1 litre with H_2O

Tris-EDTA (TE) (pH 7.4 – 8.0)

Tris standard buffer is used to resuspend and store DNA. It can be prepared by using stock solutions of 1 M Tris-HCl at pH values ranging from 7.4 to 8.0. This can be stored indefinitely at room temperature.

Component and final concentration	Amount of stock to add per 100 ml
10 mM Tris-HCl	1 ml of 1 M (pH7.4 – 8.0 at 25°C)
1 mM EDTA	200 μl of 0.5 M (pH 8.0)
H_2O	98.8 ml

10x TE, pH 8.0 (100 mM Tris-HCl, 10 mM EDTA)

	Final (1×)	Mix for 10×
Tris-HCl (pH 8.0)	10 mM	100ml of 1 M
EDTA	1 mM	20 ml of 500 mM (pH8.0)

Add distilled H_2O to 1 litre, sterilize by autoclaving. Store at room temperature.

■ ■ ■

Chapter - 2

INSTRUMENTS COMMONLY USED IN THE LABORATORY: PRINCIPLE AND PRECAUTIONS

1. Weighing Balance

Accurately measuring mass of substances is a fundamental requirement for almost all types of analyses. In the laboratory, the mass is determined by comparing the weight of an object with the weight of a set of standard masses using a balance. Since acceleration due to gravity affects both the known and unknown to the same extent, equality of weight indicates equality of mass. The terminology distinction between mass and weight is therefore, seldom made.

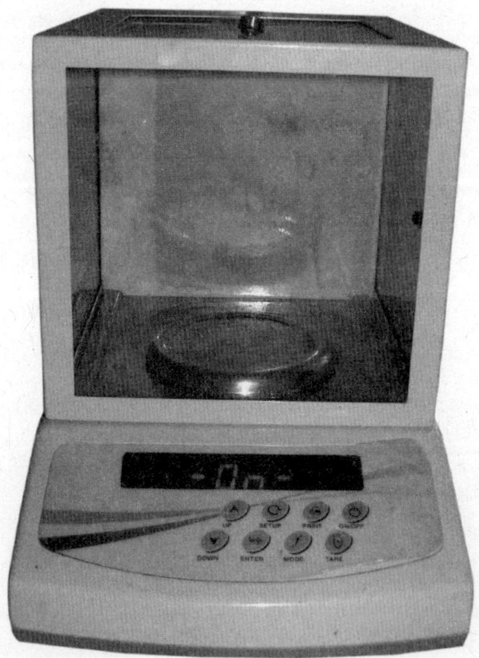

Fig. II.1 : Weighing Balance

There are several different types of balances:

Mechanical Balances: Triple beam, dual pan, primary school balances

Digital Balances: Top loading, analytical, dual range

Spring Scales: Linear, rotary dial, platform, bathroom scales.

Mechanical balances are probably the oldest form of weighing device in existence. They rely on a comparison between a known weight and the unknown sample which are "balanced" on opposite pans. Newer versions of this type of balance use a single pan for the unknown sample, which is measured by sliding weights along several calibrated beams. The capacities of these balances range from about 600 grams to about 2600 grams. The readability (or resolution) is generally one tenth of a gram. They are rugged and durable, and generally accurate enough for educational uses. Since they use moving parts to literally balance the measurement, they will eventually need to be serviced for wear and tear.

Electronic digital balances use strain gauge type load cells to measure mass. As a weight is applied to the load cell, the electric resistance of the load cell changes. By measuring this change, the mass of the sample can be accurately determined. There are many models and features to choose from, but it doesn't have to be a confusing and frustrating process if you can define your needs and match them with the proper balance.

Here are some of the things to be considered when choosing a weighing device:

Range – Also known as capacity, this is the maximum weight that the balance can measure. Putting more than the recommended mass on a digital balance can have catastrophic results. You should be careful to pick a model that exceeds the maximum weight you anticipate having to measure. On the other hand, you don't want to choose a model that has too much capacity, because cost becomes a factor when you combine a higher range with the resolution you require.

Resolution – Also known as readability, this is the increment that the balance will read in. The mechanical balances previously described have a resolution of 0.1 grams. Some of our more economical models have a resolution of 1 gram. Choose a balance that has the resolution that will meet your weighing needs.

Accuracy – Sometimes stated as linearity, this is the error factor that a balance or scale may have. It is expressed as a +/- variation from the reading that the scale or balance may give under ideal weighing conditions. Other factors such as vibration or air currents may adversely affect your readings beyond the stated accuracy.

Analytical balances, with an accuracy of ± 0.1 mg, are commonly used to determine the weight. For analysis at micro-level contaminants, balances with a precision of ± 0.01 mg may be used.

Tare

Perhaps the most widely misunderstood feature of a balance or scale is the tare function. Simply put, tare allows the user to zero the instrument to cancel the weight of a container (such as a weighing dish or plastic bag) from the reading of the instrument, thus giving the true weight of the product being tested. Without tare, it would be necessary to manually subtract the weight of the container from the displayed value. However, Tare does not increase the capacity of a scale or balance. If you are using a balance with a 100 gram maximum capacity and you place a 10 gram weighing dish on the pan and tare it, you are left with only 90 grams of capacity to do your weighing. Remember, it is an unsafe practice to overload any weighing device! Serious damage to your balance or scale will almost certainly result from overloading it. Digital balances have a push button for tare, and some mechanical balances use a tare beam or screw adjustment to rebalance the instrument back to zero. Hanging scales rely on a spring constant to make the measurement, and may have a sliding faceplate or screw adjustment for tare.

Units

Most of the balances weigh primarily in grams. The electro-nic bala-nces may also have other units as well, such as ounces, pounds, ozt (pennyweight) and Newtons. Most laboratory applications would use grams or milligrams. Force measurement in the physics lab would probably be done in Newtons. Dietetic and bathroom scales measure in ounces or pounds. Draft shields are available with many of our sensitive analytical balances to shield the weighing pan from air currents. Most balances have lock-down devices available to deter theft, and most come standard with an AC adapter for use with 110/120VAC. Many electronic balances can be battery-operated for increased portability. Some allow the balance to be interfaced with a PC. This allows the user to capture readings right onto a computer, and with some models the user can actually control the balance through their PC.

Precautions

It is of paramount importance that balances are used and maintained with great care. To avoid damage or minimize wear on the balance and obtain accurate weights adhere to the following rules:

i. Be certain that the arresting mechanism of the beam is engaged whenever the loading on the balance is being changed and when the balance is not in use.

ii. Centre the load on the pan insofar as possible.

iii. Protect the balance from corrosion. Only non-reactive glass, plastic or metal objects should be placed directly on the pan. In case of volatile or corrosive substances, take special precautions of sealing it in a weighed glass ampoule.

iv. Do not attempt to adjust or repair the balance if you are not trained to do so.

v. Keep the balance and its cases scrupulously clean. A camel's hair brush is useful for cleaning any spilled material or dust.

vi. Do not weigh an object that has been heated until it has returned to the room temperature.

vii. Do not touch a dried object with bare hands: use tongs to prevent the uptake of moisture.

viii. Always place the balance on a vibration free platform.

2. Spectrophotometer

Spectrometers are analytical instruments which disperse an emission (such as particles or radiation) according to some property of the emission (such as mass or energy) in order to measure the amount of the dispersion. This product area includes visible, infrared (IR), ultraviolet (UV), atomic absorption (AA), optical emission (OE), Raman, X-ray fluorescence (XRF) and mass spectrometers. Specific search forms are also available. Performance specifications for spectrometers include: spectral range, spectral resolution, mass range, mass resolution, and mass accuracy. Spectral range is the dispersion of the grating across the linear array, or the "size" of the spectra on the array. Spectral resolution is the width of an analytical peak at half its height as measured in nanometers (nm). This value indicates the spectrometer's ability to separate two overlapping peaks. Spectral accuracy is the agreement between a mean measured value and a true or accepted value as quantified by error. Mass range is a measurement of atomic mass that is expressed in atomic mass units (amu) or Daltons (Da). Mass resolution is the ratio of the mass divided by the change in mass over the area of two peaks.

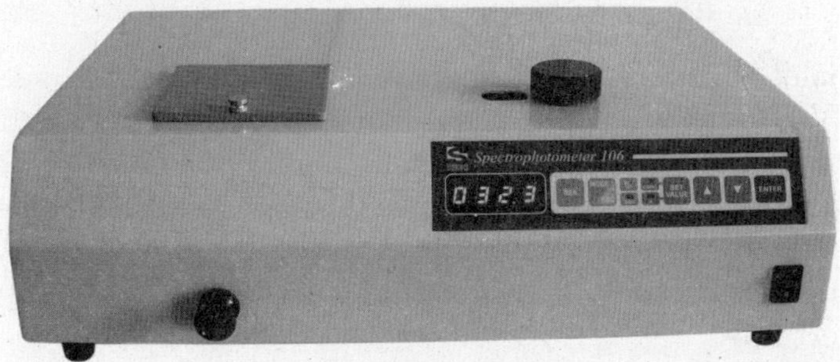

Fig. II.2 : Spectrophotometer

There are different types of spectrometers. UV and visible spectrometers measure the amount of ultraviolet (UV) and visible light transmitted or absorbed by a sample placed in the spectrometer. Infrared (IR) spectrometers measure the wavelength and intensity of the absorption of infrared light by a sample. Atomic absorption (AA) spectrometers use the absorption of light to measure the concentration of gas-phase atoms. Atomic emission (AA) and optical emission (OE) spectrometers determine analyte concentration via a quantitative measurement of the optical emission from excited atoms. Raman spectrometers are used to measure the wavelength and intensity of inelastically scattered light from molecules. X-ray solids, in which X-ray are used to excite a sample and generate secondary X-rays. Mass spectrometers separate ions by their mass-to-charge (m/z) ratios. They are used to identify compounds by the mass of one or more elements in the compound. They are also used to determine the isotopic composition of one or more elements in a compound.

Spectrophotometer is an instrument used to measure the intensity of wavelengths in a spectrum of light compared with the intensity of light from a standard source. A spectrophotometer consists of two instruments, namely a spectrometer for producing light of any selected colour (wavelength), and a photometer for measuring the intensity of light. The amount of light passing through the tube is measured by the photometer. The photometer delivers a voltage signal to a display device, normally a galvanometer. The signal changes as the amount of light absorbed by the liquid changes.

If development of colour is linked to the concentration of substance in the solution then concentration can be measured by determining the extent of absorption of light at the appropriate wavelength.

The optical density (O.D.) is directly proportional to the concentration of the coloured compound. Most spectrophotometers have a scale that reads both in O.D. (absorbance) units, which is a logarithmic scale, and in % transmittance, which is an arithmetic scale. As suggested by the above relationships, the absorbance scale is the most useful for colourimetric assays.

Ultraviolet-Visible spectroscopy or Ultraviolet-Visible spectrophotometry (UV/VIS) involves the spectroscopy of photons (spectrophotometry). It uses light in the visible and adjacent near ultraviolet (UV) and near infrared (NIR) ranges. In this region of energy, space molecules undergo electronic transitions. The method is used in a quantitative way to determine concentrations of an absorbing species in solution, using the Beer-Lambert law:

$$A = \log_{10} (I/I_0) = \varepsilon.c.L,$$

Where, A is the measured absorbance, I_0 is the intensity of the incident light at a given wavelength, I is the transmitted intensity, L the length of the cell, and c the concentration of the absorbing species. For each species and wavelength, ε is a constant known as the extinction coefficient.

When monochromatic light (light of a specific wavelength) passes through a solution there is usually a quantitative relationship (Beer's law) between the solute concentration and the intensity of the transmitted light, that is

$$I = I_0 10^{Kcl}$$

where, I sub 0 is the intensity of transmitted light using the pure solvent, I is the intensity of the transmitted light when the coloured compound is added, c is concentration of the coloured compound, 1 is the distance the light passes through the solution, and k is a constant. If the light path 1 is a constant, as is the case with a spectrophotometer, Beer's law may be written as $I/I_0 = 10^{-kc} = T$

Where k is a new constant and T is the transmittance of the solution. There is a logarithmic relationship between transmittance and the concentration of the coloured compound. Thus, logT = log 1/T = kc = optical density

Procedure for using a spectrophotometer

1. Turn the spectrophotometer on and set your wavelength. Allow to warm up for fifteen minutes.
2. Set the switch beside the readout to the transmit position or up.
3. Turn the knob on the left until the readout is zero.
4. Open the lid on the left side and insert a cuvette filled with reference.
5. Close the lid and turn the switch (next to the readout) to the absorb or down position.
6. Use the knob on the right to dial in the readout to zero.
7. Open the lid and remove the reference.
8. Put your sample in the spectrophotometer and read the number and record it in your notes.
9. Remove sample from the spectrophotometer.
10. Repeat steps 8 and 9 for each sample you have.

3. Laminar Flow Hood

A laminar Air Flow Hood is an indispensable equipment for all microbiological and tissue culture laboratories which is designed to create nearly sterile environment to prevent the contamination. These hoods function by drawing the room air through High Efficiency Particulate Air

(HEPA) filter. These filters are meant to retain 99.97% particles of 0.3u in size, thus remove almost all airborne materials which not only includes dust particles of air, but also microbes, spores, mycelia, etc.

Fig. II.3 : Laminar Flow

Filtered air is blown in smoothly through a constant laminar flow fashion over the entire work surface available at hood in parallel lines at a uniform velocity. The main body of the hood or cabinet is made up of stainless steel, it is very smooth without any gap or joint to avoid the collection of spores. It is equipped with UV light and Bunsen burner. Vertical and horizontal laminar hoods are most commonly used in labs.

Procedure for using the hood

The UV light and filters of the hood must be switched on at least 30 minutes prior to practical work to remove any particulates and to sterilize it. The surface of laminar flow hood should be thoroughly sterilized with alcohol. The objects should be properly arranged in a manner to get the full advantage of laminar flow of air. Bunsen burner is turned on and inoculation is done near the flame. After the work operation the hood must be decontaminated with alcohol or an appropriate disinfectant.

Precautions

- Keep the entry door to the lab closed when in use.
- Do not have electric fans blowing in the room as it may affects the air flow of the unit.
- Wear a closed front lab coat and gloves.
- Do not cover the air intake grill.

- Student should work well into the hood atleast 4 inches from the front grill.
- Waste and unwanted items should never enter the hood.
- Do all the calculations and labeling before entering the hood.
- Wash hands and remove the jewelry from hands and wrist before starting the work in hood.
- Avoid spraying or squirting solutions onto the HEPA filter.
- Do not place the large objects near the back of the hood as they may disrupt the laminar flow pattern of air.
- If anytime something is spilled, the work surface should be thoroughly cleaned with alcohol.
- Coughing, sneezing and talking should be minimized.

4. Autoclave

An autoclave is a device to sterilize the aqueous solutions and other objects at temperature above the boiling point of water and elevated pressure. It was invented by Chamberland in 1879.

Principle

Water boils when its vapour pressure is equal to that of the surrounding atmosphere. Under ordinary conditions water cannot be heated above 100°C in an open vessel because it results in boiling of water and steam formation. Further heating does not raise the temperature of water. However, in sealed vessels the pressure rises when heated. Hence, with increase of pressure inside the vessel, the temperature at which water boils also increases, thus killing all vegetative as well as resistant structures of microorganisms.

Fig. II.4 : Autoclave

Uses

Autoclaves are widely used in microbiological, pathological and tissue culture labs and also in medicine and metallurgy.

Operating Instructions

- Check the level of water before operating the autoclave.
- Place the material (objects) in small shallow containers.
- Cotton or glassware should be wrapped in aluminum foil.
- Close the lid and tighten the screws.
- Switch the plug on and monitor the pressure gauge until it reaches to 15 p.s.i.
- Keep it for 15 minutes, timer can also be set.
- After 15 minutes switch off the plug and wait until the chamber pressure reaches to zero again.
- Loosen the screw and wait a few minutes for steam to subside, first open the lid slowly to let the steam subside.
- Unload the items using heat resistant gloves.

Precautions

- Avoid too much loading as it prevents proper circulation of steam inside the autoclave.
- Close the lid and screw it properly.
- Open the air outlet to facilitate the escape of air from the chamber.
- Close the air outlet only when pure steam starts coming out.
- Release the pressure slowly as sudden pressure release may cause the wetting of plugs and sometimes the spilling out of liquid.
- Never open the lid if pressure has not fallen to zero. It may cause the lid to be blown off and hot liquids spilling all over.

5. Centrifuge

A centrifuge is one of the most common instruments found in science laboratories. It uses the centrifugal force to separate substances of different densities. The first successful centrifuge, a cream separator was invented in 1883 by a Swedish engineer, Carl de Laval. The ultra centrifuge was developed by Swedish chemist Theodor Svedberg in 1920 and improved by the American physicist, Jesse Beams. Centrifuges have numerous applications, *e.g.*, for isolating suspensions, separation of isotopes, for separation of blood cells from whole blood and separation of sugar from syrup etc.

Preparative rotors are used in biology for pelleting of fine particulate fractious, such as cellular organelles (mitochondria, microsomes, ribosomes) and viruses. They can also be used for gradient separations, in which the tubes are filled from top to bottom with an increasing concentration of a dense substance in solution. Sucrose gradients are typically used for separation of cellular organelles. Gradients of caesium salts are used for separation of nucleic acids. After the sample has spun at high speed for sufficient time to produce the separation, the rotor is allowed to come to a smooth stop and the gradient is gently pumped out of each tube to isolate the separated components.

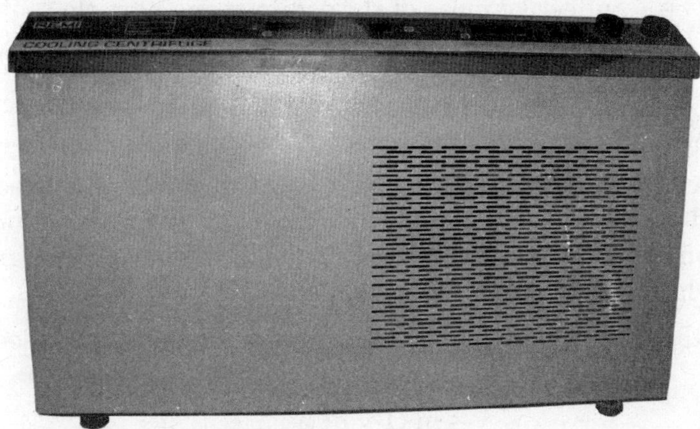

Fig. II.5 : Centrifuge

There are five types of centrifuges:

1. Tabletop / clinical / desktop or microfuge
2. Highspeed centrifuge
3. Cooling centrifuge
4. Ultracentrifuge
5. Geotechnical centrifuge (used for simulating blasts and earth-quake phenomena)

Speed

- Microcentrifuges: 15,000 rpm
- Low/High speed: 2,000 – 20,000 rpm
- Ultracentrifuges: 120,000 rpm

Principle

Centrifugation is a separation process which is used to promote accelerated settling of particles in a solid-liquid mixture leading to formation of two phases: 1. the sediment and 2. the centrifugate or centrate which is the supernatant liquid. Centrifugation separates on the basis of the particle size and density difference between the liquid and solid phases. The smaller the diameter of a centrifuge the greater the forces and accelerations exerted on the contents and the more rapidly it may be spun without breaking.

The centrifugal force, F is defined as

$$F= m\omega^2r$$

where, F = intensity of the centrifugal force

m = effective mass of the sedimenting particle

ω = angular velocity of rotation in rad/sec

r = distance of the migrating particles from the central axis of rotation

The force on a sedimenting particle increases with the velocity of the rotation and the distance of the particle from the axis of rotation. A more common form of measurement of F in terms of the earth's gravitation force, g is relative centrifugal force, RCF.

The rate of sedimentation, called the sedimentation velocity, v, is defined as

$$v = m\ (1\text{-}v\ p)\ \omega^2r\ /\ f$$

or a new term sedimentation coefficient, s is used instead

$$s = v/\ \omega^2r = m\ (1\text{-}v\ p)w^2r\ /\ f$$

the unit for s is in sec.

where, m – actual mass of particle

v – partial specific volume

p – density of the solvent

The rate of sedimentation depends upon the applied centrifugal field (g). The rate at which a particle moves downwards is proportional to centrifugation force (G) which is represented as

$$G = \omega^2r$$

ω – Angular velocity of the rotor (radians/sec.)

r – Radial distance of the particles present from the axis of rotation (cm)

Since one resolution is equal to 2 π radians,

$$\omega = \frac{2\pi . \text{r pm}}{60} \text{ rev/min.}$$

$$\omega^2 = \frac{4\pi^2 . (\text{r pm})^2}{3600}$$

By substituting the value of ω^2 in the above equation

$$\therefore \text{ G} = \frac{4\pi^2 \text{ (rpm)}^2 \text{ r}}{3600}$$

The centrifugal force is expressed as relative centrifugal field (RCF) in g units (g = 981 cm sec^{-1})

Therefore,

$$\text{RCF} = \frac{4\pi^2 \text{ (rpm)}^2 . \text{r}}{3600 \times 981} = 1.11 \times 10^{-5} \text{ (rev/min)}^2 . \text{r}$$

Thus, RCF= (1.18x 10^{-5}). $(rpm)^2$ .r

Now, net force experienced by the particles when they are centrifuged at an angular velocity of ω radian/sec. assuming particle to be spherical.

$$\text{F} = \frac{4}{3}\pi \text{r}_\text{p}^3 (\text{P}_\text{p} - \text{P}_\text{m})\ \omega^2 \text{ r}$$

Where, r_p = Particle radius

p_p = Density of particle

p_m = Density of medium

ω = angular velocity (radians/sec)

r = radial distance (in cm)

$4/3\pi r^3 p$ = is the volume os the spherical particle or radii.

If particle is spherical and rigid and it is moving at a given velocity, it will experience a frictional force which tends to retard its movement.

Frictional force $F_0 = v \times f$

Where,

v – velocity or sedimentation rate

f – frictional co-efficient of particles in the solution.

For an unhydrated spherical particle,

$$f = 6\pi\eta r_p.v$$

Where, η = Viscosity co-efficient of medium.

r_p = radius of the particle

An unhydrated spherical particle of fixed volume and density accelerates under the application of centrifugal force, the net force of the sedimentation equals the frictional force.

$$F = F_0$$

or

$$\frac{4}{3}\pi r_p^3 (P_p - Pm)\,\omega^2 r = 6\pi\eta\, r_p$$

When net force experienced by particle will become zero, it will no more accelerate and the particle achives maximum velocity at which sedimentation occurs at constant rate, so rate of sedimentation of particle,

$$w = \frac{dr}{dt} = \frac{2\, r_p^2 (P_p - P_m)}{9\eta} .\, w^2\, rp$$

Sedimentation time,

$$t = \frac{9\eta}{2\,\omega^2\, r_p^2 (P_P - P_m)} \ln \frac{r_b}{r_t}$$

Where, r_b = radial distance from axis of rotation to bottom of tube.

r_t = radial distance from axis of rotation to liquid meniscus.

Precautions

1. Proper use of the instrument and rotors within recommended limits and careful maintenance of rotors to prevent corrosion and to detect deterioration are necessary to avoid this hazard.
2. The centrifuge should always be securely anchored by use of suction cups or wheel brakes. The centrifuge should have a safety shield and a top disconnect switch. Other safety proce-dures include the following:
   i. The glass tubes should be inspected for cracks.
   ii. The metal centrifuge cups should be inspected for roughness.
   iii. The tubes should be positioned opposite each other with the same weight in each tube to balance the centrifuge. Use water-filled tubes as necessary for balance.
   iv. If the centrifuge has a lid, make sure it is closed during operation.
   v. Keep hair, loose clothing, and body parts away from the centrifuge while it is operating.
   vi. A spinning centrifuge should not be touched for any reason. Do not try to stop a centrifuge by grabbing it. Makes sure the centrifuge is completely stopped before removing test tubes from it.

vii. The centrifuge should operate vibration-free at the top speed.

viii. A running centrifuge should never be left unattended.

ix. The rotors and buckets should be kept clean.

x. Close lid during operation.

xi. Allow the rotor to come to complete stop before opening.

3. Aerosols are created by filling centrifuge tubes, removing plugs or caps from tubes after centrifugation, removing supernatant, resuspending sedimented pellets and by the very process of centrifugation. A hazard is created if a tube breaks during centrifugation. To minimize the generation of aerosols when centrifuging biohazardous material, the following procedures should be followed:
   - Use sealed tubes and safety buckets that seal with 0-rings. Before use, inspect tubes, 0-rings and buckets for cracks, chips, erosions, bits of broken glass, etc. Do not use aluminum foil or loose plastic caps to cap centrifuge tubes because it may detach or rupture during centrifugation.
   - Fill and open centrifuge tubes, rotors and accessories in a biosafety cabinet (BSC). Avoid overfilling of centrifuge tubes so that closures do not become wet.
   - Add disinfectant to the space between the tube and the bucket to disinfect material in the event of breakage during centrifugation.
   - Always balance buckets, tubes and rotors properly before centrifugation.
4. If corrosive solutions are used, be sure to wash and air dry the rotor prior to reuse. If hazardous materials (bio-hazardous, radioactive or chemical) are spilled, decontaminate the rotor before re-use. Also, be sure to clean the centrifuge before it is serviced.
5. If biohazardous samples are run in a centrifuge, an aerosol may exist in the containers. These sample containers should be opened in a biological safety cabinet. When working with bio-hazardous samples, use tubes with O-rings, centrifuge safety cups, or biosafety lids for fixed angle rotors.

Maintenance/Safe Operation

- Use safety cups, whenever possible.
- Disinfect weekly and also after all spills or breakages.
- Lubricate O-rings and rotor threads weekly.

- Do not operate the centrifuge without the rotor properly balanced.
- Do not use rotors that have been dropped.
- Contact your centrifuge rep for specific information.
- If you suspect that leakage has occurred from the centrifuge, leave the area, do not open the centrifuge for at least 30 minutes to allow aerosols to settle, then access situation while taking appropriate precautions.

6. pH Meter

The pH is negative log of the concentration of hydrogen ions. The 'p' in pH stands for 'potential'. A solution having pH 7 is said to be neutral, if the pH is less than 7 then it is acidic and if the pH is more than 7 then it is an alkaline solution. Water is an amphoteric substance serving as a proton donor (acid) or a proton acceptor (base).

$$H_2O = H^+ + OH^-$$

In pure water $[H^+] = [OH^-] = 10^{-7}$

Most of the chemical reactions in a living organism are profoundly influenced by H^+ ion concentration, where,

$$pH = -\log [H^+]$$

Principle

The various components of a pH meter are:

i. *Reference electrode:* The basic function is to maintain a constant electrical potential against which deviation can be measured. The two widely used reference electrode are calomel and Ag-AgCl electrode.

$$AgCl + e^- \longrightarrow Ag^+ + Cl^-$$

Potential can be measured by Nerst equation –

$$Eref = E^o - \frac{In\ Rt}{nf}$$

ii. *Glass electrode:* Its function is to establish an electrical potential that responds to changes in the H^+ ion activity of the solution being tested. Glass electrode consists of high resistance glass bulb at the bottom. When a pH sensitive electrode and a refere-nce electrode are placed in a solution, a galvanic cell is set up.

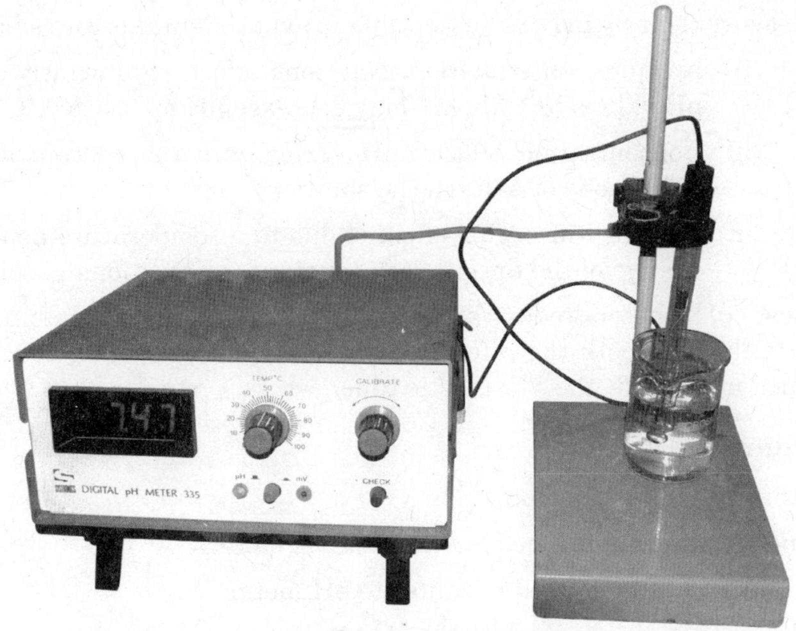

Fig. II.6 : pH Meter

The potential difference that develops between the two electrodes is measured as a voltage as given in following equation:

$$V = E\,\text{constant} + \frac{2.303\,RT}{F}\,\text{pH}$$

Where, V = voltage of the completed circuit

E constant = potential of reference electrode

R = the gas constant

T = the absolute temp.

F = the Faraday constant

The potential of this galvanic cell is the algebric sum of the two electrodes. The reference electrode and the pH sensitive electrode are related by equation

$$E_{cell} = E_{ref} + E_{glasselectrode}$$

The cell is now seen to vary only or a function of the test solution pH, if the temperature is constant.

iii. *Electrometer:* Measurement of the pH that is generated in a galvanic cell of a pH meter is done by electrometer.

Most pH meters have a single combinational electrode. Both the reference and pH-dependent electrode are present in a single glass or plastic tube.

Measurement by pH are susceptible to experimental errors such as,

(i) Na+ ion concentration: Na+ ions effect significantly at high pH values (0.1 M Na+ may decrease pH by 0.4 to 0.5.

(ii) Concentration effects: pH varies with the concentration of buffer ions or salts in the solution.

(iii) Temperature: This effect is due to a temperature dependent change of the dissociation constant (pKa) of ions in solution.

Most pH meter today are obtained as a single combinational electrode. In this, both the reference and the pH-dependent electrode are contained in a single glass or plastic tube.

Procedure

1. pH meter is switched on.
2. Knob is adjusted for measurement of pH meter.
3. Standard buffer is used to calibrate pH meter.
4. Electrode is dipped into standard buffer.
5. After reading is obtained, wash the electrode, dry it and then dip it into the solution, whose pH is to be measured.
6. Note the reading again, wash the electrode with distilled water.
7. After taking the reading, the electrode should be dipped in distilled water.

Precautions

1. The glass electrode must be washed thoroughly and kept clean.
2. The glass electrode is fragile and expensive, therefore, it must be handled with proper care and gentleness.
3. The standard buffer is used to calibrate the pH meter, it should be of pH 4, 7 and 9.
4. As the characteristics of glass-electrode change with the time, so calibration should be done at regular intervals.
5. Glass electrode is sensitive to dehydration, so it should not be dried completely, but stored in solutions suggested by manufacturers.

7. Incubator

Incubators are the most commonly used laboratory equipment, particularly in microbiology labs. These instruments are double walled chambers designed to promote the growth of microorganisms by maintaining a constant temperature within a narrow range.

Fig. II.7. : Incubator

There are many types of incubators available that offer an array of features. Advanced features may include carbon dioxide (CO_2) atmosphere, circulating fans, humidity controls, recording thermometers and alarm systems which may provide an enhanced environment for growth of certain microorganisms. They also have device to monitor current environmental conditions of the unit, or alert the user when tolerance limits are not met.

Principle

Basic to all models is a temperature control device or thermostat. The temperature range for incubators is generally from 5°C above the ambient temperature up to 99.9°C, however this will vary for different incubators. They can be adjusted below room temperatures also. The thermostat should be adjustable to obtain the desired temperature which is checked every hour to ensure that incubator is properly regulated.

Uses

Incubators are used for providing appropriate temperature for growth of microbial cultures in the laboratory.

Precautions

1. The temperature of the incubator be allowed to completely stabilize after each adjustment.
2. The door of the incubator should be opened only when necessary.

3. To avoid hot spots, items placed in the incubator (*i.e.*, culture plates, candle jars) should be arranged in a manner that allows for maximum air circulation. Avoid covering shelves completely, allowing free space in corners.
4. The interior of the incubator should be cleaned on a regular basis or immediately after the spills occur.
5. Cleaning and fumigation of incubator time to time is advisable.
6. If a CO_2 tank is required, it should be securely chained to the wall in an upright, stationary position.
7. Always refer to the manufacturer's operator's manual for specific instructions regarding use, maintenance or trouble shooting problems with the incubator.

8. Hot Air Oven

Fig. II.8 : A typical laboratory oven

Laboratory ovens are meant for high-forced volume thermal convection applications. These ovens generally provide uniform temperatures throughout. Typical sizes are from one cubic foot to 32 cubic feet (0.91 m^3) with temperatures that can be over 340°C. The outer body is made up of stainless steel. The oven has heating elements to maintain the temperature inside the oven. In many ovens temperature is controlled by digital indicator-cum-controller with safety alarm.

Principle

Laboratory oven is also one of the essential equipment of any laboratory. The oven generally consists of double walled chamber, the gap between two walls is insulated. It is heated from below by the use of heating elements which are arranged in a manner to heat the inner chamber uniformly. In microbiology lab it is used for sterilization of glassware and metal devices. The principle of oven is dry air sterilization. It kills the microorganisms by oxidizing their chemical constituents. It is less effective than moist heat sterilization.

Uses

Oven is used mainly for sterilization purpose in medical, agricultural and industrial researches. Besides, it is also used for drying, annealing, die-bond curing, polyimide baking, etc., in laboratories, hospitals and industries. It also destroys bacteria, viruses and fungi by application of dry heat. Thus ovens are used for sterilization of glasswares, soil samples, etc. Process applications for laboratory ovens can be for sterilization, drying and other industrial laboratory functions.

Precautions

1. The articles should be kept properly so that circulation of the air is not obstructed.
2. The glassware must be of good quality to tolerate the sterilization temperature.
3. The glassware should be free of moisture, otherwise it may break.
4. After the holding time is over, the glassware should not be taken out immediately to avoid the possibility of breakage.

9. Compound Microscope

A microscope is an instrument used to see objects too small for the naked eyes. This is a tool which is essential to study microorganisms. Microscopes can be separated into several different classes. One grouping is based on what interacts with the sample to generate the image, *i.e.*, light (optical microscopes), electrons (electron microscopes) or a probe (scanning probe microscopes). Alternatively, microscopes can be classed on whether they analyze the sample via a scanning point (confocal optical microscopes, scanning electron microscopes and scanning probe microscopes) or analyze the sample all at once (wide field optical microscope and transmission electron microscopes).

The most common type of microscope and the first invented is the optical microscope. It contains one or more lenses producing an enlarged image of a sample placed in the focal plane. Optical microscopes have refractive glass, occasionally of plastic or quartz, to focus light into the eye or another light detector. Mirror-based optical microscopes also operate in the same manner. Typical magnification of a light microscope, assuming visible range light, is up to 1500× with a theoretical resolution limit of around 0.2 μm or 200 nm. The magnification depends on the type of objective lense used with ocular lense. The greatest useful magnification of a microscope is that which makes the smallest visible object clearly resolvable.

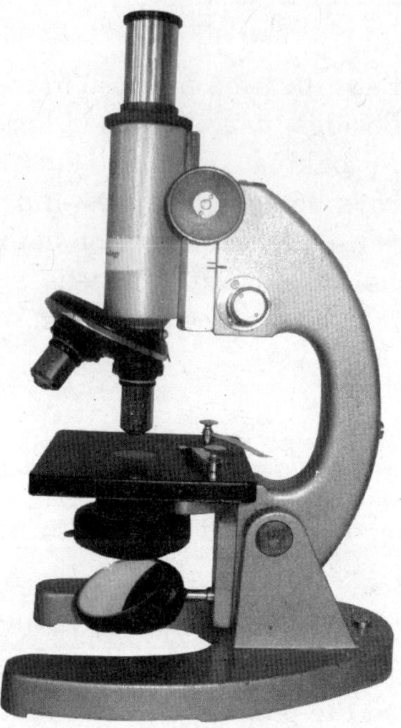

Fig. II.9 : Compound Microscope

In a simple compound microscope a light source illuminates the object to be seen. That source can be a mirror, or the instrument can itself be equipped with an illuminating device. The light passes through the object, the lens nearest to object, is the objective lens which produces an enlarged image of the object in the primary image angle. The lens that one looks into is the eyepiece which acts as a magnifier and produces an enlarged image of the image produced by the objective lens. To determine magnification, multiply the eyepiece magnification, which is usually 10×, by the magnification of the objective lens, usually 4×, 10×, 40× and 100×.

The resolving power of microscope is dependent on the wavelength of the beam used for illumination and on the optical quality of the lenses. Shorter wavelengths give better resolution. The resolving power is determined by following formula:

$$\text{Resolving power (RP)} = \frac{\lambda}{2 \times \text{NA}}$$

Where, λ = wavelength of light used (it is usually set at 550 nm);

NA = numerical aperture, which can be calculated as NA= $n \sin\theta$

n represents the refractive index of the medium through which light passes and is engraved on the barrel of objectives.

For low power objective, a common NA is 0.25, for high power objective, 0.65 and for oil immersion objective, 1.25.

$$\text{Thus, for low power objective RP} = \frac{550}{2 \times 0.25}$$

$$= 1100 \text{ nm}$$

$$= 1.1 \ \mu\text{m}$$

Similarly, it can be calculated for high power and oil immersion objectives.

Precautions

1. The microscope should always be kept covered when not in use. It should never be stored with the eyepiece removed or uncovered, since dust will collect in the body tube and it will become very difficult to clean.
2. The lens surface should never be touched with hands.
3. The lenses should be cleaned with the proper lens cleaner and lens paper before use.

10. ELISA Reader

ELISA stands for enzyme linked immunosorbent assay. In short, it is an *antibody test* or a test for *immune response* to virus, bacteria and allergens.

The test is done in an ELISA plate, also known as a *96-well plate* or *microplate*. The ELISA reader (Fig. II.10) reads the plate. An ELISA plate reader is also referred to the microplate reader, which essentially perform a number of functions including measuring fluorescence and luminescence where the chemical dyes fluoresce or emit one wavelength when exposed to light. The amount of reflection, absorption, and the colour are then used to identify and measure the amount of a substance. Initially, the ELISA plate readers were designed to measuring antibody tests, but lately, these instruments have been adapted to perform other advanced functions are aforementioned above. What makes an ELISA plate reader preferable at times is because it can measure more samples in a short time frame. Researchers use the ELISA plate readers for protein and enzyme assays.

Fig. II.10 : ELISA Reader

Principle

An ELISA reader measures and quantitates the colour differences in the wells. ELISA readers or micro plate readers perform spectrophotometry; they emit light at one wave length, and measure the amount of light absorbed and reflected by an object such as a protein. A spectrophotometer measures ultraviolet and visible light. Additionally, ELISA plate readers can also measure fluorescence and luminescence. Chemical dyes fluoresce or emit one colour or wavelength when exposed to light. The amount of reflection, absorption and the colour identify, and measure the amount of a substance.

Procedure

To use a spectrophotometer or ELISA plate reader, the molecule must be dissolved in solution. A spectrophotometer requires between 400 micro-liters and four milliliters, depending on the manufacturer and model. An ELISA plate reader needs about two to 100 micro-liters; ELISA plate readers use much less of a sample to get a result. ELISA plate readers measure more samples in a shorter period of time. A spectrophotometer measures one to six samples at a time. Typically, an ELISA plate measures 96 wells in an equivalent amount of time.

Throughput, flexibility, and cost are the top three considerations when choosing a plate reader. In pharmaceutical research, for example, throughput might be the top criterion. Regarding flexibility, a single-mode instrument, such as an absorbance reader for ELISAs, may fit the bill, but a multimode reader that offers additional capabilities, such as fluorescence may offset the additional cost outlay. Integrated software, type of plates accepted, and built-ins like incubators and plate shakers are additional considerations.

Precautions

1. Accurate measurement of samples and solutions should be added using a micropipette.
2. Readings should be taken at appropriate wavelength.
3. Wells in microtiter plates should be washed properly with distilled water.

■■■

Chapter - 3

III. EXPERIMENTS

A. MICROBIOLOGY

INTRODUCTION

Microbiology is the study of microscopic organisms which are not visible from the naked human eye. The examples are bacteria, viruses, archaea, fungi and protozoa. It not only includes the structure, classification and nomenclature of organisms but also the fundamental researches on the biochemistry, physiology, cell biology, genetics, ecology, molecular biology and clinical aspects of microorganisms. The microorganisms can range from unicellular to multicellular and prokaryotic to eukaryotic.

In 1676 Antonie van Leeuwenhoek observed the minute organisms first time under the simple self made microscope, hence considered as father of microbiology. However, Louis Pasteur was instrumental in starting industrial microbiology in 19th century and Robert Koch initiated the medical microbiology. Microbiologists traditionally relied on techniques, such as culture, staining, and microscopy for identification and study of microorganisms. However, less than 1 per cent of the microorganisms present in common environments can be cultured using current means.

Microorganisms are the world of possibilities and are playing an important role in finding the answer to key questions of present world. Applications of microbiology in medicine, pharmacy, agriculture and various industries are well known. Though many microorganisms are pathogenic, but others have played the key role in the development of life saving drugs like antibiotics and vaccines. The exploitation of microbes for the use in industrial processes includes industrial fermentation. The microorganisms are involved in industries, such as dairy, vinegar, food, alcohol, organic acids, tobacco, tea, leather tanning etc. Environmental microbiology is an important branch as microorganisms can be deployed as voracious scavengers removing pollutants, and are also used in waste-water treatment and bioremediation. Microbes also help in maintaining soil fertility and biogeochemical cycles in nature.

In view of applications of microbiology, it has become a very sought after course among students with openings in various fields. Therefore, it is worth to be well versed with practical knowledge of the field.

1.1 MICROMETERY AND MEASUREMENT OF MICROORGANISMS

Determination of size of microorganisms is important for their identification and the measurement of size of microorganism is known as micrometery. Micrometery is done with the help of ocular and stage micrometers. The ocular micrometer is a circular glass disc with 100 equally spaced divisions, marked 0–10. This is placed inside the eyepiece of microscope. The values of these divisions are not constant and vary with the magnification power of ocular and objective lens of microscope. Hence, the distance between graduations of an ocular micrometer is determined by calibrating it with a known scale, the stage micrometer.

Stage micrometer is a glass slide having generally a scale of 1 mm. This 1 mm distance is divided into 100 equally spaced divisions (10 large divisions and each large division again divided into 10 small divisions). Therefore, the value of the one division is 1/100 = 0.01 mm or 10 μm. The value of each division of stage micrometer increases with increasing magnification, *i.e.*, with high power (45x) and oil immersion (100x) objectives of microscope and 10x and 15x ocular lenses. In some cases the stage micrometer scale is of 1 cm divided into a total of 100 smaller divisions, there the value of 1 division is equivalent to 1/100 = 0.01 cm or 0.1 mm or 100 μm.

1.1.1. Calibration of Microscope by using Ocular and Stage Micrometer

Requirements

Compound microscope, ocular micrometer, stage micrometer, oil.

Procedure

For low power objective (10 × 10 magnification)

1. Remove the eyepiece, unscrew it and place the ocular micrometer in it.
2. Replace the eyepiece in the microscope.
3. Place the stage micrometer on the stage of the microscope.
4. Adjust the light so that two different scales of ocular and stage micrometer are in sharp focus.
5. Turn the ocular lens until the lines of ocular scale are parallel to those of the scale of stage.

6. Focus the superimposed scales and coincide a line at left side and find another line set which coincides by focusing towards right side of the first coinciding line.
7. Count the divisions coinciding between two links in both micrometers.
8. Take atleast 6 readings randomly by following the steps 5–7.
9. Repeat the procedure with high power (45x) and oil immersion (100×) objectives.

Observation and Results

Record your observations in the form of a Table and take the mean value of all 6 readings taken each for low, high and oil immersion objectives.

Table: Calibration of ocular for 10 × 10 magnification.

S. No.	No. of division of ocular micrometer	No. of division of stage micrometer	Value of one division of ocular micrometer (μm)
1	15	25	16.6
2			
3			
4			
5			
6			

The least count of the micrometer or the values of calibration factor can be calculated in the following way:

Scale of stage micrometer = 1 mm.

Since 100 divisions of stage micrometer = 1 mm.

1 division of stage micrometer = 1/100 = 0.01mm.

or

$\therefore$ 1 mm = 1000 μm

0.01 × 1000 = 10 μ

Hence, 1 division of stage micrometer = 10 μ.

(a) Low power objective (10×) and 10× eyepiece (10 × 10 magnification).

Suppose, 15 divisions of ocular micrometer = 25 divisions of stage micrometer.

$\therefore$ 1 division of ocular micrometer = 25/15 = 1.66 divisions of stage micrometer.

Since 1 division of stage micrometer = 10 μ.

1.66 division of stage micrometer =1.66 × 10 μ = 16.6 μ.

(b) High power objective (45×) and 10× eye piece (10 × 45 magnification).

∴ Suppose, 16 divisions of ocular micrometer = 6 divisions of stage micrometer.

1 division of ocular micrometer = 6/16 × 10 = 3.7 μ.

(c) Oil immersion objective (100×) and 10x eye piece (10 × 100 magnification).

Suppose, 14 division of ocular micrometer = 2 division of stage micrometer.

∴ 1 division of ocular micrometer = 2/14 × 10 = 1.4 μ.

It can also be directly calculated by the following formula:

$$\text{1 division of ocular micrometer } (\mu\text{m}) = \frac{\text{No. of divisions of stage micrometer}}{\text{No. of divisions of ocular micrometer}} \times 10$$

1.1.2. Measurement of Size of Micro-organism/Spore

Requirements

Slide of fungal spores/bacteria, microscope, ocular micrometer, stage micrometer.

Procedure

1. Calibrate the ocular micrometer as given in previous experiment.
2. Replace the stage micrometer by prepared slide of smut spores (any other spore or bacterial cells can also be used for preparing the slide).
3. Adjust the objective (low, high or oil immersion so that the spores/cells are clearly seen).
4. Count the number of divisions of ocular which cover the whole spore/bacterium. If the spore is not round, take length and width separately.
5. Measure the size of 10 spores by rotating the ocular lens, write down your readings in an observation Table and take the average.

Results

Determine the size of spore as follows:

Size of spore = No. of ocular micrometer divisions occupied x calibration factor, *e.g.*, if on an average the small spores occupy 3 divisions of ocular micrometer in high power and the calibration factor of ocular micrometer is 3.7 μm, the size of spore will be $3 \times 3.7 = 11.1$ μm.

If the spore is not round, *e.g.*, *Alternaria* then the length and width both should be calculated as follows:

Length: 10 divisions $\times$ 3.7 = 37 μm;

Width: 2 divisions $\times$ 3.7 = 7.4 μm;

and it is written as 37×7.4 μm.

1.2 PREPARATION OF CULTURE MEDIA

Like all other living organisms, microorganisms also need food in the form of different nutrients to grow. In nature, microbes grow as a mixed population by drawing the nutrients from their respective environments depending upon where they are growing. However, in lab all the nutrients required for the growth must be provided to them. Hence, the food base or the mixture of nutrients that support the growth of an organism is called culture medium and the growth of an organism on a medium (pl. media) is called culture.

Although all the microorganisms have the same basic requirements, such as water, carbon, oxygen, nitrogen, hydrogen, phosphorus, sulphur, etc., but due to their vast nutritional diversity as to the use of organic and inorganic compounds, the culture media may vary in form and composition depending upon the species to be cultivated. Some microorganisms also require specific growth factors, *viz.*, amino acids, vitamins, etc., for their metabolism. Therefore, the culture medium is prepared in such a way so that it has all the nutrients required by the microorganism to be grown on it. The culture media are prepared in the laboratory by weighing and dispensing the individual ingredients in water. However, ready-made media can also be procured from the market.

According to the consistency of the media, they can be liquid, more commonly known as broth or solid in which a solidifying agent, *e.g.*, agar-agar is mixed. Semi solid media, though less commonly used, can also be prepared which have custard like consistency. Based on their composition, the media can be of 3 types: (i) natural medium contains all natural products, such as urine, milk, blood, juices, animal tissues, etc., (ii) semi-synthetic medium has some known constituents, whereas, others unknown, *e.g.*, Potato Dextrose Agar, Beef Peptone Agar, Oatmeal Agar in which the exact chemical constituents present in Potato, Beef Extract and Oatmeal, respectively are not known, (iii) synthetic medium is the one in which all the chemicals used are of known composition, *e.g.*, Czapek's Dox Agar.

The synthetic media may be of following types:

(i) General Purpose Media: These can be used for routine microbiological work for large variety of bacteria and fungi.

(ii) Selective Media: They can be used only for selected microbes as they contain compounds which promote the growth of a few selected species only.

(iii) Differential Media: These media contain dyes and colour indica-tors to give a particular biochemical response (*e.g.*, MacKonky agar, EMB agar) so that it can be used for differential isolation of microorganisms.

(iv) Assay Medium: It is used for the assay of vitamins, amino acids and antibiotics.

The compositions of some of the important media are given in Appendix IV A.

1.2.1 Preparation of Liquid Culture Medium (Broth)

Generally bacteria are cultivated in broth, *i.e.*, medium lacking solidifying agent. The most common constituents of basic medium are beef extract (a beef derivative and a source of mineral salts, nitrogen and vitamins) and peptone (a semi-digested proteins).

Luria Bertani Broth

(a) Requirements

Tryptone 10 g, Yeast extract 5 g, NaCl 10 g, distilled water 1l, pH meter, heater/gas, autoclave, measuring cylinder, beaker, glass rod, 250/500 ml flasks or bottles, cotton.

(b) Procedure

1. Weigh all the constituents and dissolve them in distilled water separately.
2. Heat to dissolve the constituents properly, if necessary.
3. Make the volume upto 1 litre by adding more distilled water.
4. Adjust the pH to 7.5 ± 0.2 by adding acid (1N HCl) or alkali (1N NaOH) solution drop by drop.
5. Pour the medium in flasks/bottles up to 2/3rd capacity.

6. Prepare the cotton plugs and close the mouth of flasks/bottles. Apply aluminum foil around the neck of flasks/bottles. Brown paper can also be wrapped around using rubber band or thread.
7. Autoclave at 121°C and 15 lbs pressure for 20 minutes. In absence of autoclave, pressure cooker can also be used.
8. After the temperature subsides, remove the flasks/bottles and store for future use at 4°C.

(c) Precautions

1. Cotton plugs should not be too tight or too loose.
2. Do not pour medium over 2/3rd capacity of flasks/bottles.
3. Make sure that all the components are dissolved properly (magnetic stirrer and heating can be used for dissolving various chemicals).
4. Autoclaving should be done properly as recommended.

1.2.2 Preparation of Solid Medium

When a solidifying agent, *e.g.*, agar-agar is added to any liquid medium, it is called agar medium. Agar-agar is a complex carbohydrate produced by red algae *Gelidium* and *Gracilaria.* It makes the medium liquefiable solid which means it liquefies when heated and hardens on cooling at room temperature.

Potato Dextrose Agar Medium (PDA)

(a) Requirements

Potato tubers 200 g, Dextrose 20 g, Agar 20 g, Distilled water 1000 ml, knife, muslin cloth, heater/gas, beaker, flasks/bottles, pH meter, glass rod, cotton, measuring cylinder, autoclave.

(b) Procedure

1. Weigh potato tubers, peel the skin and make the slices.
2. Put them into beaker containing 500 ml water and boil on heater until soft.
3. Filter through muslin cloth, squeeze and collect the filtrate (potato extract).
4. Add dextrose in it and dissolve.
5. Take another 500 ml water in a beaker and melt agar-agar.
6. Finally, mix potato extract into melted agar, measure it and make up to 1000 ml.

7. Adjust the pH between 5.6 to 6.4 by adding 1 N HCl or 1 N NaOH solution drop wise.
8. Pour the medium in flasks/bottles upto 2/3rd capacity.
9. Close with cotton plug and wrap with aluminum foil/paper.
10. Autoclave at 121°C and 15 lbs pressure for 20 minutes.
11. Allow it to cool, remove the flasks/bottles and store at 4°C.
12. It can be re-melted whenever needed in future.

1.2.3 Preparation of Agar Plates and Agar Slants

Agar plates are prepared for isolation and culturing of micro-organisms in lab and agar slants are required for storing the pure culture for further use. The slanting surface of culture tubes gives more surface area for culture to grow.

(a) Requirements

Petri plates, culture tubes, funnel, solid medium, heater, laminar air flow.

(b) Procedure

Agar Plates

1. Sterilize the Petri pates at 160°C for 2 hrs in hot air oven and let them cool.
2. Melt the already prepared and sterilized medium and bring it to around 40°C.
3. Transfer the sterilized Petri plates and flasks containing medium into laminar air flow. It should be pre-sterilized by UV radiation for 30 min and surface sterilized by alcohol.
4. Pour the medium (about 15–20 ml) aseptically into the Petri plates by sliding the lid slightly. If there is more problem of contamination the pouring should be done near Bunsen burner flame (In the labs where there is no problem of contamination, the Petri plates can be poured and kept in hood to cool. Once the medium is solidified, the lid is placed. This avoids the condensation of water in Petri plates).
5. Place the lid and wait for the medium to solidify.
6. Use the agar plates immediately or store them at 4°C for future use.

Agar Slants

1. Melt the already prepared and sterilized medium.
2. Pour 8–10 ml of medium in culture tubes with the help of funnel.
3. Make the cotton plug and close the tube. However, screw capped tubes can also be used.
4. Wrap the tubes with aluminum foil or brown paper if cotton plugs were used.
5. Put them in a temperature proof basket and autoclave at 121°C and 15 lbs pressure for 20 minutes.
6. Allow the tubes to harden in a slanted position. For preparing agar deep tubes, the agar medium is allowed to harden in upright position (Fig. III.1.1).
7. Use the slants for transferring the pure cultures immediately or keep the slants for future use in refrigerator.

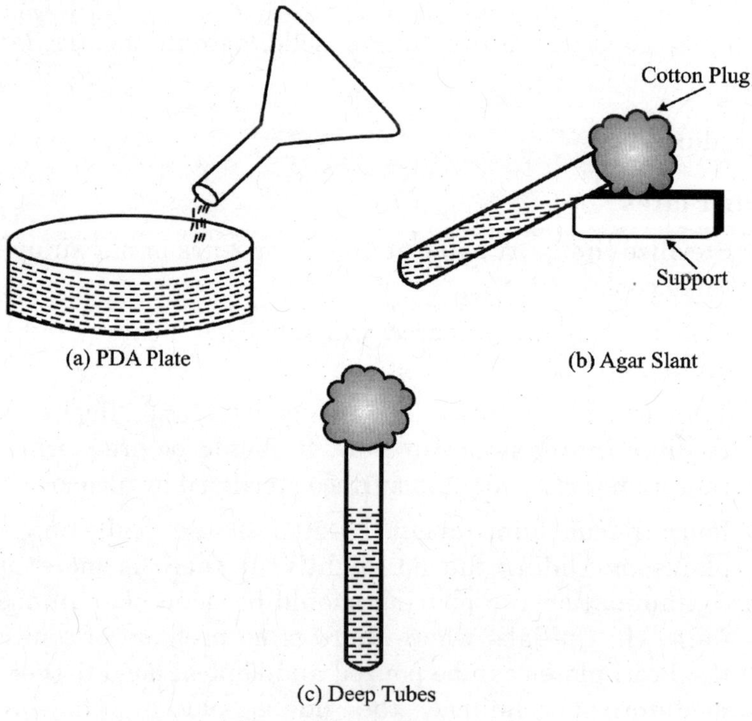

Fig. III. 1.1

(c) Precautions

1. Do not pour very hot medium in Petri plate, otherwise water will be accumulated on the lid after condensation.
2. Pour medium quickly to avoid contamination.

3. Prepare the agar Petri plates in a hood with provision for sterilization.
4. Pouring should be done near flame and close the lid as soon as possible.
5. Pour the medium only to the bottom half of the Petri plates. Medium should not touch the lid of Petri plates.
6. For slants, fill the culture tube only 2/3rd so that at the time of boiling the medium should not touch cotton plug.
7. Prepared Petri plates or slants should be stored at 4°C in refrigerator for future use.

1.2.4 Demonstration of Selective and Differential Media

1.2.4.1 Eosine Methylene Blue (EMB) medium

EMB medium is used to differentiate between enteric lactose fermenters and non-fermenters as well as colon bacilli by lactose and dyes (eosine and methylene blue) present in it. *E. coli* colonies look blue black with metallic green sheen caused by the large quantity of acid produced which precipitates the dyes onto the surface of culture. *Enterobacter aero-genes* produces thick, mucoid pink colonies on this medium. Enteric bact-eria that do not ferment lactose produce colourless colonies which because of their transparency, appear to take on the purple colour of the medium. This medium is also partially inhibitory to the growth of gram positive organisms, thus gram negative growth is more abundant.

(a) Requirements

EMB medium (composition given in Appendix IVA), 24h old *E. coli*, *Enterobacter aerogenes* and *Staphylococcus aureus* cultures, Petri dishes, inoculating loop, Bunsen burner.

(b) Procedure

1. Prepare the EMB plates as described earlier.
2. Inoculate *E. coli, E. aerogenes* and *S. aureus* on plates
3. Incubate the plates at 37°C for 2 days.

(c) Results

Gram positive *Staphylococcus* does not grow or grows only poorly on the plates, whereas, the other two gram negative bacteria grow well, *E. coli* with green metallic sheen and *E. aerogenes* with pink sheen.

1.2.4.2 Mannitol Salt Agar

It has high salt concentration, hence used for isolation of staphylo-cocci, most of the other bacteria being inhibited on it. Mannitol is used for the detection of fermenting ability which can be identified by yellow halo around staphylococci.

(a) Requirements

Mannitol Salt Agar (composition given in Appendix IVA). 24h old *Staphylococcus aureus* and *E. coli* cultures, Petri dishes, inoculating loop, Bunsen burner.

(b) Procedure

1. Prepare the Mannitol Salt Agar plates as described earlier.
2. Inoculate *Staphylococcus* and *E. coli* in the separate Petri dishes by making a single streak.
3. Incubate the plates in inverted position at 37°C for 48 h.

(c) Results

S. aureus colonies show yellow zones around them, indicating fermentation of mannitol, whereas, *E. coli* will not grow on plates.

1.2.4.3 Pikovskaya Agar

Many fungi and bacteria have the ability to solubilize inorganic phosphates. Pikovskaya medium contains the insoluble inorganic phosphate source (TCP) as its ingredient. The microbes with solubilization capability grow on it and solubilize the otherwise insoluble phosphate source, thereby forming a halo around the culture growth in the medium. This characteristic can be used to distinguish between the phosphate solubilizers from non-solubilizers.

(a) Requirements

Pikovskaya Agar, rhizosphere soil sample, Petri dishes, inoculating loop, Bunsen burner.

(b) Procedure

1. Prepare the Pikovskaya Agar plates as described earlier.
2. Collect the rhizosphere soil sample from nearby agriculture field in polythene bags and bring to the laboratory.
3. Sprinkle the soil with sterilized spatula on agar surface.
4. Incubate the plates in inverted position at 37°C for 48 h for bacteria and at 28°C for fungi.

(c) Results

The colonies showing halo around them are phosphate solubilizing bacteria/fungi.

1.3 ASEPTIC METHODS

Microorganisms are omnipresent and may be the source of contamination while cultivating a particular type of microorganism in the laboratory. Contamination of unwanted organism(s) may spoil the whole experiment. Therefore, utmost care is required while handling the microbes. The principle of aseptic technique is to remove or kill all microorganisms from the material/object. The concept of sterilization was given by Louis Pasture. Sterilization can be defined as making the object devoid of any living organism. This is an absolute term, even if 1 or 2 microbes are left alive, the object cannot be called as sterilized. However, disinfection means decreasing the load of microbes or reduction in number of viable microbes, it should not be confused with sterilization.

There are several methods of sterilization. A suitable method should be selected depending upon the material to be sterilized.

1.3.1 Physical Methods of Sterilization

Physical method is the most common method which includes sterilization through heating, radiation and filtration.

A. Heating: It is most reliable method of sterilization. At high temperature protein degradation and oxidative damage take place, thus killing the microorganisms.

a. Incineration: This method is also known as flame sterilization. This is used to sterilize inoculating needles, forceps etc. The metal part of the needle is put into the flame of Bunsen burner or spirit lamp until red hot. The needle is brought back to the normal temperature by dipping it in beaker containing alcohol or by jabbing it into agar or by simply keeping it in laminar hood for few minutes.

b. Dry heat or hot air sterilization: Hot air sterilization is done in an electric oven operating at 160–180°C for 2 hours. This is used for sterilizing glassware, syringes, forceps, scalpel, pipettes etc. High temperature kills all living organisms and viable spores due to destructive oxidation of cell contents. The glassware should be removed only after the temperature subsides.

c. Moist heat: This is more effective as heat conduction is rapid and death rate is higher as compared to dry heat. It can penetrate porous materials also. The lethal effect is due to denaturation and coagulation of proteins. This method is useful for sterilizing culture media, water, aqueous solutions, cloth, rubber etc.

(i) *Streaming steam*: Arnold steam sterilizer is used which allows the live steam to come in contact with the material or object. The temperature is maintained at 100°C for 90 min.

(ii) *Tyndallization or Intermittent sterilization*: This technique is useful for those substances which are affected by heating at high temperature. It involves exposure to steam at 100°C for 30 min for 3 successive days. First exposure kills all vegetative cells, whereas, the time gap of 24 h permits the dormant structures and resistant spores to germinate which can be killed on subsequent heatings.

(iii) *Steam under pressure or autoclaving*: This is most common mode of sterilization for culture media and aqueous solutions. Increased pressure in autoclave raises the boiling point of water. Autoclave has been described in chapter II. At 121°C temperature for 15 minutes which is obtained by increasing pressure up to 15 psi, all bacteria and spores are killed.

B. Filtration: This method is used for thermolabile media, enzymes, antibiotics, tissue extracts etc. Bacteriological filters have less than 0.45 μm pore size, small enough to prevent bacteria to pass through it. Chamberland Pasteur filter, Berkfield filter, membrane filter are some examples, however, membrane filters are most commonly used.

C. Radiation: Non-ionizing as well as ionizing radiations can also be used for sterilization. They have bactericidal effect and are lethal to DNA of microbes, *e.g.,* UV, IR, X-rays, gamma rays. Normally UV radiation is used for sterilization of laminar flow or chambers.

1.3.2 Chemical Methods of Sterilization

Many chemicals are available which can be used for freeing of an article from some or all of its burden of live microorganisms. The chemicals used for this purpose should have wide spectrum of activity, high penetration power and speedy action. They should also be cheap and easily available. Alcohol is commonly used for sterilizing working table top and hands. Other chemicals are iodine, chlorine, heavy metals, aldehyde, salts of Cu, Hg, etc., which can be used for various materials.

1.3.3 Gaseous Sterilization

Ethylene oxide is used only in special cases when sterilization is not possible by heat and other methods. The vapours are used under pressure in a special equipment. This gas is explosive and toxic to man hence great care is needed to handle it. Formalin poured over potassium permanganate creates the formaldehyde gas which has bactericidal, sporicidal and virucidal effects. It can be used to disinfect the room or laboratory.

1.4 CULTIVATION TECHNIQUES

Microorganisms are found in mixed population in nature. Hence, it becomes essential to obtain these microbes in pure form to be able to study them. A culture containing a single species is called pure culture. There are several methods of isolation of microorganisms from mixed population.

1.4.1 To isolate the Microorganisms by Streak Plate Method

(a) Principle

The objective is to produce well isolated colonies of bacteria from concentrated cell suspension. The colonies are separated by spreading the concentrated cell suspension on the surface of agar plate by using the inoculating needle, thus forming a gradient. After the proper incubation period crowded colonies are obtained at the starting point of streak (Fig. III.1.2), whereas, well separated colonies are developed at the end. These well separated colonies can be picked up and transferred to another agar plate or directly to the agar slant.

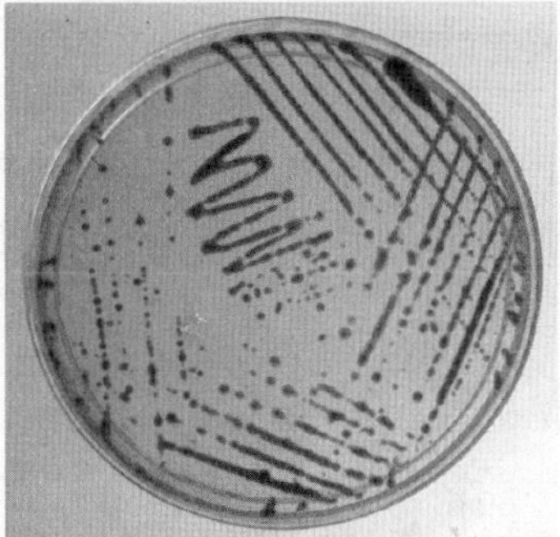

Fig. III. 1.2 : A plate showing growth of bacterial cells after streaking.

(b) Requirements

Inoculating loop, sterilized Petri plates, LB agar medium (or any other solid medium), mixed culture of bacteria, Laminar air flow.

(c) Procedure

1. Melt the agar medium, pour it into Petri plates in laminar air flow and evenly distribute the medium by gently rotating it.
2. Let the medium properly solidify in Petri plate.
3. Flame sterilize the inoculating loop and cool it.

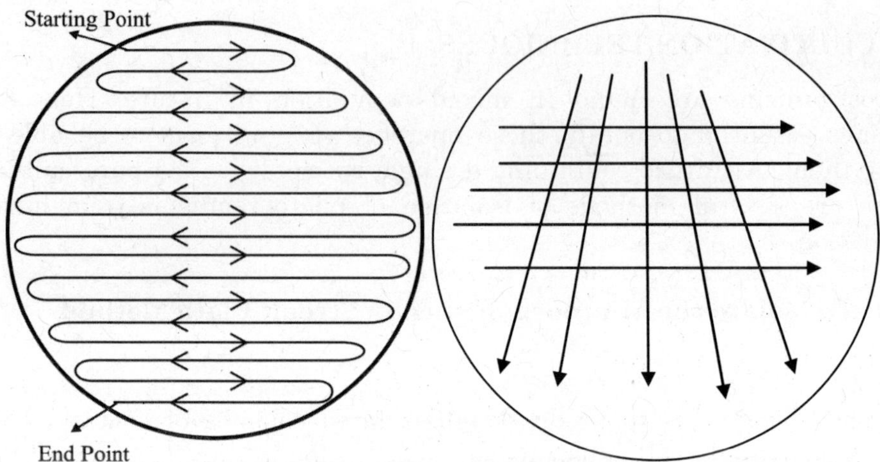

Continuous streaking Radiant Streaking

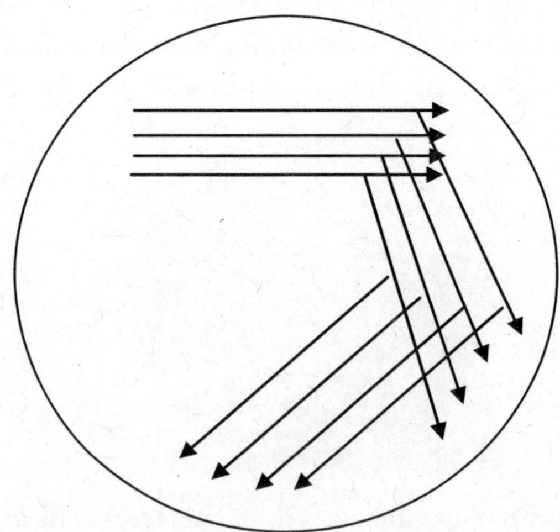

Quadrant streaking

Fig. III. 1.3 : Some popular methods of streaking

4. Dip the sterilized loop in mixed culture suspension and streak the plates following any of the methods shown in Fig-III.1.2 and

1.3 by opening the lid as minimum as possible. However, continuous streaking is the most common method used for the purpose.

5. Incubate the plate at 37°C in inverted position in an incubator.
6. Observe the plates after 24–48 h and pick up well separated colonies and transfer them to another plate or agar slants.

(c) Result

Well isolated colonies are observed at the last part of streak.

(d) Precautions

1. Strictly maintain the aseptic conditions.
2. Open the Petri plate for minimum time period to avoid contamination.
3. Do the streaking only after the agar medium has been properly solidified otherwise instead of streaking at the surface the needle may be inserted in the medium.
4. Do not touch any non-sterile object while streaking.

Note: *For isolation of fungi PDA medium can be taken, the optimum temperature for incubation is 28 ± 2°C and incubation time is 4–5 days.*

1.4.2 To Isolate the Microorganisms by Pour Plate Technique

(a) Principle

The principle of this technique is based on the fact that if the original mixed culture is too concentrated, the colonies formed by it on the plates may be overlapping. Therefore, to make them separate and countable, the original culture of microorganisms may be diluted by using liquefied agar medium. It permits a thorough distribution of cells or spores in the medium and the colonies formed after pouring this medium into Petri plates are well separated.

(b) Requirements

Mixed culture of bacteria/fungi, agar medium (LB agar for bacteria, PDA for fungi), Petri plates, inoculating loop, glass marker, water bath or hot water beaker.

(c) Procedure

1. Take 3 (or 4) sterilized tubes containing liquefied LB agar medium and mark the tubes as 1, 2 and 3.

2. Make the suspension of mixed bacterial culture or sample provided to you.
3. Transfer a loopful of suspension to tube no. 1 by sterilized inoculating loop. Shake the tube for uniform distribution of inocula. To avoid solidification of medium keep the tube in hot water bath at 40–45°C.
4. Similarly, transfer the loopful of inoculum from tube no. 1 to tube no. 2 and then from tube 2 to tube 3, each time shake the tubes vigorously (Fig. III.1.4).
5. Pour contents of each tube into sterilized Petri plates.
6. After solidifying the medium incubate the plates at 37°C in inverted position for 2 days. In case of fungi the incubation temperature would be 28°C for 4–5 days.
7. Observe the plates after 2 days for bacterial colouries.

(d) Results

First Petri plate contains numerous colonies, the number of colonies decreases with subsequent Petri plates. The last Petri plate shows well separated colonies.

(e) Precautions

1. Keep the tubes in hot water bath otherwise the medium will solidify in tubes.
2. Shake the tubes for even distribution of samples.

Note: *This is an easy method and less skilled persons can also perform this, but it has few disadvantages also. As the medium is kept at 40-45°C, hence it is not a suitable method for isolation of Psychrophilic organisms. Vigorous shaking of tubes may affect the sensitive micro-organisms.*

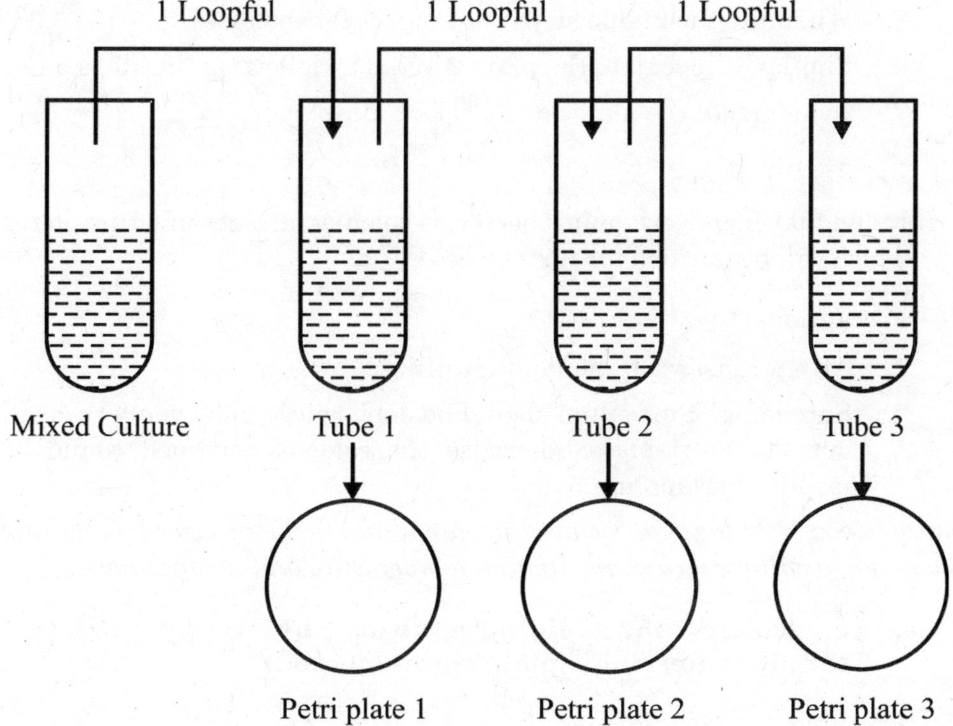

Fig. III.1.4 : Pour plate technique

1.4.3 To Isolate the Microorganisms by Spread Plate Technique

(a) Principle

In this technique, the microorganisms are spread over agar surface so that propagules are well distributed and grow into separate colonies.

(b) Requirements

LB agar, Petri plates, L shaped glass rod, Bunsen burner, mixed culture of bacteria.

(c) Procedure

1. Prepare three LB agar plates and label them.
2. Inoculate the 1st Petri plate by taking the suspension from mixed culture.
3. Sterilize the L-shaped glass rod by dipping it in alcohol and then flaming.
4. Gently spread the inoculum on the agar surface of Petri plate by using L-shaped glass rod. Use the turn table for this.

5. When the turn table stops running, cover the plate.
6. Similarly, inoculate the plate 2 and 3 by following the steps 3–5.
7. Incubate all the plates at 37°C for 2 days.

(d) Results

Individual non-overlapping bacterial colonies are obtained in plates which can be purified for further use.

(e) Precautions

1. All the glassware should be sterilized.
2. Spreading of inoculum should be done gently and evenly through out the Petri plate otherwise the colonies obtained would be mostly overlapping.

Note: *Same technique can be used for fungi and actinomycetes by selecting the suitable culture media, incubation temperature and incubation time.*

1.4.4 To Isolate the Microorganisms by Serial Dilution Technique (or viable plate count method)

(a) Principle

This is the most widely used method for isolation of bacteria, fungi and actinomycetes from soil, milk, food stuffs etc. This is also known as viable plate count method because when suspension containing propagules is cultured, each viable microorganism in it will develop into a colony. Therefore, the number of colonies hence obtained are thought to be equal to the number of viable organisms present in the sample. This makes the basis of calculating the number of microorganisms present in the original sample. The sample may be too concentrated to give rise to discrete and separate colonies, hence the series of dilutions are made to facilitate the counting of exact number. In this method, the known amount of sample is suspended into known volume of sterilized water or buffer solution in a series, so that dilution can be made and all the microorganisms/propagules may develop into individual colony. The sample size taken, *i.e.* generally 10 ml/ 10 g or 1 ml/ 1g is suspended into 90 ml or 9 ml of sterile water to keep the total volume 100 or 10 ml. For making the series of dilution 10 ml or 1 ml is transferred to another flask/tube containing 90 ml or 9 ml water. This process is followed until required dilution is made as 10^{-2}, 10^{-3}, 10^{-4}, 10^{-5}, 10^{-6} and so on. Finally, 1 ml aliquot of these dilutions are spread over agar Petri plates and incubated (Fig. III.1.5).

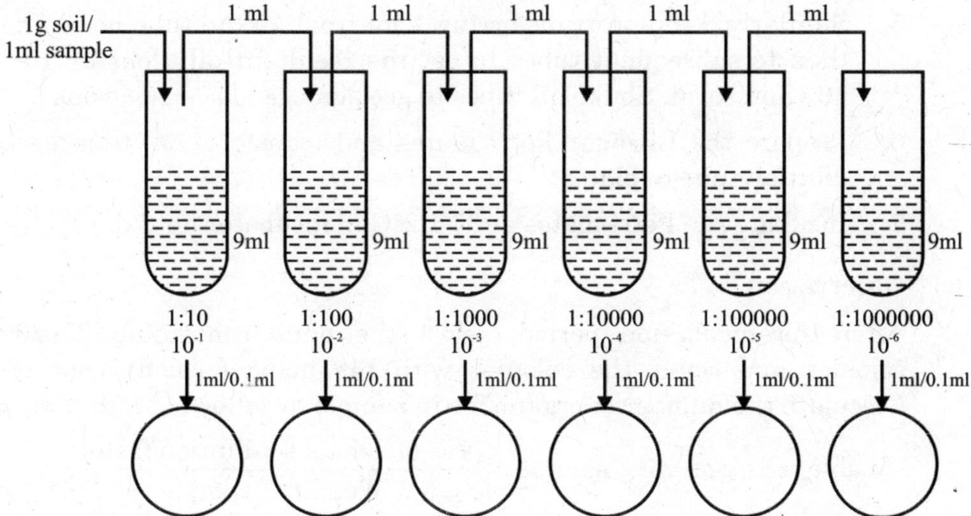

Fig. III. 1.5 : Isolation of microorganisms from sample using dilution plate technique.

The number of colonies appearing in plates are calculated by following formula:

$$\text{No. of cells / ml or g of sample} = \frac{\text{No. of colonies} \times \text{dilution factor}}{\text{Weight of sample}}$$

Note: *The dilution series can also be made by mixing 10 g/10 ml sample into 90 ml water in flasks, and then transferring 10 ml from 10^{-1} dilution flask to subsequent flasks.*

(a) Requirements

LB agar medium for isolation of bacteria, pipettes, Erlenmeyer flasks (250 ml capacity) or tubes, incubator.

(b) Procedure

1. Collect the sample for microbial analysis.
2. Name the tubes containing 9 ml sterile water as 1, 2, 3.... and so on with marker.
3. With a sterile pipette remove 1 ml or weigh 1 g sample (soil, sewage, food material etc.) and transfer to tube no. 1. Vortex the tube. This tube contains 10 times diluted sample, *i.e.,* 1:10 = 1/10 = 10^{-1} dilution.
4. Remove 1 ml from the tube no. 1 (*i.e.,* 10^{-1} dilution) to tube no. 2 containing 9 ml water. Vortex it. This tube will contain 1:100 dilution ($1/10 \times 1/10 = 1/100 = 10^{-2}$).

5. Similarly, keep on transferring 1 ml from 10^{-2} to tube no. 3 and then to subsequent tubes to get the resultant dilutions as 10^{-3}, 10^{-4} and so on. Shake all tubes to get homogenous suspension.
6. Prepare the LB agar Petri plates and transfer 1 ml from each dilution tube to plates.
7. Incubate the Petri plates at $37 \pm 2°C$ in incubator for 2 days.

(c) Observation

After the incubation period, select the plate containing 20–200 colonies and count the colonies with the help of colony counter. Calculate the number of organisms in sample by following formula:

$$\text{Viable cells/g or ml of sample} = \frac{\text{No. of colonies} \times \text{dilution factor}}{\text{Weight of sample}}$$

Sample calculation: Suppose in 10^{-6} dilution, 100 colonies were present in Petri plate and sample was 10 g, then

$$\text{cell/g of sample} = 100 \times \frac{10^6}{10}$$

$$= 10 \times 10^6 = 10^7$$

$$\text{Here dilution factor} = 1/10^{-6} = 10^6$$

Hence, the cells present in the original sample would be 10^7 cells/g

In research experiements, the Petri plates are taken in triplicate for each dilution. In that case the number of colonies is the average of colonies in 3 plates.

Note: *This technique can be efficiently used for bacteria, fungi and actinomycetes. However, the suitable media should be selected accordingly. For enumeration of bacteria in soil generally 10^{-4} to 10^{-7} dilution is suitable, for fungi 10^{-2} to 10^{-4} and for actinomycetes it is 10^{-3} to 10^{-6} according to their proportion in sample. For fungi, the media should be supplemented with streptomycin (30 mg/l) and for actinomycetes with aureomycin (30 mg/l) to avoid bacteria.*

The density of bacteria in soil is around 10^8-10^9 colony forming units (cfu). Besides, 10^7 to 10^9 actinomycetes and 10^6 fungal and protozoan propagules are present in per gram of dry soil.

(d) Precautions

1. If the sample is soil, composite sample should be taken collecting the soil randomly from various parts of field.

2. Each dilution tube must be vortexed before transferring the aliquot to next tube.
3. All the glassware should be properly sterilized.
4. Antibiotics should be added in melted medium cooled to 40–45°C just before pouring into Petri plates.

1.5 PRESERVATION AND MAINTENANCE OF CULTURE

Microorganisms have to be maintained and preserved for future use. The maintenance of viability of cultures is very important and requires great care. The cultures must be transferred at regular intervals to prevent drying and mutation. This process is known as subculturing. The cultures are maintained at low temperature to allow minimum metabolism thus increasing their life. In laboratories, stock cultures are prepared and preserved and for day to day routine work, working cultures are prepared to decrease the contamination.

1.5.1 Maintenance by Sub-culturing

This is most common method of maintaining the cultures in laboratories for short period yet a labour intensive method.

(a) Requirements

Nutrient media (LB agar for bacterial cultures and PDA for fungal cultures), culture tubes, inoculating needle, Bunsen burner.

(b) Procedure

1. Make the agar slants. For making the slants culture tubes with cotton plugs or tubes with screw caps can be used.
2. Inoculate the culture on the surface of slant.
3. Incubate it at appropriate temperature, *e.g.*, 37± 2°C for bacteria and 28±2°C for fungi.
4. Once the cultures are grown on slants, store at 4°C in refrigerator.
5. Do the subculturing every 3–6 months depending upon the growth of culture.

(c) Precautions

1. Minimum incubation should be allowed to prevent accumulation of metabolic products in medium.
2. Retransfer the culture before it looses its viability.

1.5.2 To Prepare Agar Slants with Mineral Oil (Paraffin Method)

This is simple and economical method and can be used for all common microorganisms. This method increases the longevity of bacterial and fungal cultures upto several years due to slow metabolic activity and prevention of dehydration.

(a) Requirements

Nutrient media (LB agar for bacterial cultures and PDA for fungal cultures), culture tubes, mineral oil/paraffin, inoculating needle, Bunsen burner.

(b) Procedure

1. Sterilize mineral oil/liquid paraffin/medical paraffin of 0.03 – 0.09 g specific gravity.
2. Prepare the agar slants and culture the microorganism.
3. Once the organism has grown, cover the agar slant with a layer of oil or paraffin, atleast about ½ inch above the tip of the slant.
4. Store the slant upright at room temperature or in fridge for long term maintenance.
5. For reviving the culture, draw a loopful of culture under oil, drain off excess oil and streak on agar plate.
6. Incubate the plate at 37 ± 2°C for 2–3 days for bacterial cultures and at 28 ± 2°C for 4–5 days for fungal cultures.

(c) Precautions

1. Make the tube in duplicate.
2. Keep the tubes upright.

1.5.3 Storage in Soil

The viability of pure culture of soil microbes can be maintained in soil medium.

(a) Requirements

Nutrient media (LB agar for bacterial cultures and PDA for fungal cultures), soil, dextrose, culture tubes, inoculating needle, Bunsen burner.

(b) Procedure

1. Collect fertile soil, pass through 2 mm sieve and fill in tubes.
2. Saturate the soil with 1% dextrose solution (This step is optional).

3. Autocalve the tubes for 2 successive days.
4. Suspend the fungal spores/bacterial cells of fully grown culture in 1 ml sterile water.
5. Pour this suspension into soil tube.
6. Allow the culture to grow in tubes at room temperature for 10 days.
7. Store in refrigerator.
8. Revive the culture by sprinkling the soil on suitable agar plate and incubating it at appropriate temperature according to the need of microorganism.

(c) Precautions

1. Prepare atleast 2 tubes of each culture.
2. Properly sterilize the soil before using.

1.5.4 To Store the Cultures in Liquid Nitrogen

This method is used for long term preservation of microorganisms. The principle of this method is to freeze the cells in liquid nitrogen at temperature of –196°C to bring the cells to zero metabolism. Some stabilizing agents (cryoprotectants) like glycerol or dimethyl sulfoxide (DMSO), that prevent the formation of ice crystals which may cause injury to cells, are also added.

(a) Requirements

Nutrient media (LB agar for bacterial cultures and PDA for fungal cultures), glycerol/DMSO, liquid nitrogen, Petri plates, cryotube, glass marker, Bunsen burner.

(b) Procedure

1. Prepare the plates and grow the cultures.
2. Add stabilizing agent (4 ml of 10% glycerol or 5% DMSO) on growing culture plate.
3. Scrape the culture and mix to make cell suspension.
4. Add 0.5 ml of cell suspension in cryotubes.
5. Label the tube and transfer it to liquid nitrogen.
6. Store the tube in liquid nitrogen.
7. To retrieve the culture, thaw the tube at room temperature for 2–3 hours and then transfer the culture on suitable medium.
8. Grow the bacterial culture at 37°C for 24–48 h or fungal culture at 28°C for a week.

(c) Precautions

1. Prepare the tubes in duplicate.
2. A suitable cryoprotectant should be added.

1.5.5 To Store the Culture at -70°C

This method is recommended for long term storage of cultures in deep freezers. This is very simple and easy to handle yet effective method.

Requirements

Nutrient media (LB agar for bacterial cultures and PDA for fungal cultures), glycerol/DMSO, Petri plates, tube with screw cap, glass marker, Bunsen burner.

(a) Procedure

1. Grow the cells on suitable medium on Petri plates.
2. Add cryo-protectant, either glycerol or DMSO as given for liquid nitrogen method.
3. Transfer 0.5 ml cell suspension in small tube with a screw cap.
4. Store in a box at -70°C deep freezer.

(b) Precautions

1. Prepare the tubes in duplicate.
2. Continuous electric supply to deep freezer is ensured.

1.5.6 To Preserve the Cultures by Lyophilization (Freeze drying)

The principle of lyophilization is that the culture is frozen and then the water is removed by sublimation or evaporation. Since the cells are in dry condition, hence saved from the injuries caused by crystallization of liquid. Also, the organism maintained at zero metabolic state in absence of liquid can retain the viability for 30–50 years. In this technique, the culture is rapidly frozen at –70°C, dehydrated by vacuum, sealed and stored at 4°C in refrigerator. This technique is widely used to preserve bacteria, yeast and fungi but usually not recommended for algae and protozoa. This is the most preferred method of preservation in National and International Culture Collection. The transportation of lyophilized ampoules is also very convenient.

(a) Requirements

Sterilization proof slightly constricted at the mouth glass ampoules (110 × 6 mm), pipettes, cotton, lyophilizer, sterilized skim milk.

(b) Procedure

1. Grow the microbial cells on Petri plates containing a suitable medium.
2. Apply 4 ml of 10 percent skim milk over the culture and make the suspension of cells.
3. Sterilize the glass ampoules in an oven at 160°C for 2 h.
4. Transfer 0.1–0.2 ml of culture suspension in pre-sterilized ampoule. Place coded number label inside the ampoule and close it with cotton wool plug.
5. Keep the ampoules at –7°C in freezer.
6. Connect the ampoule to a hole in the centrifuge head of freeze drying apparatus or lyophilizer at –5°C, start the vacuum pump for about 5 h. This process is known as primary drying and results in rapid dehydration and freezing of contents of ampoules.
7. Push the cotton plug completely into the ampoule.
8. The neck of ampoule is drawn out and the ampoule is given secondary dehydration for 12 h in order to get the final product as highly powdered substance.
9. Seal the ampoule under vacuum and store away from light preferably at 4°C.
10. For reconstitution of culture, flame the ampoule neck, make a mark on ampoule using a glass cutter just around the equator of the cotton wool plug and apply pressure at the scratch mark by glass rod to crash it.
11. Remove the plug carefully using forcep.
12. Add about 0.5 ml of sterilized broth into the ampoule.
13. Keep it for rehydration of culture for few minutes.
14. Transfer the rehydarated culture to a flask or tube having appropriate broth using sterile pipette.
15. Incubate it at desired temperature as per the requirement of culture.

(c) Precautions

1. It is particularly important to have high initial count of cells on pre-drying medium. It has been observed that the cultures which have passed the logarithmic growth phase survive better than the cultures harvested in their active growth phase.

2. All possible care should be taken to maintain aseptic conditions while transferring the culture suspension to ampoules.
3. The time between filling the ampoules and freeze drying operation should be kept at minimum to avoid further growth or germination of spores.
4. For many species several sub culturing may be required before they regain their original characteristics.
5. More care should be taken for freezing pathogenic species.

1.6 STAINING TECHNIQUES

Generally bacterial cells are small and transparent and do not form colourful spores as fungi; hence cannot be seen properly under the simple microscope. This problem can be overcome by colouring the cells using selective dyes to differentiate them from their surroundings. This also helps to observe the morphological characteristics of the cell as shape and size and with special stains even the nature of cell wall, flagella, endospore and cell inclusions can also be studies. The dyes which are used for colouring the bacteria for this purpose are known as stains and the process of colouring is known as staining.

There can be natural and synthetic dyes, but the ones used for bacteria are all synthetic products of coal-tar or aniline type. Chemically, a dye or stain can be defined as an organic compound containing a benzene ring plus a chromophore and auxochrome group. Chromophore group provides the colouring property to the compounds and are called chromogens, whereas, the auxochrome group furnishes salt forming properties to the compound. Thus, the dyes or stains can be of two types, the acidic and basic. The acidic dyes are anionic which ionize to give the dye portion of molecule a negative electric charge. They are the coloured salts of acids, most commonly Na salts, but sometimes salts of K, Ca, Mg etc. and are used to stain the cytoplasmic components of the cells which are more alkaline in nature, *e.g.*, eosin, acid fuchsin, picric acid. Basic dyes on the other hand ionize to impart to the dye portion of the molecule, a positive electric charge. They are salts of colour bases and are usually chloride, sulphate or acetate. The basic dyes are used to stain negatively charged components of the cell, such as neuclic acid, *e.g.*, crystal violet, safranin, methylene blue. However, neutral stains can be prepared by mixing acidic and basic dyes.

There are two kinds of staining procedures, **simple** and **differential**. Simple stains employ a single dye to colour the cell or structure within the cell. They are easily prepared, less subjected to deterioration and give rapid results with one application. These stains can have basic or acid dye *e.g.* methylene blue, crystal violet, corbol fuchsin. Differential stains on the contrary require more than one dye and are used to distinguish between the otherwise morphologically indistinguishable bacteria by targeting their chemical differences, thus are more complicated, *e.g.*, Gram's stain, Ziehl-Neelsen stain, etc. For visualization of structures, such as flagella, capsule, spores, nuclei, food granules, etc., some special stains called structural stains are needed which are more or less differential stains. The principle is based on the fact that the particular structure shows a different degree of affinity for the dye from the rest of the cell structures which makes the differentiation possible.

1.6.1 Preparation of Bacterial Smear and Fixation

Before staining, a thin film of bacteria is prepared on glass slide in order to spread and fix the organisms to be observed. Heat fixing denatures proteins preventing autolysis and helps in adhering cells on slide. Some chemicals like methanol can also be used for this purpose.

(a) Requirements

Culture of bacteria in broth or agar medium, glass slides, inoculating loop, burner.

(b) Procedure

1. Take a clean glass slide and dry it.
2. Flame the inoculating loop and transfer a drop of bacterial culture from broth. In case of culture grown on agar medium place a drop of sterile water along with a small bit of culture and mix it well.
3. Spread the culture on the slide with wire loop of inoculating needle to form a thin layer or film. Another glass slide can also be used for spreading the culture. For this keep the second slide at 30 degree angle on first slide and gently push it on slide surface facilitating the spreading of culture.
4. Allow the film to dry at room temperature.
5. Fix the film by passing the slide through the flame of burner quickly 2–3 times. Chemical fixation can also be done for this purpose. Apply few drops of 5 percent formalin or acid alcohol or methanol to the film, hold it for few minutes and let the smear dry.

(c) Result

The smear is ready for staining.

(d) Precautions

1. The slide should be totally clean.
2. Do not prepare a thick or dense smear, it will not give good resolution of cells.
3. Do not heat too much, it may cause distortion of normal shape of bacteria.
4. If the smear is too thin to be located under oil immersion objective, mark the border of smear with glass marker.

1.6.2 Simple Staining of Bacteria

The simple staining is performed to study cell shape, size and arrangements of bacterial cell. It is also known as monochrome staining as only one stain is used. Basic stains are used for this purpose.

(a) Requirements

24 hour old culture of cocci (*Spaphylococcus aureus*), bacilli *(Bacillus subtilis*), spiral *(Spirillum* sp.*)* bacteria inoculating loop, Bunsun burner, glass slide, any basic dye, *e.g.*, methylene blue (0.5%), crystal violet (1%), carbol fuchsin (10 times diluted), microscope.

(b) Procedure

1. Prepare a smear as described earlier.
2. Flood the smear with about 5–6 drops of any one of the stains.
3. Allow the stain to remain on slide for one minute.
4. Gently wash the excess of stain with tap water.
5. Dry the slide with blotting paper.
6. Examine the slide under an oil immersion objective of microscope with or without cover slip (if preparing the permanent slide by mounting it by DPX or Canada balsam, then apply the coverslip).

(c) Observation

Note the shape, size and arrangement of cells. Bacteria stain a deep blue if methylene blue stain is used. The colour of cells is different with different stains.

(d) Results

The shape of *Staphylococcus aureus* cells is spherical, *Bacillus subtilis* is rod, whereas *Spirillum* is spiral.

1.6.3 Negative Staining of Bacteria

Negative staining is the technique in which the background is stained and the cell remains colourless. This is done by using acidic stains like sodium eosinate, nigrosine and India ink. Since the overall charge of bacterial cell is negative, so the negatively charged dye is repelled, the result being the bacterial cells remaining colourless. Instead the dye is deposited around the cells, colouring the background. However, this technique distinguishes the bacterial cells in dark background, indicating their shape, granules, endospore and size. It can also show cell breakage, if any. Here no heat fixing is done so cells are never distorted. Moreover, even capsulated bacteria which are difficult to stain can be observed.

(a) Requirements

24 hour old culture of bacteria (*Staphylococcus*), nigrosine solution, glass slide, inoculating loop, Bunsen burner and microscope.

(b) Procedure

1. Place a drop of 7 percent nigrosine solution on a slide.
2. Transfer a loopful culture and mix gently.
3. Place a second slide at a 30° angle on 1st slide and spread the stained culture across the entire surface of 1st slide by pushing the 2nd on the 1st one.
4. Allow the smear to dry at room temperature.

(c) Observation

Note the shape and size of the cells.

(d) Results

Colourless bacteria are observed against dark background.

(e) Precautions

1. Culture should be freshly grown.
2. The thickness of the smear decreases gradually towards the other side of the slide.
3. Do not heat the smear.

1.6.4 Gram Staining of Bacteria

The most commonly used differential stain is Gram stain described by a Danish physician Christian Gram in 1884. The Gram reaction is determined by colouring the cells by crystal violet and iodine which is followed by decolourization step using a solvent. The cells of some bacteria loose the stain and are known as Gram-negative, whereas, the other cells resist the decolourizing solvent and remain coloured are called as Gram-positive. At the end counter staining is done by another stain of different colour to make the colourless cells visible again. Generally alcohol is used for decolourizing the colour as it serves as lipid solvent and dehydrating agent for protein.

The difference in staining responses of two groups of bacteria is due to some basic differences in their cell wall chemistry. The cell envelope of Gram-negative bacteria is complex structure which contains thin layer of peptidoglycan (cell wall), but high lipid content in outer membrane resulting in formation of large pores in cell wall due to dissolving of higher amount of lipid by the solvent, alcohol. These pores do not close on dehydration of cell wall proteins. This facilitates the leakage of crystal violet-iodine complex resulting in decolourization. On the contrary, Gram-positive bacteria have thick cell wall mainly composed of protein and cross-linked mucopeptides, hence small pores are formed which are readily closed by dehydration caused by alcohol. Thus, the loss of stain complex is not allowed, thereby, cells remaining violet.

Steps in Gram staining

S. No.	Steps	Stains/ solvent	Gram-positive	Gram-negative
1	1st stain applied	Crystal violet	Cells become violet	Cells become violet
2	2nd stain applied	Iodine solution	CV-I complex formed, cells remained violet	CV-I complex formed, cells remained violet
3	Decolourization	Alcohol	Stain complex stays within the cells. Cells remain violet	Stain complex extra-cted out of the cells, cells become colourless
4	Counter stain	Safranin	No effect, cells remain violet	Colourless cells are stained and become red

(a) Requirements

24 h old culture of *E. coli* or freshly set curd, slide, crystal violet, Gram's iodine solution, 95 percent ethyl alcohol, safranin, Bunsen burner, microscope.

(b) Procedure

1. Prepare a smear of *E. coli* or curd, air dry it.
2. Stain with crystal violet for 30 seconds.
3. Wash off the excess stain with tap water and dry it.
4. Flood the slide with iodine solution for 30 seconds.
5. Wash the slide with tap water to remove the excess stain.
6. Decolourize with 95 percent ethyl alcohol until no more colour flows from smear (at this step Gram-positive bacteria will not be affected, whereas Gram-negative ones will loose the colour completely).
7. Wash again with tap water.
8. Counter stain it with safranin for 45 seconds, rinse with water and dry it.

(c) Observation

1. Observe under oil immersion objective.
2. See carefully the Gram's reaction and identify bacteria according to the colour stain.

(d) Results

E. coli smear appear red, hence it is Gram-negative, whereas, smear made from curd remains violet in colour indicating Gram-positive bacteria in it.

(e) Precautions

1. The smear should be thin because if the film is too thick Gram-negative bacteria may seem Gram-positive.
2. Culture of bacteria should be young to avoid misleading results.

1.6.5 Acid-fast Staining

Presence of waxy substances like mycolic acid in bacteria sometimes interferes with staining. Species of *Mycobacterium* and actinomycetes belong to this category. Acid-fast is a differential stain and has diagnostic value in medical microbiology. Paul Ehrlich developed it in 1882 and Franz Ziehl and Friedrich Neelsen modified it. Bacteria are classified as acid-fast if they retain the colour of primary stain even after washing with strong acid. Non-acid-fast bacteria loose their stain, hence should take the colour of counter stain.

(a) Requirements

24 h old culture of *Mycobacterium*, slide, Ziehl-Neelsen carbolfuchsin, hydrochloric acid, ethanol, methylene blue, Bunsen burner, microscope.

(b) Procedure

1. Make the smear of *Mycobacterium*.
2. Cover with tissue paper.
3. Flood slide with carbol-fuchsin, the primary stain, for 2 minutes while heating with steam or heating on hot plate.
4. Remove paper cover, decolourize slide with a mixture of hydrochloric acid and ethanol.
5. Counter stain with methylene blue.

(c) Results

Mycobacterium cells retain the colour and stain red in blue tissue.

(d) Precaution

1. *Mycobacterium* being pathogenic, care should be taken to handle it.

1.6.6 Endospore Staining

Some bacteria produce dormant structure such as endospore to survive the unfavourable conditions. A single bacterium forms a single spore within the cell, which is surrounded by impermeable spore envelope. The spores come out of the cell by disintegrating the cell wall, survive the extreme conditions and germinate to their active vegetative state on return of favourable conditions.

Endospores are formed by few genera only, *e.g.*, *Bacillus* and *Clostridium*, the most common soil inhabitants. In this method malachite green penetrates the spores, but vegetative portion of cell do not accept this stain and can be removed by washing, thus spores look green and rest of the cell red due to counter stain.

1.6.6.1 Endospore staining by malachite green

(a) Requirements

48h old culture of *Bacillus cereus*, Malachite green (5% aqueous), safranin (0.5%), glass slides, inoculating loop, Bunsen burner, microscope.

(b) Procedure

1. Make a smear of *Bacillus*, air dry and fix it by heating through a flame 20 times.
2. Flood the smear with malachite green. Steam for 5 minutes by adding more stain to smear.
3. Wash the slide under tap water gently.
4. Counter stain with safranin for 30 seconds.
5. Wash with water, blot and air dry.

(c) Observation

Examine the slide under oil-immersion objective of microscope. Green spores can be observed. Also, look for the position and size of endospores.

(d) Results

In *Bacillus cereus* the cells are rod shaped containing green coloured centrally located endospore.

1.6.6.2 Endospore staining by Dorner's method

This is a negative method of staining endospores. In this method hot carbol fuchsin penetrates the endospore, staining them red whereas, nigrosine acts as decolourizing agent as well as negative stain for the smear. The negative charge, thus repelled by the overall negative charge of bacterial cell, leads to unstained cells in dark background.

(a) Requirements

Carbol-fuchsin, safranin, broth culture of *Bacillus* sp., glass slide, test tube, Bunsen burner, water bath, inoculating needle.

(b) Procedure

1. Take a little portion of *Bacillus* culture suspension in sterile test tube.
2. Add equal quantity of carbol fuchsin.
3. Boil in water bath for 10 minutes.
4. Transfer 2 loopful of the suspension to slide and add a small drop of nigrosine solution, mix thoroughly and spread in a thin film with the edge of glass slide.
5. Dry in air.

(c) Observation

Observe the slide under microscope. Endospore will stain red and vegetative cell will remain unstained.

(d) Results

Red coloured endospores are visible against dark background.

Note: *Instead of Bacillus, Clostridium tetani culture can also be used to see terminal endospores, but this species being pathogenic, is not preferred for class room practicals.*

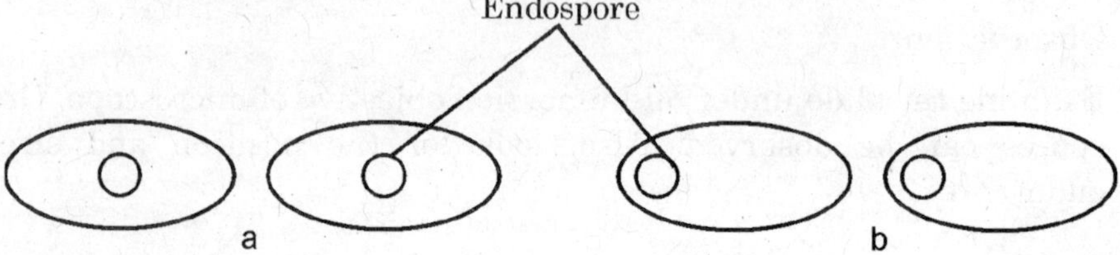

Fig III 1.6 : a. Endospores a. Centrally located in *Bacillus cereus*
b. Terminally located in *Clostridium tetani*

1.6.7 Capsule Staining to Detect the Capsule or Slime Layer in Bacteria

Capsule slime layer is an additional layer on the cell surface of certain bacterial species formed due to exopolysaccharides or sometimes exopolypeptides deposition. This is a loose, more or less amorphous layer. The term capsule is restricted to a well defined persistent layer, whereas, slime layer forms loose mass present more diffusedly around the cell.

The capsule is not an integral part of the cell, it is rather dispensable. Hence, any dye which stains bacterial cell will not stain the capsule, for this special staining technique is used. Capsule staining is a diagnostic method for detecting pneumonia and other disease causing pathogens which are encapsulated.

(a) Requirements

24 h culture of *Enterobacter aerogenes* or any capsule forming bacteria, congo red, acid alcohol, acid fuchsin, inoculating loop, glass slide, microscope.

(b) Procedure

1. Place the culture on glass slide and prepare the smear.
2. Apply congo red stain and air dry.
3. Fix the smear with acid-alcohol for 15 seconds.

4. Wash with water.
5. Flood the slide with acid fuchsin for 60 seconds.
6. Wash with water and blot dry.

(c) Observation

Examine under microscope and look for the colour of capsule and cell.

(d) Results

In encapsulated bacterial cells, capsule appears colourless, cells are stained red in colour and the background is dark blue.

(e) Precautions

1. Do not heat fix the smears.
2. Capsulated bacteria are generally pathogens so take all the precautions needed to handle pathogen.

1.6.8 Flagella Staining

Motility in bacteria is due to thread like filamentous appendages protruding from the cells. Bacterial flagella are 70 μm in length and 0.01 – 0.02 μm in diameter, hence too thin to see under ordinary light microscope without special staining meant for flagella. In order to make these thin flagella visible first flagella are coated with a mordant to increase their diameter and then applying some special stain.

1.6.8.1 Lee's method

(a) Requirements

Culture of *Pseudomonas fluorescens*, Lee's stain, glass slides, osmic acid, burner, Petri plates, inoculating loop, Canada balsam.

(b) Procedure

1. Put a drop of motile bacterial culture on slide.
2. Kill the bacteria in osmic acid (2%) fumes for 1 min. (Put the osmic acid in a closed Petri plate and invert the slide containing the culture for 1 min.
3. Air dry it.
4. Flood the slide with water for 1 min., dry with mild heat.
5. Flood the slide with freshly prepared stain by mixing solution A and B for 1 min.
6. Drain off the stain and expose the slide to air for 30 seconds.

7. Wash with water and dry it.
8. Mount in Canada balsam.

(c) Observation

Observe the slide under oil immersion objective. Try to locate the position of flagella.

(d) Results

Pseudomonas fluorescens contains lophotrichous flagella.

1.6.8.2 Leifson's method

(a) Requirements

Bacterial culture (*Pseudomonas fluorescens*), Leifson's stain, burner, inoculating loop, microscope, glass slides.

(b) Procedure

1. Take clean glass slide and warm through a flame of burner.
2. Remove the loopful of cell suspension from culture and put it at one end of clean slide.
3. Allow the drop to spread on slide by tilting it. Air dry it.
4. Mark the boundary line by a marker.
5. Flood the stain on smear and leave the slide for 10–15 min.
6. Wash gently with tap water and air dry.
7. Examine under microscope in 100×

(c) Results

The cells and flagella retain the red colour.

(d) Precautions

1. To prepare a clean slide, immerse it in dichromate solution, rub gently with a clean cloth, wash with water, rinse in alcohol (95%) and wipe with cheese cloth.
2. Suitable bacterial species should be selected and grown on suitable culture media.
3. Flagella are easily detached from the cell so require careful handling.
4. For Lee's method, carefully apply the osmic acid as the fumes are injurious.
5. In Leifson's method do not heat the smear.

1.6.9 To Test the Viability of Bacteria by Staining

Certain changes take place in bacterial cells when they die making different staining response. This is made the basis of distinction between living and dead cells.

(a) Requirements

2–3 week old culture of bacteria, Loeffer's methylene blue solution, dilute carbol fuschin, slide, burner, blotting paper, microscope.

(b) Procedure

1. Prepare the smear and heat fix.
2. Add methylene blue on smear for 10 min.
3. Wash the slide under tap water. Smear becomes pale blue.
4. Flood with carbol fuschin for a short time.
5. Wash with tap water and blot dry.

(c) Observation

Examine the slide under oil immersion.

(d) Results

Living bacteria appear purple, whereas dead bacteria appear red. As for spores, living one shows pink colour and dead spores stain blue.

1.6.10 Staining of Fungi

The stain used for colouring fungal mycelium and spores is lactophenol cotton blue. It stains the fungal cytoplasm and provides a light blue background. The constituents of this stain are cotton blue, which is a dye, lactic acid which acts as cleaning agent, phenol is a fungicide and glycerine makes the semi permanent slide. For all routine works related to fungal identification lactophenol cotton blue stain commonly called as mounting fluid, is used. For microscopic observation the fungal culture is teased on slide with needles, mounted with stain and examined after placing cover slip on it.

(a) Requirements

Young culture of *Rhizopus, Aspergillus* or *Alternaria*, mounting fluid (lactophenol cotton blue), slides, cover slips, needles, burner, microscope.

(b) Procedure

1. Place a drop of mounting fluid on a slide.
2. Take hyphal mat from the edge of the colony and transfer it into the drop of fluid.
3. Tease the fungal mat with the help of needles to spread the hyphae.
4. Place the coverslip on it taking care to avoid air bubbles.

(c) Sealing of slides

1. Remove air bubbles, if any, by gentle heating or adding more mounting fluid.
2. Remove the excess of stain by alcohol (70%) cotton swab.
3. Apply a layer of nail polish or euperol around the margins of coverslip to seal it.
4. Dry overnight.

(d) Observation

For fungi, low as well as high power objectives can be used. Look for hyphae, sporangia, sterigmata, sporangiophore/conidiophores, spores/conidia, etc. Use oil immersion objective for the tiny structure such as spores.

(e) Result

Fungal cytoplasm is stained blue, whereas the cell wall remains unstained.

(f) Precautions

1. Air bubbles should be removed.
2. A gentle pressure by thumb on coverslip helps in spreading the hyphae and making it more visible.
3. Take the fungal culture from the margins of the young colony growing on agar to avoid too many spores.
4. In case of overgrown culture, spores can be washed in alcohol (70%) in a watch glass.

1.6.11 Staining of Arbuscular Mycorrhiza (AM)

Mycorrhiza is a symbiotic fungi found in association with roots of higher plants. One of the most common type of mycorrhizae, arbuscular mycorrhiza, previously known as vesicular arbuscular mycorrhiza (VAM) form dichotomously branched haustoria within the host cell (arbuscles).

In addition, the fungal hyphae also bear a swollen thick walled structure, called vesicles. These are quite wide spread in plant kingdom. AM infection is observed in cleared roots, hence, the roots are first preserved, squashed and then stained.

(a) Requirements

Plants roots (freshly collected), FAA (Formalin-acetic acid-alcohol, 5:5:90 v/v), KOH (10%), mounting fluid (Chlorohydrate 20 g, gum arabic 20 g, glycerine 20 ml, glucose syrup 3 ml, basic fuschin 10 drops prepared by dissolving 0.3 g in ethanol (95%), water 35 ml), staining solution (equal volume of lactic acid (80%), glycerine and distilled water with chlorazol black E (0.1%)), water bath, slides.

(b) Procedure

1. Wash the roots thoroughly to remove adhereing soil particles and cut 1 cm long pieces.
2. Fix them in FAA overnight.
3. Remove the root pieces and wash several times in tap water.
4. Transfer them in KOH solution.
5. Simmer the solution to 90 percent for about 2-3 hours.
6. Decant KOH, rinse with several changes of tap water and finally with deionized water.
7. Stain the roots for 1 h at 90°C.
8. Put the stained roots in glycerine overnight. At this stage roots can be kept for future use.
9. Take a clean glassslide and mount the roots using mounting fluid.
10. Gently press the slide by thumb to make the squash.

(c) Observation and Result

Observe the slide under the microscope and look for hyphae, VAM spores, vesicles and arbuscules.

1.7 ISOLATION AND ENUMERATION OF MICROBES

Microorganisms are ubiquitous in nature. They can be aerobic and anaerobic according to their demand for O_2 and saprophytic or parasitic according to their nutritional requirements. They are found in mixed form in environment hence, several techniques are prescribed to isolate an individual species in pure culture.

1.7.1 Isolation and Enumeration of Fungi from Soil

Soil is the reservoir for microorganisms. There is no perfect single method to isolate all types of fungi from the soil, hence, more than one techniques should be employed for isolation purpose. Dilution plate and soil plate methods are the two most important methods. Dilution plate method has already been described in section 1.4.4 in detail. The other method is known as Warcup method on the name of the microbiologist who described it in 1950. This method is easier to use and less time consuming.

1.7.1.1 Isolation of fungi by dilution plate method

(a) Requirements

Soil sample, pipettes, Erlenmeyer flasks, streptomycin, PDA plates.

(b) Procedure

1. Collect the soil randomly in sterilized polythene bags, make composite sample by mixing the soil samples aseptically.
2. Sterilize 90 ml water in 4 flasks.
3. Weigh 10 g of soil and transfer in first flask and mark it as 1:10 dilution (10/100 = 1:10)
4. Make the serial dilution as describes in section 1.4.4.
5. Pour 1 ml soil suspension from each flask into agar plates (triplicate for each dilution) supplemented with streptomycin @ 30 mg/l. Gently spread the suspension on the solid surface of medium by rotating the plate.
6. Incubate the plates at 28 ± 2°C for 4–7 days.

(c) Observation

Examine the differently appearing fungi on plates, calculate the number of each type of colonies in all plates and record your observation in a table form. Also, purify and identify the colonies.

S. No.	Fungi	No. of colonies						Average No. of colonies
		10^{-2}			10^{-3}			
1	Blackish green (*Alternaria*)	8	6	8	4	4	3	
2	Black (*Aspergillus niger*)							
3								
4								
Total no. of colonies								

Calculate the frequency, occurrence and colony forming units of fungi by following formulae:

$$\text{Percent frequency} = \frac{\text{No. of plates in which a particular type of colony occurs}}{\text{Total no. of plates studies}} \times 100$$

$$\text{Percent Occurrence} = \frac{\text{Total no. of colonies of a species in the plate}}{\text{Total no. of colonies of all the species in all the plates}} \times 100$$

$$\text{Colony forming units (cfu) per g dry soil} = \frac{\text{Average no. of colonies} \times \text{Volume of sterile water}}{\text{Dry weight of soil*} \times \text{dilution factor}}$$

**For dry weight of soil put 10 g soil in an oven in Petri dish at 80°C for 24 h and weigh the soil.*

(d) Results

Various fungal species are isolated, calculated and studied.

1.7.1.2 Isolation of fungi by Warcup method

(a) Requirements

Soil sample, Sabouraud agar medium, streptomycin, burner.

(b) Procedure

1. Collect the soil samples randomly and make a composite mixture by mixing samples aseptically.
2. Transfer 0.5 to 1 g of air dried soil in Petri plates in triplicate with the help of sterile transfer needle (sample can also be sprinkled on the medium after it has solidified in the Petri plate).
3. Add 15–20 ml of melted and cooled (45°C) medium supplemented with streptomycin to each Petri plate.
4. Gently rotate the plate for evenly distributing the soil particles.
5. Allow the plates to solidify.
6. Incubate the plates at 28±2°C in an inverted position for a week.

(c) Observation

Colonies start appearing after 3 days, some slow growing fungi appear after a week (or even till 15 days). Purify various differently appearing colonies and study them.

(d) Results

Various fungal species are isolated and studied.

(e) Precautions

1. Plates should be taken in triplicates.
2. Aseptic conditions should be maintained from the soil collection to pouring step.
3. Plates should be incubated at inverted position.

1.7.2 Isolation of Seed Microflora

Seeds are infested by various types of microorganisms in field as well as storage. The microbes may remain on the surface of seeds or may be seed borne. For isolation of seed borne microflora, seeds are surface sterilized to eliminate the surface microflora.

1.7.2.1 Agar plate method (ISTA, 1966)

(a) Requirements

Wild or cultivated seeds, sodium hypochlorite solution (1%), sterile distilled water, PDA for fungi, nutrient agar for bacteria.

(b) Procedure

1. Collect the seeds in sterile polythene bags.
2. Surface sterilize the seeds with sodium hypochlorite solution (1%) for 2–5 minutes and wash these seeds with water 4–5 times. This is done for internally seed borne microorganisms. For externally associated microbes this step is skipped.
3. Prepare the plates of PDA (with streptomycin)/nutrient agar in triplicate.
4. Place 10 seeds per plate at equidistance.
5. Incubate the plates in an incubator at 28±2°C and 37±2°C for fungi and bacteria, respectively.
6. Purify and identify different fungal and bacterial colonies.

(c) Results

Different appearing fungal colonies grow on PDA plates and different bacterial colonies grow on nutrient agar plates.

1.7.2.2 Blotter method

(a) Requirements

Blotter paper, Petri plates, seeds, sterile distilled water.

(b) Procedure

1. Cut the blotting paper in the shape of Petri plate and moist the paper with water.
2. Sterilize the Petri plates.
3. Place 10 seeds on blotter as in previous experiment.
4. Incubate the plates in an incubator as given in 1.7.2.1.
5. Examine the microbial colonies, purify and identify them.

(c) Results

Different fungal and bacterial colonies are obtained, purified and identified.

1.7.3 Isolation of Aeromycoflora

Air is a carrier for particulate matters and microorganisms. The number and kind of microbes present in a particular place and region vary because of sources of contamination in the environment, *e.g.*, in hospitals air may have more pathogens spread due to coughing and sneezing. Circulation of dust particles, pollens, etc., from the surface of earth by air current is also a major source of air contamination.

For sampling of air two most widely used methods are gravity slide and settle plate techniques. In gravity slide technique, the glass slide containing an adhesive like vaseline or petroleum jelly is exposed into air to trap the spores, whereas in settle plate technique Petri dishes containing agar medium are exposed.

(a) Requirements

Petri plates, PDA or Czapek's Dox Agar medium, (for bacteria, nutrient agar medium can be used).

(b) Procedure

1. Pour melted, cooled (45°C) PDA or Czapek's Dox Agar supplemented with streptomycin in sterilized Petri plates and allow the

plates to solidify. Do not use streptomycin for bacterial isolations.

2. Remove the lid and expose the Petri plates in triplicate to air for 5–10 minutes in laboratory and outside the department.
3. Cover the plates and incubate them at 28 ± 2°C for 4–5 days.

(c) Observations

Observe the fungal colonies appearing on plates and compare the number and type of colonies obtained from air of laboratory and outside. Percent frequency and percent occurrence of each type of colony may also be calculated. If you have done the experiment for bacteria also, the comparison between the number of bacterial and fungal colonies can also be made.

(d) Results

Calculate the number and type of fungi. The number may be more in air in the laboratory than in the air outside.

1.7.4 Isolation of Arbuscular Mycorrhizal (AM) Spores from the Soil

Mycorrhiza is the symbiotic association found in nature between fungi and roots of higher plants. AM fungi were previously known as vesicular arbuscular mycorrhiza (VAM). A number of mycorrhizal fungi such as *Glomus, Endogone, Gigaspora, Glaziella, Sclerocystis, Entroph-ospora,* etc. reside in soil which in presence of suitable host, colonize the roots. There are various methods of isolating the AM fungal spores, but sieving and partition methods are widely used.

1.7.4.1 Wet-sieving method

(a) Requirements

Soil sample, sieve set, beaker, Bunsen burner, water.

(b) Procedure

1. Mix 100 g soil in 1 litre of water and allow the soil to settle.
2. Decant the soil through 710 μm sieve first to remove large organic matter and roots.
3. Again wash coarse particles and roots on the sieve with water and collect soil suspension.
4. Pass the soil suspension thus obtained through 250, 75 and 45 μm sieve.

5. Collect the entire residue on 45 μm sieve.
6. Wash the residue well with water and collect the spores in watch glass.

(c) Observation and Results

Examine under the microscope and identify the genera with the help of literature available in lab.

1.7.4.2 Floatation method

(a) Requirements

Soil sample, sucrose solution (20, 40, 60%), fine sieve, blender, centrifuge.

(b) Procedure

1. Collect the soil sample and take 20 g out of it.
2. Blend it at high speed for 1-2 minutes to make the spores detach from roots, or clay particles.
3. Pass these contents through a fine sieve and wash the colloidal material thoroughly with a strong stream of water.
4. Add 10 ml of sucrose (20%) into a centrifuge tube, followed by 10 ml of 40% and then 60% sucrose into the bottom of tube.
5. Add 10–15 ml of blended sieving onto the surface of 20% sucrose layer.
6. Centrifuge at 3000 rpm for 3 min.
7. Remove the debris accumulated at the interface of 20–40% and 40–60% sucrose.
8. Gently wash the spores present on fine sieve with a strong stream of water to remove the sucrose.
9. Collect the spores.

(c) Observation and Results

AM spores are found at the interface of 20–40 percent and 40–60 percent of sucrose that has been removed. Observe the size, shape and colour of spores and identify with the help of literature available.

1.7.5 Isolation and Enumeration of Microorganisms from Rhizosphere

Rhizosphere is the region of soil immediately surrounding the roots of plant. This region extends a few mm from the surface of roots. This zone has increased microbial population and activity due to rich food material provided by the added sloughed off portions of root tissues and root exudates. Root exudates are composed of sugar, amino acids, vita-mins and other growth factors which influence the population of micro-organisms inhabiting this region. The microorganisms in rhizosphere region exceed by a factor of 10 in comparison to neighbouring region devoid of plant roots, known as non-rhizosphere region. The rhizosphere to soil ratio (R : S) is called as rhizosphere effect.

(a) Requirements

Freshly collected root sample, Petri plates, sterile polythene bags, Czapek's Dox Agar, Nutrient Agar, Starch Casein Agar media, sterile pipettes, Bunsen burner, glass marker, sterile water.

(b) Procedure

1. Prepare Czapek's Dox agar plates for fungi, nutrient agar plates for bacteria and starch casein plates for actinomycetes in triplicate.
2. Collect the roots and soil from rhizosphere and non-rhizosphere regions in separate polythene bags.
3. Separate rhizosphere soil from roots in Petri plates using sterile spatula or brush.
4. Add 10 mg of rhizosphere soil in 90 ml water sterilized in flask.
5. Make the series of dilutions (up to 10^{-6} dilutions) in flasks as described in section 1.4.4 and mark them.
6. Transfer 1 ml aliquots of soil suspension from dilution 10^{-3} on Czapek's Dox agar plates, 1 ml from 10^{-4}, 10^{-5} dilutions on starch casein plates and 1 ml from 10^{-5}, 10^{-6} dilution on nutrient agar plates.
7. Repeat the same procedure for non-rhizosphere soil.
8. Incubate the Czapek's Dox and starch casein plates at 28°C and nutrient agar plates at 37°C for 7, 15 and 2 days, respectively.

(c) Observations

Observe the plates after their respective incubation periods, record the number of colonies per plate, identify the colonies and compare the rhizosphere and non-rhizosphere microflora.

Microbial count of rhizosphere and non-rhizosphere soils g^{-1} is calculated by the following formula:

$$\text{No. of microorganisms/g of soil} = \frac{\text{No. of colonies/plate} \times \text{Dilution factor}}{\text{Dry weight of soil}}$$

$$\text{Rhizosphere effect} = \frac{\text{No. of microoranisms in rhizosphere soil}}{\text{No. of microorganisms in non - rhizosphere soil}}$$

(d) Results

Microorganisms are more in rhizosphere region. Greater rhizosphere effect is observed in case of bacteria (10 to 100) than with fungi and actinomycetes, whereas, with regard to protozoa and algae, there are hardly any changes in its value.

1.7.6 Isolation of Microorganisms from Rhizoplane

Rhizoplane microorganisms are those which are intimately associated with root surfaces.

(a) Requirements

Freshly collected root sample, Petri plates, sterile polythene bags, Czapek's Dox agar for fungi, Nutrient agar for bacteria, Starch casein agar for actinomycetes, sterile pipettes, Bunsen burner, glass marker, sterile water.

(b) Procedure

1. Collect the root sample.
2. Take root pieces with soil attached to them in sterilized flask containing 100 ml water, shake well.
3. Serially wash roots 5–6 times with sterilized water until clear root surface is seen.
4. Transfer the root pieces on all 3 media plates in triplicate for fungi, bacteria and actinomycetes.
5. Incubate the plates as the experiment 1.7.5.

(c) Observation and Results

Observe the colonies on plates and count the number.

1.7.7 Isolation of Protozoa from Soil

Protozoa are single celled eukaryotes that commonly show the characteristics usually associated with animals, most notably mobility and heterotrophy, and play a key role in maintaining the ecosystem. Many pathogenic and non-pathogenic species of protozoa have been identified.

Free living non-pathogenic species are found in soil and are critical link in food chain serving as a nutrient source for higher organisms.

(a) Requirements

Soil sample, water-agar medium, young culture of *Aeromonas* sp., Petri plates, pipettes, glass bottle, inoculating loop.

(b) Procedure

1. Prepare sterile water agar plates.
2. Spread a thick suspension of *Aeromonas* in the form of a circular patch of 0.5 cm diameter at 3 points.
3. Add around 50 mg of soil on the centre of each circle.
4. Moist the soil with a few drops of sterilized water.
5. Incubate the plates at room temperature for about 2 weeks.
6. Keep on adding water to soil, as and when required during the incubation period.

(c) Observations

Examine the plates under low power objective of microscope.

(d) Results

Protozoans are visible on agar surface and can be recognized by their movement through flagella or cilia. Isolate, purify and identify them.

1.7.8 Isolation of Cyanobacteria

Cyanobacteria, previously called as blue green algae, are prokaryotic organisms with oxygenic photosynthesis and biological nitrogen fixation capability. They occur in sea and fresh water as well as in soil. Many genera have been recorded from paddy fields where they make nitrogen rich environment. Many of these cyanobacteria are now grown at large scale to produce biofertilizer.

(a) Requirements

Soil or water sample from paddy field, Pringsheim's broth, conical flasks, pipettes, illuminated growth chamber, bottles or polythene bags, inoculating loop, Bunsen burner.

(b) Procedures

1. Collect soil or water in sterilized polythene bags/bottles.
2. Pour the Pringsheim's broth (composition given at appendix IV A) in 100 ml flasks and sterilize.

3. Prepare the series of dilutions (up to 10^{-7}) taking soil or water sample as described in experiment no. 1.4.4.
4. Pour 1 ml aliquot from each dilution into another series of flasks containing 20 ml broth.
5. Incubate the flasks at 30°C for 2–3 weeks in an illuminated growth chamber.
6. Pick up individual colonies and purify on fresh broth.

(c) Observations

Various coloured colonies with mucilage appear in broth which can be identified with the help of literature.

(d) Results

Cyanobacteria which are commonly found in rice fields are *Anabaena, Cylindrospermum, Nostoc, Scytonema, Aulosira, Tolypothrix, Westiella* etc.

1.7.9. Isolation of Aquatic Fungi by Baiting Method

Many aquatic fungi occur in water ecosystem which grow on litter of plants and animals and degrade them. These fungi can be isolated by baiting technique. The common baits used are hemp seeds, hard green apple, small pear, loquat, plum, tomato, grapes, green banana, mango, sweet corn kernels, wax paper, vegetable chips, plant leaves, human hair, etc.

(a) Requirements

Bait (any of the above), nylon bags, sampling bottles, water sample, Petri dishes.

(b) Procedure

1. To collect the sample place the bait in nylon bag and suspend it in river.
2. Withdraw the samples, *i.e.*, baits at regular intervals and bring to the laboratory.
3. Wash the samples in sterilized water.
4. Transfer the washed baits to sterile Petri plates containing sterile water and fresh baits.
5. Incubate in dark at 5–20°C for 2 weeks.

(c) Observations

Examine the colonized baits for fungal growth. Purify and identify the fungi.

(d) Results

Some of the common genera found in water are *Allomyces, Aphanomyces, Olpidium, Pythium, Saprolegnia* etc.

1.7.10 Isolation of Yeast

(a) Requirements

Petri plates, YPS medium (composition given in appendix IVA), flour.

(b) Procedure

1. Mix flour and sterile water in the ratio of 1:10 (w/v).
2. Prepare the plates of YPS medium.
3. Transfer 1 ml from it into prepared Petri plates.
4. Incubate the plates at $20 \pm 2°C$.
5. Yeast colonies will appear after a few days.

(c) Results

Vegetative cells of yeasts showing budding can be seen under the microscope. Asci and ascospores can also be observed.

1.7.11 Isolation of Rhizobia from the Root Nodules

Rhizobium is a symbiotic bacterium forming root nodules in leguminous crops. It is gram negative in gram shows reaction and fixes nitrogen, thus can be exploited as biofertilizer.

(a) Requirements

Root nodules, Yeast extract mannitol (YEM) agar medium, $HgCl_2$ (0.1%), ethyl alcohol (70%), pipettes, inoculating needles, Petri dishes, Bunsen burner, sterilized polythene bags, glass rod.

(b) Procedure

1. Uproot the leguminous plant and bring it to laboratory in sterilized polythene bag.
2. Wash the roots with tap water thoroughly to remove the soil.
3. Select healthy pink, intact nodules and wash them with water.
4. Surface sterilize the nodules in $HgCl_2$ for 5 min.

5. Wash the nodules with sterile water 4 times to remove the traces of $HgCl_2$.
6. Dip the nodules in ethyl alcohol (70%) for 3 min.
7. Again wash the nodules with sterile water.
8. Prepare the YEM agar plates.
9. Crush the nodules aseptically with glass rod.
10. Prepare the dilutions in tubes/flasks up to 10^{-5} dilution.
11. Spread 1 ml suspension from each dilution on YEM agar plates.
12. Incubate the plates at 28°C for 4 days.

(c) Observations and Results

Observe *Rhizobium* colonies which appear gummy, translucent or opaque. Perform the gram staining for identification. Another test to identify the bacterium is to stain with corbol fuschin to see β hydroxy-butyrate granules within the cells.

(d) Precautions

1. Healthy nodules (pink in colour) should be selected.
2. Surface sterilization and subsequent washings should be done properly.

Note: *Rhizobium is also found in soil as free living bacterium, hence can be isolated from collecting the agriculture soil of leguminous crops and performing the dilution plate technique thereafter.*

1.7.12 Isolation and Cultivation of Anaerobic Bacteria

Strict anaerobes require special media and techniques for their growth. A number of methods are available; two of them are described below:

1.7.12.1 Candle jar method

Requirements

Bell jar, Vaseline, candle, bacterial cultures (*Bacillus* sp., *Clostridium* sp.), Nutrient agar, Petri plates, inoculating needle.

(a) Procedure

1. Prepare the Petri plates with nutrient agar medium as described in experiment 1.2.3.
2. Streak the plates with each culture in duplicate.
3. Fix a lighted candle in it.

4. Seal the bottom of bell jar to glass plate with vaseline tightly (Fig. III.1.7).
5. Once the flame is extinguished place this assembly at 37°C in incubator for 2 days.
6. Examine the growth of bacteria in Petri plates.

(b) Results

Clostridium, which is an anaerobic bacterium will show the growth, whereas, *Bacillus* which is aerobic will not grow.

1.7.12.2 Gaspak anaerobic jar method

The principle of this method is the production of water because of the reaction of H_2 generated in the jar and O_2 available in it. Thus, the anaerobic condition is created which can be ensured by methylene blue strip.

(a) Requirements

Gaspak anaerobic jar assembly, palladium pellets, methylene blue, bacterial culture of *Clostridium* sp., Nutrient agar, Petri plates, inoculating needle.

(b) Procedure

1. Prepare the nutrient agar plates as described in experiment 1.2.3.
2. Inoculate the plates with bacterial culture.
3. Keep the plates in the jar having Gaspak indicator strip of methylene blue and gas pak generator envelopes.
4. Pull the strip halfway so that the colour change can be observed.
5. Cut the envelope and add 10 ml of tap water into the open envelope (Fig. III.1.8).
6. Close the whole assembly and incubate at 37°C.
7. Check the jar after 2–3 hours to see whether indicator strip has totally lost its blue colour which shows absence of O_2.
8. Incubate it for 2 days.

(c) Results

Anaerobic bacteria will show the growth.

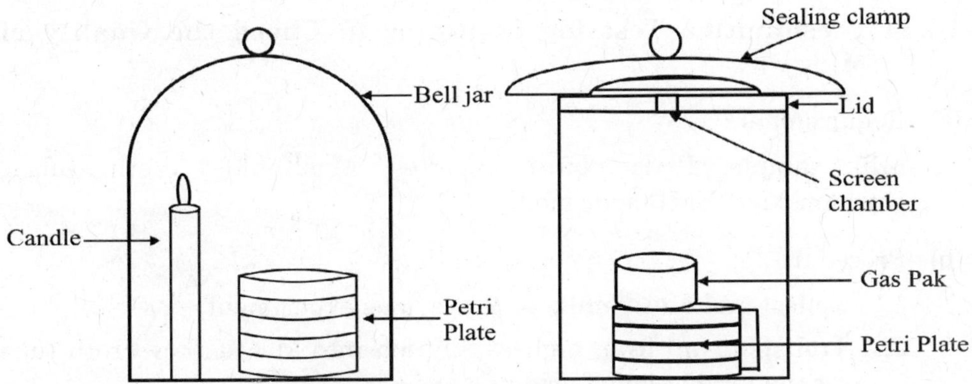

Fig III.1.7 : Candle jar method for cultivation of anaerobic bacteria
Fig III.1.8 : Gaspak anaerobic jar for cultivation of anaerobic bacteria

1.7.13 Analysis and Enumeration of Bacteria in Milk

Milk is a highly perishable food and can be the source of contamin-ation of various pathogenic microorganisms. American Public Health Association (APHA) has recommended the standard plate count as the official method of analysis of milk.

(a) Requirements

Milk sample, Petri plates, Nutrient agar, pipettes, colony counter.

(b) Procedure

1. Collect raw milk sample.
2. Dilute the milk sample in sterile water up to 10^{-6} as described in 1.4.4.
3. Prepare the nutrient agar plates and transfer 1 ml from each dilution to agar plates in duplicate.
4. Incubate the plates in inverted position at 32°C for 2 days.
5. Count the number in those Petri plates which contain 20–200 bacterial colonies.
6. Calculate the number of organisms/ml of milk sample using the formula given in serial dilution experiment.

(c) Results

Bacterial colonies will grow on plates. Record the number of bacteria/ ml of sample.

1.7.14 Presumptive Test for Coliforms to Check the Quality of Milk

(a) Requirements

Milk sample, Petri plates, pipettes, MacKonkey broth, tubes, Durham tube, EMB agar medium.

(b) Procedure

1. Collect and dilute milk as in previous experiment.
2. Transfer 1 ml from highest dilution into MacKonkey broth tube for the production of acid and gas.
3. Place a single Durham tube in inverted position inside the broth tube and incubate for 1 day at 37°C.
4. Observe the acid and bubble production.
5. For confirmation streak the diluted milk from positive broth tube onto EMB agar plates.
6. Incubate the plates at 37°C for 2 days.
7. Examine the colonies with greenish metallic sheen.

(c) Observation and Results

Change in pH of broth and bubbles in Durham tube will show the acid and gas production, respectively and coliform bacteria with greenish sheen will grow in EMB agar plates.

1.7.15 Microbiological Examination of Cheese

(a) Requirements

Cheese samples, Bromocresol purple lactose agar, Petri plates, pipettes, Skim milk agar, methylene blue.

(b) Procedure

1. Collect samples of cheese of different ages.
2. Take 1 g of each sample and grind in sterile mortar and pestle.
3. Prepare the series of dilutions from this.
4. Prepare Bromocresol purple lactose agar plates.
5. Transfer 1 ml from each dilution to plates.
6. Incubate the plates at 37°C for 2 days.
7. For proteolytic organisms inoculate 1 ml of 10^{-3} sample into plates containing skim milk agar, incubate as above.
8. For gas producers follow the method given in previous experiment.

(c) Observation and Results

Count the number of microorganisms g^{-1} of cheese. Prepare smear, stain with methylene blue and examine under microscope. Record the type of bacterial cells. If proteolytic organism is present clear zone will be formed and production of gas will confirm the presence of gas formers in sample.

1.7.16 Microbiological Examination of Butter

Butter is not an ideal medium for the growth of microorganisms, however, undesirable microorganisms, such as yeasts and moulds, proteolytic and lipolytic organisms and coliforms may arise and give objectionable odour and colour of butter. The following satisfactory levels are recommended for creamery butter: psychrophiles <1000/ g, coli-aerogenes organisms <10/ g and yeasts and moulds <50/ g.

(a) Requirements

Old butter sample, Ringer's solution, plates containing selective media, *viz.*, malt extract agar, milk agar, butter fat agar and violet red bile agar, pipettes.

(b) Procedure

1. Collect the sample in sterilized container and melt by placing it in water bath at 45°C.
2. Add Ringer's solution quarter strength with agar (0.1%).
3. Dilute the sample up to 10^{-4}.
4. Pour 1 ml from each dilution on plates containing malt extract agar, milk agar, butterfat agar and violet red bile agar.
5. Incubate malt extract agar plates at 25°C for 5 days, butter fat agar and milk agar plates at 30°C for 3 days and violet red agar plates at 30°C for 2 days.

(c) Observations

Yeasts and moulds will grow on malt extract agar plates, colonies of proteolytic organisms will be surrounded by clear zone on milk agar, lipolytic organisms will also show clear zones due to hydrolysis of fat on butter fat agar, whereas, coliforms, if any will appear in violet red bile agar plates.

(d) Results

Record the number of microbial types per g of butter.

1.8 COUNTING OF CELLS/SPORES

Counting of cells/spores is required in various experiments. It can be done by direct or indirect methods. For direct method, Petroff Hausser counting chamber, electronic cell counter and are used for counting the number of microorganisms on the slide under the micro-scope. For indirect methods, dry weight, cellular constituents, oxygen update, CO_2 production, turbidimetric measurement, membrane filter count or plate count are determined. The last two methods are used for counting the viable cells, whereas all other methods measure viable as well as non viable cells.

1.8.1 Measurement of Cells/Spores by Counting Chamber (Hemocytometer procedure)

Hemocytometer was originally meant to count blood cells, but this is also an excellent method for determining the number of sporangia, zoospores, spores and conidia in various fungi. One entire grid on standard s with Neubauer rulings can be seen at 40× (40× objective). The main divisions separate the grid into 9 large squares. Each square has a surface area of one square mm, and the depth of the chamber is 0.1 mm. Thus, the entire counting grid lies under a volume of 0.9 mm-cubed. Each square of the haemocytometer (with cover slip in place) represents a total volume of 1.0 mm^3 or 10^{-4} cm^3. Since 1 cm^3 = 1 ml, the subsequent cell concentration per ml can be determined.

(a) Requirements

Hemocytometer, cell/spore suspension, pipette, microscope.

(b) Procedure

1. Prepare a cell/spore/conidial suspension by filtering the fungal culture suspension through muslin cloth to get rid of mycelium.
2. Prepare a haemocytometer for use by carefully cleaning all surfaces of the and cover slip.
3. Centre the cover slip on the haemocytometer.
4. Pipette approximately 50 µl (this volume will vary slightly with the brand of haemocytometer) of the spore suspension into one of the two counting chambers using a clean pipette tip (Fig. III.1.9).
5. Fill the chambers slowly and steadily avoiding injecting bubbles into the chambers.

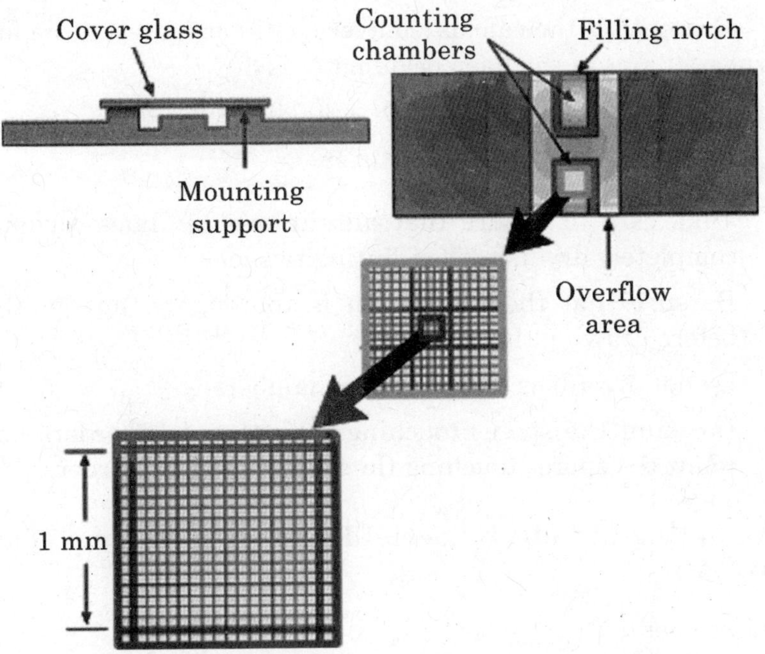

Fig. III.1.9 : Counting chamber of hemocytomer

6. Count all the spores in each of the four corner squares labeled A to D in Fig. III.10.

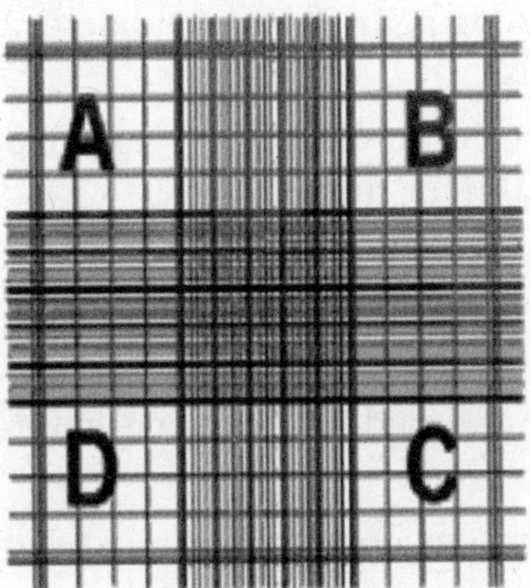

Fig. III.1.10 : Enlarged view of counting chamber.

Determining the spore count

1. Calculate the total spores counted in the four corner squares.
2. Calculate the spore count using the following equation:

$$\text{Spores/ml} = (n) \times 10^4$$

where, n = the average cell count of the four corner squares counted.

Example: If the calculated average (n) of spores in the four 1 mm corner squares of the haemocytometer is 30

$$\text{spores/ml} = 30 \times 10^4 = 300{,}000 \text{ spores/ml}$$

(c) Precautions

1. Take care to ensure that all surfaces of haemocytometer are completely dry using non-linting tissue.
2. Be sure that the suspension is thoroughly, but gently, mixed before drawing the samples.
3. Do not overfill or underfill the chambers.
4. Do count the spores touching the top or left borders but do not count the spores touching the bottom or right borders.

1.8.2 Counting of Cells by Serial Dilution Technique (Viable cell count)

(a) Requirements

Bacterial or fungal suspension, Petri plates, pipettes, Nutrient agar / PDA media, sterile water, flasks, L shaped glass rod.

(b) Procedure

1. Prepare nutrient agar plates for bacterial cells and PDA plates for fungal cells.
2. Prepare the series of dilution as described in 1.4.4. up to 10^{-7} for bacteria and 10^{-4} for fungus.
3. Mark the flasks as 10^{-1}, 10^{-2}, 10^{-3} and so on.
4. Transfer 1 ml of suspension to Petri plates in triplicate (alternatively, 1 ml suspension can be transferred to sterile empty Petri plate and then medium can be poured on it, rotated and solidified).
5. Spread the suspension with L shaped sterile glass rod.
6. Incubate the plates in inverted position at 37°C for bacteria for 2 days and 28°C for fungi for 5 days.

(c) Observation

Examine the type of colonies appearing on the Petri plates and count the no. of colonies by Quebec colony counter. Fungal colonies can be counted by naked eyes.

(d) Results

Calculate the number of cells per ml of original sample by following formula:

$$\text{Cells/ml} = \frac{\text{Average no. of colonies}}{\text{Amount plated} \times \text{dilution}}$$

Here, amount plated may be 1 ml or 0.1 ml and dilution may be 10^{-1}/10^{-2}/10^{-3} and so on.

1.8.3 Counting of Bacterial Population Spectrophotometrically

The broth becomes turbid with the growth of bacteria which can be used as an indicator of bacterial density as turbidity is directly proport-ional to the number of cells. Turbidity can be quantified with spectropho-tometer by measuring the light transmitted or absorbed directly through the sample.

The density of a cell suspension is expressed as absorbance or optical density (OD), a value derived from the percentage of transmission.

(a) Requirements

Test tubes, culture of *Bacillus* sp., pipette, Bunsen burner, spectro-photometer, nutrient broth.

(b) Procedure

1. Prepare 12 tubes by filling nutrient broth.
2. Inoculate 0.1 ml of culture in only 7 of the tubes.

The use of spectrophotometer is given in the Instrumentation Chapter II. OD will increase with increasing turbidity. You need to create a standard curve of OD versus colony counts first. Make different dilutions of a culture. Take aliquots to do colony counts, and take the OD readings. Once you have all the data, plot the standard curve. From now on, all you have to read is the OD of a culture, you will get the estimation of bacterial number, without waiting a couple of days for the colony to show up. Incubate the tubes at 37°C for 2 days.

(c) Observation

Record the OD of the incubated tubes at 0, 2, 4, 6, 8, 12, and 24 h. against un-incubated tubes (blanks) using spectrophotometer.

(d) Results

Plot the graph of OD against time period.

1.9. MICROBIAL GROWTH

Unicellular organisms are said to be growing when their cells multiply and they increase in number. When bacteria are inoculated in suitable medium and incubated under the conducive conditions, they grow very fast and in batch culture show a sigmoid curve.

Four stages can be clearly observed in batch culture: **Lag Phase** is the time gap before the multiplication starts in nutrient broth. Once bacteria are adjusted in new environment they start consuming nutrients and multiply and this phase is called as **Log Phase**. In this phase cells divide steadily at a constant rate and show the exponential growth. Then the growth becomes stagnant as the nutrients in broth are utilized by growing cells. This phase is known as **Stationary phase** in which the regeneration of new cells is equivalent to the number of cells dying. If new medium is not added the nutrients are diminished and the toxic metabolites are accumulated as a result the growth is retarded. Thus **Decline Phase** is characterized by death of bacterial cells.

1.9.1 Determination of Bacterial Growth Curve by Spectrophotometric Method

The time in which the cell number is doubled is called the doubling time or generation time. The average generation time is 30–60 minutes under optimum conditions. *E. coli* grows very rapidly and has generation time of 20 minutes only, whereas, *Mycobacterium* has longest generation time of 10–30 days.

(a) Requirements

Pure culture of *E. coli* (10–12 hours old), Nutrient broth, Erlenmeyer flask, incubator shaker, cuvettes.

(b) Procedure

1. Activate *E. coli* culture for 10–12 hours.
2. Autoclave 100 ml of medium in 250 ml Erlenmeyer flasks in duplicate.
3. Inoculate the medium with 1 ml of activated culture.
4. Withdraw 2 ml of broth in a cuvette in laminar air hood and record initial OD.
5. Place the flask in incubator shaker at 150 rpm and 37°C for 24 hours.
6. Withdraw 2 ml of broth at every 30 minutes and measure OD at 600 nm wavelength. Uninoculated medium is used as blank.
7. Plot the graph between optical density versus incubation time.

(c) Result

A sigmoid growth curve of *E. coli* is obtained.

Note: *The graph can also be plotted using logarithms of bacterial number versus time.*

Generation time (g) can be calculated by following formula:

$$g = \frac{t \log^2}{\log^b - \log^a}$$

Where,

t = time of incubation.

a = number of cells in inoculum.

b = number of cells at the time of final recording.

log = logarithm to the bases 10.

1.9.2 Measurement of Fungal Growth by Colony Diameter Method

Most fungal species make observable colonies in Petri dish. One of the methods of measuring fungal growth is to measure the colony at right angles to each other and average is calculated. Cells divide exponentially and form colony in solid medium. The colony grows rapidly at its edges while the cells at centre grow more slowly or begin to die because they get smaller quantities of nutrients and are also exposed to wastes. Therefore, in contrast to bacteria all the phases of growth curve occur simultaneously in fungal colonies.

(a) Requirements

Culture of *Aspergillus* sp. or any other fungi, Czapek's Dox agar plates, cork borer, scale, burner.

(b) Procedure

1. Inoculate *Aspergillus* culture in the centre of Czapek's Dox Agar plate and incubate the plate at 28 ± 2°C for 3–4 days.
2. Sterilize the cork borer and cut the agar discs from the edge of growing colony. Transfer one disc to the centre of each of 3 plates.
3. Incubate the plates at 28 ± 2°C in inverted position for 10–12 days.

(c) Observation

Examine the plates every day and measure the colony diameter in all the plates. Take two readings at right angle to each other and calculate the mean values. Plot a graph.

(d) Result

A linear growth of *Aspergillus* is observed in the graph.

1.9.3 Measurement of Fungal Growth by Dry Weight of Mycelium

(a) Requirements

Culture of *Aspergillus* species (or any other fungi) Czapek's Dox agar and broth, cork borer, 100 ml conical flasks, Whatman filter paper, funnel, burner, oven.

(b) Procedure

1. Prepare a Czapek's Dox agar plate.
2. Inoculate the culture of *Aspergillus* in the centre of plate and incubate it at 28 ± 2°C.
3. Prepare the flasks in triplicate by pouring 30 ml of Czapek's Dox broth and sterilize them.
4. Cut the mycelial discs of 5 mm diameter from the edges of the growing colony in agar plate.
5. Inoculate the disc in each flask and incubate them at 28 ± 2°C for 5–10 days depending upon the growth rate of test fungus.
6. Filter the contents of flasks through a pre-weighed Whatman filter paper to obtain the mycelial mat.
7. Dry the filter paper with mycelial mat at 80°C for 1 day.

(c) Observation and Results

Calculate the dry weight and growth rate by following formulae:

Dry weight = (Weight of filter paper + Mycelial mat) – Weight of filter paper

Growth rate/day = Dry weight/Days of incubation

Note: *For inoculation of Czapek's Dox broth flasks 1 ml of spore suspension in sterilized water prepared from pre-grown culture can also be taken.*

For measuring the growth increments in dry mycelial weight, start the experiment with 20 flasks, inoculate them as above and incubate at 28 ± 2°C. Take out 2 flasks each consecutive day and filter the mycelial mat, oven dry and calculate the dry weight. Repeat the process until 10 days. Prepare a graph illustrating the growth increments in dry wt. of mycelium.

1.9.4 Estimation of Biomass

For estimation of biomass two methods of direct measurement can be carried out:

Dry cell weight: It provides the simple and crude means of quantifying the cell mass through centrifugation or filtration. The net result of dry cell weight analysis may be ±30% or more from actual cell mass in fermentor. Besides, it reflects both dead and live cell mass.

Packed cell volume: This is a rapid method employed by industries to follow the kinetics of fermentation. Packed cell volume measured by conical graduated centrifuge tubes depends not only on time but also on the amount of non cellular solids, the culture morphology and osmolarity of the medium.

1.9.4.1 Dry cell weight estimation

(a) Requirements

Culture of *S. cerevisiae*/*A. niger*/*E. coli*, graduated Eppendroff tubes, MYE (Malt extract yeast extract)/Czapek's Dox media/LB broth.

(b) Procedure

1. Prepare MYE broth, dispense 100 ml of it in 250 ml Erlenmeyer flask and sterilize.
2. Inoculate the flask with *S. cerevisae* pure culture and incubate in an incubator shaker for 24 h.
3. Weigh empty graduated Eppendroff tubes.
4. Transfer 1.5 ml culture broth to the tubes in triplicate from the flasks.
5. Keep the flask again in incubator shaker.
6. Centrifuge the tubes, discard the supernatant and wash the pellet with distilled water.
7. Place the tubes in an oven to dry and weigh the tubes with pellet.
8. The steps 3–7 are repeated after every 24 h for 3 days.

(c) Observations

Calculate the dry cell weight by subtracting the weight of empty tubes from final reading.

S. No.	Sampling time	Wt. of empty Eppendroff tubes (g)	Wt. of Eppendroff tubes + pellet (g)	Dry cell wt. g/1.5/2 ml of culture medium	Average
1					
2	24 h				
3					
4					
5	48 h				
6					
7					
8	72 h				
9					

(d) Result

Dry cell weight/100 ml medium is calculated.

1.9.4.2 Packed cell volume (PCV) determination

(a) Requirements

Same as above (1.9.4.1).

(b) Procedure

1. Prepare MYE broth dispense 100 ml of it in 250 ml Erlenmeyer flask and sterilize.
2. Inoculate the flask with *S. cerevisae* pure culture and incubate in an incubator shaker for 24 h.
3. Take 1.5 ml of culture broth from the flask in graduated Eppendroff tubes in triplicate and centrifuge at 5000 rpm for 15 min.
4. Discard the supernatant, wash the pellet and centrifuge.
5. Measure the volume of remaining pellet using graduation on the tubes.
6. Repeat the procedure for 3 consecutive days.

(c) Observations and Results

S. No.	Sampling time	Initial volume (ml)	Packed cell volume (ml)	PCV (ml of pellet/ volume of culture medium)	PCV (ml of pellet/ 100 ml of culture medium)	Average
1		1.5				
2	24 h	1.5				
3		1.5				
4		1.5				
5	48 h	1.5				
6		1.5				
7		1.5				
8	72 h	1.5				
9		1.5				

(d) Precautions

1. All possible aseptic conditions should be maintained.
2. Culture should be shaken well before sampling, so that aliquot represents the whole culture.

1.9.5 Effect of Temperature on Growth of Fungi/Bacteria

Temperature is an important factor that influences the growth of a microorganism. The temperature, lower or higher than the optimum range slows down the metabolism of the organism by affecting its enzyme system. Therefore, it is essential to know the optimum temperature for the growth of an organism. Depending upon the temperature range which they can tolerate, microorganisms can be divided into mesophiles (25–40°C), psychrophiles (–2 to 20°C) and thermophiles (40–80°C).

(a) Requirements

Fully grown pure culture of fungi (*e.g., Aspergillus*/*Alternaria*) or bacteria (*E. coli*), Nutrient agar/broth or Czapek Dox agar/broth, Bunsen burner, inoculating loop.

(b) Procedure

1. Prepare 24 flasks of Czapek's broth for fungi or nutrient broth for bacteria.

2. Inoculate each flask (18) with 1 ml of spore/bacterial cell suspension made in sterilized water from previously grown culture on agar plates.
3. Incubate the flasks in triplicate in incubators set at 5, 15, 25, 35, 45, 55°C. For 5°C (or 4°C) refrigerator can be used. Uninoculated flasks (one) will serve as control for each temperature.
4. Observe the flasks after 6 days in case of fungi and on 2nd day in case of bacteria.
5. Estimate the dry weight of mycelial mat for fungi and turbidity for bacteria by spectrophotometric method as described in earlier protocols.

(c) Observation

Make an observation table and enter all the readings.

S.No.	Temperature (°C)	Dry Weight	Average dry weight
1			
2	5		
3			
1			
2	15		
3			
1			
2	25		
3			

(d) Results

Determine the optimum range of temperature and classify whether the organism is psychrophilic, mesophilic or thermophilic.

To make the experiment simpler, the flasks can be observed by naked eyes for presence of mycelium and spores (for fungi) and turbidity (for bacteria) and note down as follows:

Maximum growth	+++
Moderate growth	++
Slight growth	+
No growth	–

This experiment can also be performed in agar plates by streaking the test organism on it. At the end the qualitative estimation of growth can be done by naked eyes as above.

(e) Precautions

Incubators must be pre-set at appropriate temperatures.

1.9.6 Effect of pH on Growth

Each enzyme system of an organism has a particular pH range at which it can function. The pH values below and above at which an organism fails to grow are respectively termed as minimum and maximum hydrogen ion concentrations. Each species has the ability to grow within a specific pH range, that may be broad or limited, in which the growth occurs at maximum rate. For fungi the optimum pH range is 5.5 to 6.5 and for bacteria 6.7 to 7.5, however, the generalization can not be made.

Many microbes produce organic acids and alkali which change the pH of the medium and influence adversely on the growth of organism. Thus to maintain equilibrium and to neutralize these acids and alkali, buffers are added. Peptones, proteins and amino acids act as natural buffers due to their amphoteric nature.

(a) Requirements

Cultures of fungi (*Aspergillus*/*Penicillium*) or bacteria (*E. coli*/ *Staphylococcus*), 100 ml Erlenmeyer flasks, buffered broth (Czapek's Dox for fungi and Nutrient broth for bacteria), inoculating loop, Bunsen burner, pipettes.

(b) Procedure

1. Adjust the pH of the broth (30 ml) to 2.5, 3.5, 4.5, 5.5, 6.5, 7.5, 8.5 and 9.5 in triplicate flasks and sterilize. Instead of flasks, tubes can also be taken.
2. Inoculate the culture by transferring 1 ml to each flask (or 0.1 ml to each tube).
3. Incubate the flasks at 37°C for bacterial cultures for 2 days. For fungi the tubes can be incubated at 28°C for 5 days.

(c) Observation and Results

Record the growth of organism qualitatively or quantitatively in an observation Table as done for experiment for effect of temperature (section 1.9.5).

(d) Precautions

1. pH should be adjusted carefully.
2. pH is usually changed a little after sterilization, hence for more accuracy (as required for research experiments), it should be

standardized before main experiment or pH should be adjusted after sterilization.

1.9.7 Determination of Antibiotic Sensitivity by Disc Method

Antibiotics are life saving drugs which are used to control the bact-erial infections. The antibiotic resistant genes are present on plasmids in bacteria which are transferred from one cell to another through conjug-ation thus increasing the antibiotic resistance in bacteria. The antibiotic sensitivity test helps in selecting the appropriate line of treatment. The effectiveness is based on size of inhibition zone.

(a) Requirements

Petri plates, Nutrient agar, swab, bacterial culture (*E. coli*), antibiotic discs, forceps, incubator.

(b) Procedure

1. Prepare the nutrient agar plates.
2. Inoculate the bacterial culture with the help of sterilized swab to make bacterial lawn on entire Petri plate.
3. Place 2 ready made antibiotic discs and 1 control disc without any antibiotic with sterilized forceps equidistant on agar surface.
4. Use as many antibiotics as available on different plates.
5. Incubate the plates for 2 days at 37°C.
6. Observe the plates and measure the diameter of inhibition zone.

(c) Observation

Enter the size of zone of inhibition with respect to the type of antibiotic disc in mm in observation Table.

(d) Results

Inhibition zone around the discs shows the sensitivity of bacteria towards the respective antibiotic. Size of zone indicates its capability to inhibit growth of bacteria.

Note: *If readymade antibiotic the discs are not available, prepare the solution of known concentration of an antibiotic in sterile distilled water and dip the disc of Whatman filter paper no. 1 in it.*

1.10 FERMENTATION TECHNOLOGY

Fermentation process is the basis of industrial microbiology. Originally, fermentation referred to the bubbling observed when sugar and starchy materials underwent a transformation to yield alcoholic beverages. Pasture applied the term to those anaerobic reactions through which microorganisms gained energy for growth in the absence of oxygen. Through the years this word has gained new meanings and today it is applied to both the anaerobic and aerobic metabolic activities of microorganisms in which specific chemical changes are brought about in an organic substrate. In fact, from an industrial microbiology standpoint the meaning is yet broader, and includes almost any process mediated by or involving microorganisms in which a product of economic value accrues.

In recent years, many fermentations of potential commercial value have been studied extensively, but only a few are actually practiced. Alcohol, acetone-butanol, lactic acid and antibiotic fermentations are some common examples of type of fermentations mostly used in various industries.

1.10.1 Estimation of Lactic Acid in Curd

The lactic acid fermentation has been one of the important methods of food preservation for untold centuries. Fermented pickles and dairy products constitute a vital part of the human diet in many parts of the globe. The preservation and concomitant transformation of flavour and texture of milk by fermentation had been known since antiquity. Milk is the normal habitat of lactic acid bacteria which may cause spontaneous souring. For sophisticated control, modern industrial processes utilize specially prepared lactic acid bacteria as starter culture in the manufacture of fermented dairy products. *Streptococcus* and *Lactobacillus* are common organisms needed to set curd or 'Dahi'. Temperature and pH are the crucial factors for maximum yield. The biochemical changes involve the conversion of lactose of milk to lactic acid by enzymatic action.

(a) Requirements

Curd, beaker, burettes, phenolphthalein solution (10 g of phenolphthalein, 1 ml methanol and make up to 10 ml distilled water), 0.02 N NaOH, muslin cloth.

(b) Procedure

1. Take 20 ml curd and make three samples by diluting it to different concentrations and label them as I, II, and III, respectively.
2. Solution I contains well mixed curd, II contains small quantity of water mixed in curd (2.5 ml of water : 2.5 ml of curd) and solution III contains high amount of water in curd (1.25 ml of curd : 3.75 ml water).

3. Filter the samples of curd through muslin cloth.
4. Take 5 ml filtrate in a conical flask from each sample and titrate against 0.02 N NaOH solution after adding few drops of phenolphthalein as indicator in the filtrate.
5. Titrate until the end point, *i.e.*, light pink colour appears.
6. Finally calculate the normality and strength of lactic acid.

(c) Observations and Results

Sample	Volume of NaOH used (ml)			Average (ml)
	I	II	III	
1	19	21	20	20
2				
3				

$$N_1V_1 = N_2V_2$$

$$N_2 = N_1V_1 / V_2$$

$$= 0.02 \times 20 / 5$$

$$N_2 = 0.08 \text{ N}$$

$$\text{Strength} = N_2 \times \text{molecular weight}$$

$$= 0.08 \times 90$$

$$= 7.20 \text{ g/litre}$$

The strength of lactic acid will decrease with dilution of samples.

(d) Precautions

1. The base should be added drop wise.
2. Flasks should be shaken constantly.

1.10.2 Production and Estimation of Citric Acid by *Aspergillus niger*

Citric acid is widely used by the food, pharmaceutical and chemical industries as a major substrate for the production of a variety of products. The demand for citric acid is increasing day by day. When applied to appropriate mass balances, it is possible to predict the utilization of substrates and the yield of individual products. Fermentation media for citric acid biosynthesis should consist of substrates necessary for the growth of microorganism, primarily the carbon, nitrogen and phosphorus sources. The basic substrates for citric acid fermentation using submerged technique of fermentation are beet or cane molasses.

Many moulds have the ability to produce citric acid from sugar, however, *Aspergillus niger* is most widely used species in industries due to its high yielding strains, fairly uniform biochemical properties and production of low amount of byproducts such as oxalic acid.

(a) Requirements

Czapek's Dox agar, citric acid producing strain of *A. niger*, Erlenmeyer flasks of 250 ml capacity, pipette, burette, TCA (5%), anhydrous acetic anhydride, pyridine, standard solution of citric acid, screw cap tubes.

(b) Procedure

(I) Production of citric acid

1. Prepare the medium, put 100 ml into 5 flasks and sterilize.
2. To activate the culture inoculate one flask with spores of *A. niger* and incubate it at 28°C for 2 days.
3. Inoculate the remaining 4 flasks from the activated culture and incubate at 28°C for 5 days.
4. Filter the mycelial mat using funnel containing filter paper.
5. Collect the filtrate in flasks and use further for estimation of citric acid.

(II) Estimation of citric acid

1. Take 0.5 ml of unknown sample in duplicate tubes and make the volume up to 1 ml by adding TCA (5%).
2. To prepare the standard graph, take 0.01, 0.02, 0.03 ...... 0.10 ml of standard in different screw cap tubes.
3. Make the volume of each tube up to 1ml by adding TCA (5%).
4. Prepare the blank with 1 ml of TCA (5%) (Instead of 5% TCA distilled water can also be taken to make up the volume of tubes and for blank).
5. Add 8 ml of anhydrous acetic anhydride to all the tubes and screw cap tightly.
6. Place the tubes in water bath at 60°C for 10 min.
7. Cool the tubes, add 1 ml of pyridine to each tube, re-stopper them quickly and transfer on ice bath for 40 min.
8. Read the absorbance at 420 nm and plot the graph between absorbance and amount of citric acid in standard solutions.
9. Calculate the amount of citric acid in sample by standard graph.

Observation Table

S. No.	Volume of citric acid solution (ml)	Volume of 5% TCA (ml)	Amount of citric acid μg/ml	Incubation				OD at 420 nm
1.	0.0	1.0	0	Add 8 ml of anhydrous acetic anhydride to each tube	Place the tubes in water bath for 10 min.	Add 1 ml of pyridine to each tube on cooling.	Place in ice bath for 40 min.	0.00
2	0.01	0.99	40					
3	0.02	0.98	80					
4	0.03							
5	0.04							
6	0.05							
7	0.06							
8	0.07							
9	0.08							
10	0.09							
11	0.1							
12	0.5 (unknown sample)	0.95	–					

(c) Results

The amount of citric acid present in 1 ml is estimated by standard graph.

(d) Precautions

1. Make sure that the strain of *A. niger* selected for the experiment is good citric acid producer.
2. Transfer of inoculum to medium should be done aseptically.
3. Great precautions should be taken while adding anhydrous acetic acid and pyridine, and heating at 60°C should also be done carefully.

1.10.3 Estimation of Growth and Product Yield During Bioconversion of Tannic Acid to Gallic Acid by *Aspergillus niger* MTCC 282

Gallic acid (3,4,5-trihydroxybenzoic acid) is an organic substance occurring in many plants either as a free molecule or as part of tannic acid molecule. Gallic acid is extensively used as an ingredient of developer in photography and printing inks. It also serves as a precursor for the commercial production of an anti-microbial drug trimethoprim, a food – preservative propyl gallate and some dyestuffs.

Various microbial species are well known for transformation or bioconversion of substrates into industrially important compounds through enzymatic reactions. Tannin acyl hydrolase produced by *A. niger* MTCC-282 converts tannic acid into gallic acid. Gallic acid formed can be estimated spectrophotometrically. Residual tannic acid and gallic acid formed are determined by using two wavelengths estimation method, where 254.6 nm and 293.8 nm refers to the wavelengths showing maximum difference in the absorbance of tannic acid and gallic acid.

$$\underset{\text{(Penta galolyl glucose)}}{\text{Tannic acid}} \xrightarrow{\text{tannase}} \underset{\text{(3,4,5 trihydroxy benzoic acid)}}{\text{Gallic acid}} + \text{Glucose}$$

(b) Requirements

A. niger MTCC 282, Czapek's Dox medium, tannic acid, flasks, pipette.

(c) Procedure

1. Grow *A. niger* MTCC 282 in Czapek's Dox medium (containing 0.1% tannic acid) for 24–48 hrs. at 28–30°C in incubator shaker and use it as inoculum.
2. Sterilize the tannic acid stock solution and add to the bioconversion flasks in triplicate containing Czapek's Dox medium to get the final concentration of tannic acid (5%).
3. Inoculate the bioconversion flasks with the already prepared inoculum of *A. niger* and incubate at 28 ± 2°C in an incubator shaker.
4. Prepare a control without inoculating with fungus.
5. Draw the samples after 0, 24, 48 and 72 hrs from each flask.
6. The samples were subjected to biomass and product analysis.

Observation Table

Time	Sample	OD at 254.6 nm	OD at 293.8 nm	Concentration of Tannic acid (µg/ml)	Concentration of Gallic acid (µg/ml)
0 min	1				
	2				
	3				
24 h	1				
	2				
	3				

Determination of tannic acid and gallic acid

1. Filter the samples through filter paper.
2. Dilute the filtrate to 1:1,000.
3. Take OD at 254.6 nm and 293.8 nm against distilled water as blank.
4. Calculate the concentration of tannic acid and gallic acid in µg/ml using the following formula:

$$\text{Tannic acid } (\mu g/ml) = 34.41 \times A_{293.8} - 6.98 \times A_{254.6}$$

$$\text{Gallic acid } (\mu g/ml) = 21.77 \times A_{254.6} - 17.17 \times A_{293.8}$$

(d) Observations and Result

Calculate the percent conversion of tannic acid to gallic acid. Give the optimum time of incubation for maximum conversion percentage.

(e) Precautions

1. Sampling should be done at regular intervals.
2. The dilution of the sample should be done carefully.
3. Aseptic conditions should be maintained.

1.10.4 Determination of Alcohol Production in the Fermented Broth

(a) Principle

Fermentation is a metabolic pathway by which cells catabolize, or break down a carbon source to produce energy. Fermentation refers to catabolic processes where organic molecules, such as sugars or amino acids, are broken down to produce energy without the use of a membrane-bound electron transport chain. Depending upon the organism, fermentation can occur in the presence (aerobic) and/or in the absence (anaerobic) of oxygen. Fermentation pathways produce byproducts such as carbon dioxide, ethanol (alcohol), or organic acids (lactic acid or acetic acid, for example).

Under aerobic condition acids are produced and there is increase in biomass. Under anaerobic condition yeasts produce alcohol. In order to maximize the conversion of carbohydrate to alcohol, the medium is initially aerated and after proper growth aeration is stopped. Thus the condition becomes anaerobic and the biomass converts the remaining carbohydrate to ethanol. Most alcoholic beverages are manufactured by the metabolic activity of *Saccharomyces cerevisae*.

This method is based on the procedure given by Caputi *et al.* (1968). The estimation method for ethanol is based on the complete oxidation of ethanol by dichromate in presence of sulphuric acid with the formation of acetic acid. The green colour produced during the reaction is due to the formation of chromate ions.

Reaction

$$2Cr_2O_7^{-2} + 3C_2H_5OH + 16\ H^+ \longrightarrow 4\ Cr^{+3} + 3CH_3COOH + 11\ H_2O$$

(yellow colour) (green colour)

(b) Requirements

- YPD (Yeast Peptone Dextrose medium).
- *Saccharomyces cerevisae* MTCC 36, inoculating needle, beaker, $K_2Cr_2O_7$ solution, absolute alcohol.

Potassium dichromate solution ($K_2Cr_2O_7$)

Disolve 34 gm of Potassium Dichromate in 500 ml distilled water and make the volume to 1000 ml by adding 325 ml of conc. H_2SO_4 by stirring and keep the flask in ice bucket.

(c) Procedure

1. Dispense the YPD medium in the flasks and sterilize.
2. Inoculate the flasks with 24 h old inoculum raised in the same medium.
3. Keep three flasks in static conditions (mouth properly sealed to create crude anaerobic condition) and three in stirred condition (with cotton plug to keep air passing into the flask) at 30°C.
4. Draw the samples from each flask after 24, 48 and 72 h for analysis of biomass and alcohol production.
5. Examine both aerobic and anaerobic media containing flasks and subject them to quantitative estimation of alcohol.

Preparation of Ethanol standard

1. Take 0, 10, 20, 40, 60, 80, 100, 120 and 140 μl of absolute alcohol in different test tubes.
2. Make the total volume up to 5 ml with Potassium Dichromate solution.
3. Incubate the tubes at 60°C for 20 minutes.
4. Take the absorbance at 540 nm against a blank.

Ethanol estimation in culture

1. Take 1 ml culture from both aerobic and anaerobic cultures in separate tubes.
2. Make volume up to 3 ml with distilled water.
3. Add 5 ml of $K_2Cr_2O_7$ solution in each tube.
4. Incubate the tubes at 60°C for 20 min.
5. Record the absorbance at 540 nm against a blank.

(d) Observation Table

Incubation period (hrs.)	Sample	OD at 540 nm	Alcohol produced (µl) by 1 ml of sample (anaerobic culture)	Alcohol produced (µl) by 1 ml sample (aerobic cultures)
24				
48				
72				

(e) Result

More alcohol is produced in anaerobic flasks than in aerobic ones.

(f) Precautions

1. Aseptic conditions should be maintained.
2. Temperature of both the aerobic and anaerobic flasks must be same.
3. Preparation of potassium dichromate should be done carefully.
4. Chilled sulphuric acid should be used and acid should be added to water to avoid heating of solution.
5. Anaerobic condition should be properly maintained.

1.10.5 Alpha-amylase Biosynthesis and Measurement of its Activity

(a) Principle

The α–amylase [E.C. 3.2.1.1, (1,4- α-D glucanolydrolase)] is an exoenzyme that hydrolyses alpha-bonds of large alpha-linked polysaccha-rides such as starch and glycogen, yielding glucose and maltose. It is widespread among living organisms. In the digestive system of humans and many other mammals, an alpha-amylase called ptyalin is produced by the salivary glands, whereas pancreatic amylase is secreted by the pancreas into the small intestine. It is also present in seeds containing starch as a food reserve. Amylase production is naturally found in some bacterial species, while it is well known in fungi. Amylase commercially produced from various species of Aspergilli is used in the initial steps in several fermentation processes to convert starch to fermentable sugar in food industries. They are also used to partially predigest food for young children, to clarify fruit juices and to manufacture laundry detergents.

The ability of alpha amylase to degrade starch is used as criteria for the determination of amylase production by microbes. The estimation of alpha amylase is carried out by srtarch-iodine method. After the reaction with the substrate, the unreacted substrate concentration is measured (Yoo *et al.*, 1987).

(b) Requirements

Young culture of *Bacillus cereus* (MTCC 1308), growth medium for *B. cereus* (see appendix IV A), starch solution, 1N HCl, phosphate buffer, I_2 solution.

Reagents

1. Starch solution: 1% soluble starch in 0.1M phosphate buffer, pH-6.9.
2. Phosphate buffer :
   (a) 0.2 M NaH_2PO_4 (27.8 g in 1 litre).
   (b) 0.2 M Na_2HPO_4 (53.65g in 1 litre).
   Mix 45 ml of (a) and 55 ml of (b), and then dilute to a total of 200 ml.
3. 1N HCl–44.5 ml concentrated HCl is made up to 500 ml with distilled water.
4. Enzyme extract.
5. I_2 solution:
   (stock) – 500 mg I_2 + 5g KI made up to 100 ml.
   (working) – 1 ml stock made up to 100 ml.

(c) Procedure

Enzyme biosynthesis

1. Activate *B. cereus* in 50 ml broth.
2. Prepare and sterilize the medium for culturing *B. cereus* in four 250 ml flasks.

3. Inoculate 3 flasks with *B. cereus* and use the fourth one as control.
4. Incubate the flasks at 37°C.
5. Collect the sample of 0 h (3 ml).
6. Similarly, collect the samples at different incubation periods such as 24, 48, 72 h.
7. Keep these samples in freezer in order to prevent inactivation of enzyme that has been produced.
8. Centrifuge all the samples in a cooling centrifuge and use the supernatant for measuring the enzyme activity.

Enzyme activity determination

1. Take 5 ml of buffered starch and 0.5 ml of suitably diluted sample.
2. Incubate the mixture at 90°C for 10 min.
3. Add 4.5 ml of 1 N HCl and vortex.
4. Mix 0.5 ml of the above mixture with 5 ml of working I_2 solution.
5. Measure the intensity of blue colour produced at 640 nm.

Control: Prepare the control with enzyme and substrate blank.

Substrate blank: 5 ml starch + 0.5 ml distilled water and 4.5 ml 1 N HCl + 5 ml I_2 solution.

Enzyme blank: 5 ml starch + 4.5 ml 1 N HCl + 0.1, 0.2 ml enzyme made upto 0.5 ml with distilled water + 0.5 ml I_2 solution.

(d) Observation Table

Time	Samp	Enz. (ml)	D.W. (ml)	Buffered sol. (ml)	Incubation at 90°C for 10 min.	1N HCL	0.5 ml of this mixture is taken	Sub. (ml)	OD (540 nm)	Average for 1 ml sample

Calculation

$$\% \text{ activity (units/ml)} = \frac{D[R_0 - R]}{R_0} \times 100$$

D = dilution factor of the enzyme

R_0 = absorbance in the control tube in the absence of enzyme

R = absorbance of the experimental tube

$\frac{R_0 - R}{R_0}$ = indicates the linearity of the enzyme activity.

Enzyme activity: one unit of enzyme is defined as the amount of enzyme needed to digest 50 mg of soluble starch in 10 min.

(e) Precautions

1. All the steps should be carried out aseptically.
2. Samples should be drawn at fixed intervals.
3. Starch should be freshly prepared.

1.11 MUSHROOM CULTIVATION

Mushrooms are edible fungi belonging to order Agaricales of Basidiomycetes. They are quite nutritious, low in calories, but rich in minerals and a good source of proteins. Due to their food value and flavour, mushrooms have been cultivated for long. The common edible commercially grown mushrooms are white button (*Agaricus bisporus*), paddy straw (*Volvariella volvacea*), oyster (*Pleurotus sajor-caju*), shiitake (*Lentinus edodes*) etc.

Production of mushroom has already crossed 5 million metric tonnes annually in the world and is expected to reach around 7 million metric tonnes in this decade. India has been known world over for its exotic mushrooms and the cultivation of mushroom, especially the button mushroom is fast picking up in India. The current production level is around 50,000 tonnes. For the successful cultivation of any mushroom, the spawn is required which is commonly known as seed of mushroom. Spawn is a pure culture of the mycelium grown on the medium. Commercially spawn production is done in controlled environmental conditions on sterilized cereal grains.

1.11.1 Production of Spawn for White Button Mushroom (*Agaricus bisporus*)

(a) Requirements

Pure culture of *Agaricus* (It can also be obtained from NCMRT, Solan), cereal grains (wheat/rye/sorghum), gypsum, calcium carbonate (chalk powder), milk bottles or polypropylene bags, cotton, autoclave, laminar flow, incubator, Bunsen burner, water.

(b) Procedure

Substrate preparation

1. Take 1 kg of good quality wheat seeds and boil in water in a container for 20–30 minutes.

2. Remove the seeds when they become soft and spread evenly on a cotton cloth to drain out the water.
3. Mix 3 percent calcium carbonate (30g/ kg of grains) and gypsum for adjusting pH and to prevent the grains to stick together.
4. Fill 250 g of grain in clean and dried bottle of 500 ml capacity (or polypropylene bags) and plug the mouth of the bottle tightly with non absorbent cotton.
5. Sterilize the bottle. After cooling transfer the bottles to inoculation chamber.

Pure culture Preparation

1. Select a well grown mushroom with membrane covering the gills.
2. Take a small bit of gill portion using forceps.
3. Inoculate on PDA slants (pre prepared) and incubate at 25°C in an incubator.
4. The mycelium covers the entire surface in a few weeks time.
5. Alternatively, the spores can also be collected from well developed fruiting body by 'spore mapping technique' and then the spores are inoculated on PDA slants and incubated.

Multiplication using mother spawn

1. Prepare the bottles with substrate and sterilize them.
2. Take 10–20 days old well grown mother spawn.
3. Stir the spawn using sterilized forceps to get the individual grains with fungal growth.
4. Transfer few grains with mycelia growth into substrate bottle under aseptic condition and plug it with cotton.
5. Place the bottles at 25–30°C.
6. Inspect the bottles regularly and discard the contaminated bottles immediately.

(c) Results

Within 15–20 days of inoculation, mycelia growth covers the entire substrate and the spawn is ready for use.

(d) Precautions

1. Use healthy and non-infected whole grains.
2. Check the master spawn for contamination and discard such bottles.

1.11.2 Cultivation of White Button Mushroom

(a) Requirements

Compost, wooden box, tray, wooden board, calcium carbonate, water, casing soil, news paper.

A. Preparation of compost

Compost is an ideal medium for mycelia growth. Basic mushroom compost is made up of wheat, straw, horse manure and gypsum. There are a variety of optional ingredients that may be added. Besides straw, molasses, brewer's grains, cotton seed meal can also serve as carbohy-drate source. Poultry, pig and sheep manure are another alternatives to horse manure for N_2 source.

Constituents of synthetic compost

Chopped wheat straw	300.0 kg
Wheat bran	15.0 kg
Calcium ammonium nitrate	6.0 kg
Superphosphate	7.5 kg
Urea	2.4 kg
Potassium sulphate	3.0 kg
Gypsum	30.0 kg
Saw dust	10-12 kg

Procedure

1. Spray the water on sawdust and mix all the ingredients except wheat straw and leave overnight.
2. Spread the wheat straw on floor and soak it in water.
3. Spread the mixture of all other ingredients over wheat straw and mix.
4. Fill the mixture in a wooden box and let it decompose for a month.
5. Turn the compost pile starting from 6th day to obtain uniform fermentation. Watering is to be done in first two turnings.

Observation

Dark brown coloured compost which is free of ammonia is prepared.

B. Filling of tray beds

1. Mix calcium carbonate (3 kg) to prepared compost.
2. Fill the tray by compost.

3. Compress the compost by pressing it by a wooden board leaving 1 cm space at the top of the tray.

C. Spawning

Spawn can be prepared taking all the precautions or purchased from any government or non-government agency.

1. Spread the spawn on half filled tray with compost and then fill the tray completely with compost. Again the spawning is done and it is pressed by fingers.
2. Cover the trays with any sheet or news paper.
3. Sprinkle water on news paper to maintain humidity.
4. Spray water regularly twice a day to keep it humid throughout the cropping period.
5. Maintain the temperature of the room between 24–25°C for 12 – 15 days, at least until the mycelium is grown all over the tray beds.

Observation

White cottony mycelium is formed over the bed

D. Casing

The compost is covered by a thin layer of soil or soil like material after the mycelium is visible in compost (spawn run). In India a number of mixtures of soil have been recommended such as:

(i) Rotten cow dung and light soil (3:1 ratio).

(ii) Soil and sand (1:1 ratio).

(iii) Farm yard manure and gravel (4:1 ratio).

(iv) Farm yard manure and loam (1:1 ratio).

(v) Soil peat mixture (2:1 ratio).

(vi) Spent compost, sand and slaked lime (4:1:1 ratio).

The last combination is most commonly used.

Preparation of casing soil

1. Prepare the soil mixture with spent compost, sand and slake lime in 4:1:1 ratio.
2. Spray the mixture with Nematicide (Nemagoan) @ 5 ml / cbm.
3. Leave the material in a pile under a tree shade for a year.
4. Turn at every four months.

5. Sieve the spent compost after a year to remove undecomposed lumps.

Observation

Observe the casing soil for black colour and pH between 8–8.5. Adjust pH by addition of lime, if needed.

Sterilization of casing soil: Sterilize the casing material either by formalin, methyl bromide or vapour or by heating. Temperature is raised to 70–75°C for 6 h. Sterilization kills harmful fungi, nematodes and insects.

Casing the beds

1. Remove the newspaper sheets from the spawned beds after 3 weeks.
2. Gently press the compost with the help of wooden board.
3. Fill the beds with 2–2.5 cm thick layer of sterilized casing material.
4. Spray the beds with water using a sprayer with a fine nozzle.

Observations

After 5–20 days mushroom crop is expected. At 15°C it generally takes 7–8 days to come to the button stage from the first appearance of pin-head. There may be many flushes or crops with the gap of 8–10 days.

Harvesting and storage: The right stage for harvesting is the one when the cap is still tight to its stalk. Mushrooms are harvested by holding the cap with forefingers slightly pressed against the soil and twisted out and collected in wooden box. Mushrooms are highly perishable commodity and turn brown rapidly, hence should be kept at 4°C in a refrigerator.

Precautions

1. For preparation of wheat straw compost, at least 300 kg pile is maintained so that sufficient heat is generated. Moreover, to maintain aerobic conditions, the compost pile should not be hard pressed.
2. Many turnings of compost are advisable to avoid ammonia smell.
3. At watering time humidity should not be more than 70-80%.
4. Air circulation should be maintained.

B. BIOCHEMISTRY

INTRODUCTION

Biochemistry is the application of chemistry to the study of biological processes at the cellular and molecular level. It is the study of the structure and function of biological molecules such as proteins, nucleic acids, carbohydrates and lipids. Biochemistry is also used to describe techniques suited to understanding the interactions and functions of biological molecules, including traditional techniques such as Western blotting, co-immunoprecipitation, and chromatography methods. It emerged as a distinct discipline around the beginning of the 20th century when scientists combined chemistry, physiology, and biology to investigate the chemistry of living systems. Biochemistry has become the foundation for understanding all biological processes. It has provided explanations for the causes of many diseases in humans, animals and plants.

2.1 CARBOHYDRATES

Carbohydrates are defined as aldehydes or ketones derivatives of polyhydric alcohols and are known as aldoses or ketoses. They perform diverse structural and metabolic roles in animals, plants and microorganism. For example glycogen and starch functions as storage poly-saccharides in animals and plants, respectively whereas glucose is amongst major source of energy.

Carbohydrates are classified as Monosaccharides, Disaccharides, Oligosaccharides and Polysaccharides.

Monosaccharides

Monosaccharides, or simple sugars, are polyhydroxy aldehydes (aldose) and ketones (ketoses). Sugars are classified by the number of carbon atoms in their hydrocarbon chain. This number may vary from three to seven or more, but the most common and the most significant sugars are the pentoses, which have five carbons, and the hexoses, which have six carbons. Sugars contain asymmetric carbon atoms and are optically active. The form found in living organisms is the D (dextro) or right-handed isomer. In aqueous solution, the monosaccharides exist primarily as closed ring structures. Different ring structures are possible for the same monosaccharide. These structures are distinguished by Greek letters such as alpha and beta. The most common monosaccharide is the hexose glucose, $C_6H_{12}O_6$. Other common hexoses include galactose, fructose, and mannose. The most significant pentoses are ribose and 2-deoxyribose, which have a crucial role in the chemistry of nucleotides.

Oligosaccharides

Oligosaccharides are polymers of monosaccharides containing two to ten residues. Oligosacchaarides are obtained upon hydrolysis of polysaccharides. When two monosaccharide molecules react or condense with each other, losing a water molecule in the process, the product is a disaccharide. Several disaccharides are common in nature, the most well-known being sucrose. Sucrose is a disaccharide of glucose and fructose. Other disaccharides which are important in nature are lactose or milk sugar, which is a disaccharide of glucose and galactose, and two disaccharides of glucose with itself, maltose and cellobiose, which differ in the joining of the two glucose rings.

Polysaccharides

A polysaccharide is a polymer of one of the monosaccharides. All of the well-known polysaccharides found in nature are polymers of glucose. Several familiar materials are polymers of glucose: starch, which is poly-alpha-glucose; cellulose, which is poly-beta-glucose; and glycogen, which is a differently branched form of poly-alpha-glucose. Cellulose is used to form plant structures since it is not soluble in water, while plants use starch for energy storage. Animals store energy as glycogen or, for longer-term storage, convert glycogen and other carbohydrates to lipids.

2.1.1 Qualitative test for Carbohydrates

Quantitative Tests are based on specific colour reactions typical for their group. These can be categorized into three groups.

(a) Based on the reducing property of the aldehyde or ketone group.

(b) Based on the dehydration of neighbouring hydroxyl groups by strong and treating the result compound with various chemicals to form coloured products.

(c) Based on preparation of derivatives, which are of course, of applicability to pure compounds only.

2.1.1.1 Molisch's Test

This is a test for all carbohydrates.

(a) Principle

Conc. H_2SO_4 hydrolyses glycosidic bonds to give monosaccharides. These monosaccharides in the presence of acid get dehydrated to form furfural and its derivatives. These products then combine with sulphonated α-napthol to give purple complex.

$$\text{D-Glucose} \xrightarrow{\text{Conc. } H_2SO_4} \text{Hydroxy methyl furfural} \xrightarrow{\alpha\text{-naphtol}} \text{Purple or violet complex}$$

(b) Reagents

1. Conc. H_2SO_4
2. α-napthol (prepared fresh in 5% w/v ethanol)

(c) Procedure

Add 2-3 drops of α-napthol solution to 2 ml of the test solution. Carefully pipette 1 ml conc. H_2SO_4 along the sides of the test tube.

(d) Observation

Two distinct layers are formed. Appearance of purple colour at the junction of the two layers indicates the presence of carbohy-drates.

(e) Precautions

1. Being unstable in nature, α-napthol solution should be prepared fresh.
2. Care must be taken in adding conc. H_2SO_4 along the sides of the test tube.

2.1.1.2 Iodine Test

(a) Principle

Starch gives blue colour complex with iodine, while glycogen reacts to form reddish brown complex.

(b) Reagents

Iodine solution: Prepare 0.005 N iodine solution in 3% (w/v) potassium iodide solution.

(c) Procedure

Add 4–5 drops of iodine to 1 ml of sample in a test tube. Note the colour of the product.

(d) Observation

Appearance of dark blue or reddish brown colour.

2.1.1.3 Barfoed's Test

(a) Principle

This is a test for monosaccharide. It can also be used for distinguishing monosaccharides from reducing disaccharides. Monosaccharides usually react in about 1–2 min while the reducing disaccharides take 10–12 min to get hydrolysed and afterwards react with the reagent. The colour obtained is brick red due to formation of cuprous oxide.

Cupric acetate + $2H_2O$ → 2 Acetic acid + cupric hydroxide

Cupric hydroxide → cupric oxide + H_2O

D - Glucose + 2 cupric oxide → D - Gluconic acid + cuprous oxide (red ppt.)

Glucose reacts with cupric oxide to form Gluconic acid and cuprous oxide

Reagents: Barfoed's reagent by dissolving 13.3 g of copper acetate in 200 ml water and add 1.8 ml of glacial acetic acid to it.

(b) Procedure

Take 2 ml of Barfoed's solution in a test tube and add 1 ml of sample solution. Keep the test tube in a boiling water bath.

(c) Observation

Note the time taken for the formation of brick red colour.

2.1.1.4. Seliwanoff's Test

(a) Principle

In the presence of conc. HCl, ketoses undergo dehydration to yield furfural derivatives more rapidly than aldoses. These derivatives form complexes with resorcinol to yield deep red colour. It is a timed colour reaction specific for ketoses.

$$\text{D-Fructose} \xrightarrow{\text{Conc. HCl}} \begin{matrix}\text{Hydroxy}\\\text{methyl}\\\text{furfural}\end{matrix} + 3H_2O \xrightarrow{\text{Resorcinol}} \begin{matrix}\text{Red}\\\text{colour}\\\text{product}\end{matrix}$$

Reagent: Seliwanoff's reagent [0.05% (w/v) resorcinol in 3N HCl]

(a) Procedure

To 2 ml of Seliwanoff's reagent add 2 drops of test solution and heat the mixture to just boiling.

(b) Observation

A deep red colour indicates ketoses.

2.1.1.5 Fehling's Test

(a) Principle

This is the test for detection of reducing sugars. The blue alkaline cupric hydroxide present in Fehling's solution, when heated in presence of the sample is reduced to yellow brown or red cuprous oxide and gets precipitated.

Cupric acetate + $2H_2O$ → 2 Acetic acid + cupric hydroxide

Cupric hydroxide → Cupric oxide + H_2O

D - Glucose + 2 cupric oxide → D - Gluconic acid + cuprous oxide (red ppt.)

(b) *Reagents*

1. Fehling's solution A: Dissolve 35 g of $CuSO_4.5H_2O$ in water and make volume to 500 ml.
2. Fehling's solution B: Dissolve 120 g of KOH and 173 g Na-K tartrate in water and make the volume to 500 ml.

(c) Procedure

Mix equal volumes (1 ml each) of Fehling's solution A and B to 1 ml of sample solution. Boil for 10 min in boiling water bath.

(d) Observation

Yellow brown or red precipitate is formed.

2.1.1.6 Benedict's Test

(a) Principle

This test is also for the detection of reducing sugars. Reducing sugars reduce cupric hydroxide in alkaline solution to red coloured cuprous oxide. Yellow to green colour precipitate is formed depending on the sugar concentration.

$$Na_2CO_3 + 2H_2O \rightarrow 2NaOH + H_2CO_3$$

$$2NaOh + CuSO_4 \rightarrow Cu(OH)_2 + Na_2SO_4$$

$$Cu(OH)_2 \rightarrow CuO + H_2O$$

$$D\text{ - Glucose} + 2CuO \rightarrow D\text{ - Gluconic acid} + Cu_2O$$

(b) *Reagents*

Benedict's reagent: Dissolve 173 g sodium citrate and 100 g of sodium carbonate in about 500 ml water. Heat to dissolve the salts and filter. Dissolve 173 g copper sulphate in about 100 ml and add it to the above solution. Stir well and make up the volume to 1 litre with water.

(c) Procedure

Add 5 drops of test solution to 2 ml of Benedict's reagent. Boil for 5 min in a water bath, cool the solution.

(d) Observation

Formation of red, yellow or green colour precipitate indicates presence of reducing sugar.

2.1.1.7 Bial's Test

(a) Principle

This test is specific for pentoses. Furfural in the acid medium complexes with orcinol in presence of ferric ion to give a blue-green coloured complex.

$$\text{Ribose} \xrightarrow[-3H_2O]{\text{Conc. HCl}} \text{Furfural} \xrightarrow{\text{Orcinol}} \text{Blue - green complex}$$

Reagent: Dissolve 1.5 g orcinol in 500 ml of conc. HCl and add 20–30 drops of 10% ferric chloride.

(b) Procedure

To about 4 ml of Bial's reagent add 1–2 ml of test solution. Heat in a boiling water bath and then cool.

(c) Observation

A green colour or precipitate will appear.

2.1.1.8 Osazone Test

(a) Principle

The ketones and aldoses react with phenylhydrazine to produce phenylhydrazone. This in turn reacts with two more molecules of phenyl-hydrazine to form the osazone crystals upon cooling.

$$\text{Aldoses} + \text{Phenylhydrazine} \rightarrow \text{Phenylhydrazone} + H_2O$$

Phenylhydrazone + 2 Phenylhydrazine → Osazone + Phenylamine + NH_3

(b) *Reagent*

1. Phenylhydrazine hydrochloride.
2. Sodium acetate.
3. Glacial acetic acid.

(c) Procedure

To 0.5 g of phenylhydrazine hydrochloride add 0.1 g of sodium acetate and 10 drops of test solution and heat for about half an hour on a boiling water bath. Allow the tube to cool slowly.

(d) Observation

Examine the osazone crystals under microscope. Glucose, fructose and mannose produce needle shaped yellow osazone crystals. Maltose produces flower shaped crystals and for lactose mushroom shaped crystals are formed.

2.1.1.9 Test for non-reducing sugars

(a) Principle

Reducing sugars are formed from non-reducing sugars after hydrolysis with conc. acid. These resulting reducing sugars can be detected by Fehling or Benedict's test.

(b) *Reagents*

1. Benedict's / Fehling's reagent A and B
2. Conc. HCl

$$\text{Sucrose (Non - reducing sugar)} + \text{conc. HCl} \xrightarrow{\text{Heat}} \text{Glucose} + \text{Fructose}$$

$$\text{Glucose} + \text{CuO (from Fehling's test)} \rightarrow \text{D - Gluconic acid} + Cu_2O$$

(c) Procedure

Add 5 drops of conc. HCl to 5 ml of test solution. Heat for 5 min on a boiling water bath.

(d) Observation

Appearance of red or yellow colour indicates non-reducing sugar.

2.1.2 Quantitative Test for Carbohydrates

2.1.2.1 Determination of total carbohydrate by Anthrone reagent

The Anthrone method is an example of a colourimetric method of determining the concentration of the total sugars in a sample. Sugars react with the anthrone reagent under acidic conditions to yield a blue-green colour. There is a linear relationship between the absorbance and the amount of sugar that was present in the original sample. This method determines both reducing and non-reducing sugars because of the presence of the strongly oxidizing sulfuric acid. Like the other methods it is non-stochiometric and therefore it is necessary to prepare a calibration curve using a series of standards of known carbohydrate concentration.

(a) Principle

Carbohydrates are first hydrolysed into simple sugars using dilute hydrochloric acid. In hot acidic medium glucose is dehydrated to hydroxy-methyl furfural. This compound forms with anthrone a green coloured product with an absorption maximum at 630 nm.

(b) *Reagents*

1. Anthrone Reagent: Dissolve 200mg anthrone in 100mL of ice cold 95% H_2SO_4. Prepare fresh before use.
2. 2.5 N-HCl.
3. Standard Glucose: Stock – 1 mg/ml.
4. Working standard – 10ml of stock diluted to 100ml with distilled water. Store refrigerated.

(a) Procedure

A. Preparation of sample (Unknown concentration)

1. Weigh 100 mg of the sample and hydrolyse by keeping it in a boiling water bath for 3 hours with 5 ml 2.5N HCl.
2. Cool to room temp and add to it solid sodium carbonate to neutralize.
3. Make up the volume to 100 ml and centrifuge. Collect the supernatant and take aliquots for analysis.
4. Dissolve the residue sugars in 10 ml of distilled water.
5. Pipette out aliquots of 0.1 and 0.2 ml in separate test tubes.

B. Standard curve to determine the unknown concentration of Sample:

1. Prepare the standards by taking 0, 0.2, 0.4, 0.6, 0.8 and 1 ml of the working standard.

2. Make up the volume to 1 ml in all the tubes including the sample tubes by adding distilled water.
3. Cool the contents of all the tubes on ice. Then add 4 ml of ice-cold anthrone reagent.
4. Heat for eight minutes in a boiling water bath, Cool rapidly.
5. Observe the green to dark green colour and take the absorbance at 630nm.
6. Draw a standard graph and from the graph calculate the amount of carbohydrate.

(d) Precautions

1. Spectrophotometer should be kept ON 15 min before use.
2. Pipetting should be done carefully.

2.1.2.2 Determination of reducing sugar by Nelson-Somogyi method

Reducing sugars have reducing property due to presence of potential aldehyde or keto group. Some of the examples are– Glucose, Galactose, Lactose and maltose. The Nelson-Somogyi method is one of the classical and widely used methods for quantitative determination of reducing sugar.

Note: The copper tartrate reagent offers definite advantage in determination of glucose using Nelson-Somogyi's method:

a. It is equally suitable for the iodometric and colourimetric determination of cuprous oxide.
b. A single reagent is suitable for macro and micro analysis.
c. In lowest range of sugar concentration the reagent yields more consistent and reproducible results.

The iodide and iodate reagent use have been omitted because these interfere with molybdate colour reagents.

Before the use of arsenomolybdate reagent phosphomolybdate reagent were used. Satisfactory proportionality was found between colour density and glucose taken over a wide range of values. However, all the phosphomolybdate reagents lacked the desired stability of colour. Therefore, arsenomolybdate reagent was used which gives satisfactory stability and reproducibility of colour.

(a) Principle

The reducing sugar when heated with alkaline copper tartrate reduce the copper from cupric to cuprous state and thus cuprous oxide is formed. When cuprous oxide is treated with hetropolymolybdate complex (arsenomolydate complex), the cuprous [Cu(I)] ions are oxidised back to cupric [Cu(II)] ions and heteropolymolybdate complex is reduced to give characteristic blue colour by forming molybdenum blue. Finally the absorption of the hetero-polymolybdenum blue is measured at 620 nm.

Note: Na_2SO_4 is included in Cu reagent to prevent reoxidation of cuprous oxide by atmospheric oxygen. The blue colour of heteropolymo-lybdate arises because the near colourless phosphomolybdate anion $PMo_{12}O_{40}^{3-}$, can accept more electrons (be reduced) to form an intensely coloured mixed valence complex.

(b) Requirements

1. Alkaline copper tartrate
   a. Reagent A: Dissolve 2.5 g of anhydrous Na_2CO_3, 2 g of sodium hydrogencarbonate ($NaHCO_3$), 2.5 g of sodium potassium tartrate, 20 g anhydrous sodium sulphate in 20 ml water and make up to 100 ml.
   b. Reagent B: Dissolve 15 g of $CuSO_4$ in small volume of distilled water. Add 1–2 drops of conc. H_2SO_4 and make upto 100ml.
   c. Reagent C: Prepare by mixing Reagent A and B. Mix 4 ml of Reagent B and 96 ml of Reagent A before use.
2. Arsenomolybdate colour reagent: Dissolve 2.5 g of ammonium molybdate in 45 ml of water. Add 2.5 ml of conc. H_2SO_4 and mix well. Then add 0.3 g disodium hydrogen arsenate dissolved in 25 ml water. Mix the above two solutions well and incubate at 37°C for 24 hr to 48 hr.
3. Standard glucose solution.
   Stock: 100 mg in 100 ml distilled water (1 mg / 1 ml).
4. Working standard solution: 10 ml of stock diluted to 100 ml distilled water (100 µg / ml).

(c) Procedure

A. Preparation of sample (Unknown concentration)
   1. Weigh 100 mg of the sample and extract the sugars with hot ethanol (80%). Repeat it twice with 5 ml ethanol each time.
   2. Collect the supernatant and evaporate it by keeping it on a water bath at 80°C.

3. Dissolve the residue sugars in 10 ml of distilled water.
4. Pipette out aliquots of 0.1 and 0.2 ml to separate test tubes.

B. Standard curve to determine the Unknown concentration of Sample:

1. Pipette out 0.2, 0.4, 0.6, 0.8 and 1 ml of the working standard solution into a series of test tubes.
2. Make up the volume in both sample and standard tubes to 2 ml with distilled water.
3. Prepare blank with 2 ml distilled water.
4. Add 1 ml of alkaline copper tartarate reagent to each tube.
5. Keep the tubes in boiling water for 10 min.
6. Cool the tubes and then add 1 ml of arsenomolybdate reagent (Reagent D) to all the tubes.
7. Make up the volume in each tube to 10 ml by adding 6 ml of distilled water, keep them for 10 min.
8. Read the absorbance at 620 nm and prepare a standard curve.
9. From the standard curve calculate the amount of sugars present in the sample.

(d) Precautions

1. Reagent C should be prepared just before use.
2. Incubation should be carried out at 37°C for 24 to 48 hrs.
3. Spectrophotometer should be on 15 min before use.
4. Pipetting should be done carefully.

2.1.2.3 Determination of Cellulose

In the analysis of plant materials and woods, one of the chief constituents to be determined is cellulose, which is important since it forms the structural framework of the tissue. The determination of cellulose, however, is for various reasons not always satisfactory. The methods of determination of cellulose in general use do not give the same product; consequently the results obtained are not always directly comparable. The method selected may be decided by the purpose for which the cellulose is required. As it occurs naturally, cellulose is always found in the closest association with other constituents of the cellwall, and these have, by various treatments, to be removed. Cellulose is composed of glucose units joined together in the form of the repeating units of the disaccharide cellobiose with numerous cross linkages.

(a) Principle

Cellulose undergoes acetolysis with acetic/nitric reagent forming acetylated cello-dextrins which get dissolved and hydrolyzed to form glucose molecules on treatment with H_2SO_4 (67%). This glucose mole-cule is dehydrated to form hydroxymethyl furfural which forms green coloured product with anthrone and the absorbance is measured at 630 nm.

(b) *Reagents*

1. Anthrone reagent: Dissolve 200 mg anthrone in 100 ml concentrated sulphuric acid. This should be prepared fresh and chilled for 2 h before use.
2. Acetic/Nitric reagent: Mix 150 ml of acetic acid (80%) and 15 ml of concentrated nitric acid.
3. Sulphuric acid (67%).

(c) Procedure

1. Add 3 ml acetic/nitric reagent to a known amount of the sample (0.5 g or 1 g) in a boiling test tube and vortex it.
2. Place the tube in a water-bath at 100°C for 30 min.
3. Cool and then centrifuge the contents for 15–20 min. Discard the supernatant.
4. Wash the residue with distilled water.
5. Add 10 mL of sulphuric acid (67%) and allow it to stand for 1 h.
6. Dilute 1 mL of the above solution to 100 mL. To 1 ml of this diluted solution, add 10 mL of anthrone reagent and mix well.
7. Heat the tubes in a boiling water-bath for 10 min, cool and measure the colour at 630 nm.
8. Prepare a blank with anthrone reagent and distilled water.
9. Prepare a standard curve corresponding to 40–200 μg of cellulose and calculate the amount of cellulose in the sample.

2.2 AMINO ACIDS AND PROTEINS

Introduction

Proteins are the major building blocks of all living cells and are made up of amino acids. Amino acids are critical to life, and have many functions in metabolism. These are molecules containing an amine group, a carboxylic acid group and a side chain that varies between different amino acids. These molecules contain the key elements of Carbon,

Hydrogen, Oxygen, and Nitrogen. These molecules are particularly important in biochemistry, where this term refers to alpha-amino acids with the general formula $H_2NCHRCOOH$, where R is an organic substituent.

Amino acids are joined covalently by peptide bonds to yield proteins (including enzymes and hormones; structural, transport, and contractile elements; and molecules of special biological activity. Conceptually, protein structure can be considered at four levels. The primary structure consists of a sequence of amino acids linked together by covalent peptide bonds and includes any disulfide bonds. Secondary structure refers to regular, recurring arrangements in space of adjacent amino acids in polypeptide chain. There are few common types of secondary structure, the most prominent being the α–helix and the β conformation. Tertiary structure is the complete three–dimensional structure of the polypeptide; it refers to the spatial relationship among all amino acids in a polypeptide. Proteins with several polypeptide chains have one more level of structure: quaternary structure, which refers to the spatial relationship of the polypeptides, or subunits, within the protein.

Proteins interact with other molecules–small molecules, other proteins, nucleic acids, and so forth–and these interactions are the basis of biological function.

Forces Influencing Protein Structure

Several different kinds of non-covalent interactions are of vital importance in protein structure. Hydrogen bonds, hydrophobic interactions, electrostatic bonds, and van-der Waals forces are all non-covalent in nature, yet are extremely important influences on protein conformations. The stabilization free energies afforded by each of these interactions may be highly dependent on the local environment within the protein.

(i) Hydrogen Bonds

Hydrogen bonds are generally made wherever possible within a given protein structure. In most protein structures that have been examined to date, component atoms of the peptide backbone tend to form hydrogen bonds with one another. Furthermore, side chains capable of forming H bonds are usually located on the protein surface and form such bonds primarily with the water solvent. Although each hydrogen bond may contribute an average of only about 12 kJ/mol in stabilization energy for the protein structure, the number of H-bonds formed in the typical protein is very large. For example, in α-helices, the C=O and N-H groups of every residue participate in H bonds. The importance of H bonds in protein structure cannot be overstated.

(ii) Hydrophobic Interactions

Hydrophobic interactions form because non-polar side chains of amino acids and other non-polar solutes prefer to cluster in a non-polar environment rather than to intercalate in a polar solvent such as water. The forming of hydrophobic bonds minimizes the interaction of non-polar residues with water and is therefore highly favourable. Such clustering is entropically driven. The side chains of the amino acids in the interior or core of the protein structure are almost exclusively hydrophobic. Polar amino acids are almost never found in the interior of a protein, but the protein surface may consist of both polar and non-polar residues.

(iii) Electrostatic Interactions

Ionic interactions arise either as electrostatic attractions between opposite charges or repulsions between like charges. Amino acid side chains can carry positive charges, as in the case of lysine, arginine, and histidine, or negative charges, as in aspartate and glutamate. In addition, the NH_2-terminal and COOH-terminal residues of a protein or peptide chain usually exist in ionized states and carry positive or negative charges, respectively. All of these may experience electrostatic interacti-ons in a protein structure. Charged residues are normally located on the protein surface, where they may interact optimally with the water solvent.

It is energetically unfavourable for an ionized residue to be located in the hydrophobic core of the protein. Electrostatic interactions between charged groups on a protein surface are often complicated by the presence of salts in the solution. For example, the ability of a positively charged lysine to attract a nearby negative glutamate may be weakened by dissolved NaCl.

The Na^+ and Cl^- ions are highly mobile, compact units of charge, compared to the amino acid side chains, and thus compete effectively for charged sites on the protein. In this manner, electrostatic interactions among amino acid residues on protein surfaces may be damped out by high concentrations of salts. Nevertheless, these interactions are important for protein stability.

(iv) Van der Waals Interactions

Both attractive forces and repulsive forces are included in van der Waals interactions.

The attractive forces are due to primarily instantaneous dipole-induced dipole interactions that arise because of fluctuations in the electron charge distributions of adjacent non-bonded atoms.

Individual van der Waals interactions are weak ones (with stabilization energies of 4.0 to 1.2 kJ/mol), but many such interactions occur in a typical protein, and, by sheer force of numbers, they can represent a significant contribution to the stability of a protein.

Proteins can be precipitated by dehydration or neutralization of polar groups.

1. *Precipitation by neutral salts (salting out):* This phenomenon is explained on the basis of dehydration of protein molecules by salts. This causes increased protein-protein interaction, resulting in molecular aggregation and precipitation. The amount of salt required for protein precipitation depends on the size (molecular weight) of the protein molecule. In general, the higher is the protein molecular weight, the lower is the salt required for precipitation. This is the reason why serum globulins are precipitated by half saturation with ammonium sulphate, while albumin is precipitated by full saturation.
2. *Precipitation by salts of heavy metals:* Heavy metal ions like Pb^{+2}, Hg^{+2}, Fe^{+2}, Zn^{+2}, Cd^{+2} cause precipitation of proteins. These metals being positively charged, when added to protein solution (negatively charged) in alkaline medium result in precipitate formation.
3. *Precipitation by anionic or alkaloid reagents:* Proteins can be precipitated by trichloroacetic acid, sulphosalicylic acid, phosphotungstic acids, etc. By addition of these acids, the proteins existing as cations are precipitated to produce protein-sulphosalicylate, protein tungstate etc.
4. *Precipitation by organic solvents:* Organic solvents like alcohol dehydrate the protein molecule by removing the water envelop and cause precipitation.

2.2.1 Colour Reaction of Proteins/Amino Acids

Tests for Amino Acids

Amino Acid Solution: Prepare individual amino acid (0.1% w/v) in distilled water. If the sample does not dissolve add a few drops of 6 N HCl.

Ninhydrin test

(a) Principle

Ninhydrin test is given by the amino acids due to formation of a complex called Rheumann's purple. It is formed by condensation of two molecules of ninhydrin with one molecule of ammonia.

(b) Reagents

Ninhydrin reagent: Ninhydrin (2% w/v) in acetone.

(c) Procedure

To 2 ml of the amino acid solution taken in a test tube, add a few drops of ninhydrin reagent and vortex the contents. Place the test tube in a boiling water bath for 5 min allow it to cool and add NaOH.

(d) Observation

Appearance of purple to blue colour indicates the presence of α–amino acids. In case of imino acid, a yellow colour is observed, for example, proline.

Xanthoproteic test

(a) Principle

The aromatic groups of either the free amino acids (tyrosine, tryptophan) or proteins undergo nitration to give nitro-derivatives that are yellow in colour in acidic medium. At alkaline pH, the colour changes to orange due to the ionization of the phenolic group.

(b) Reagents

Sodium hydroxide (40% w/v); Conc. Nitric acid.

(c) Procedure

To 2 ml of the amino acid or protein solution taken in a test tube, add 1 ml of nitric acid and excess of NaOH (40%). Boil the contents for few minutes and cool.

(d) Observation

In acid medium, the solution is yellow and upon treatment with alkali, the colour changes to orange.

Note: (i) Phenylalanine gives a weak positive reaction for this test. (ii) aliphatic amino acids give negative xanthoproteic test. (iii) In case of proteins, initially there is precipitation, which goes into solution upon heating.

Ehrlich's test

(a) Principle

This is another test for Tryptophan. The indole ring of tryptophan reacts with p-dimethylaminobenzaldehyde under acidic condition to give a purple colour.

(b) Reagent

p-dimethylaminobenzaldehyde (10% w/v) in hydrochloric acid.

(c) Procedure

To 1ml of the amino acid solution taken in a test tube, add few drops of Ehrlich's reagent and vortex the contents.

(d) Observation

Appearance of a purple colour indicates the presence of tryptophan.

Note: In case, the colour does not appear, add 1 or 2 drops of conc. hydrochloric acid and mix. Other indole derivatives also respond to this test.

Hopkins Cole test

(a) Principle

The indole ring of tryptophan reacts with aldehydes in acidic condition to yield purple or violet coloured ring at the junction of two liquids.

(b) Reagent

Glacial acetic acid, conc. H_2SO_4

(c) Procedure

Mix 2 ml of the test solution with 2 ml acetic acid in a test tube, vortex. Then carefully, add 2 ml conc. H_2SO_4 along the side of the test tube, keeping the tube in an inclined position (do not shake the test tube, while adding the acid).

(d) Observation

A purple-violet ring appears at the junction of the amino acid solution and the conc. sulphuric acid. The formation of this coloured ring confirms the presence of tryptophan.

Lead sulphide test

(a) Principle

This test is given by sulphur containing amino acids, such as cysteine and cystine. The amino acid upon boiling with sodium hydroxide (hot alkali) yield sodium sulphide. This is due to partial conversion of the organic sulphur to inorganic sulphide, which can be detected by precipitating it to lead sulphide, using lead acetate solution.

(b) Reagents

Lead acetate solution (10% w/v); Sodium hydroxide solution (40% w/v) solution.

(c) Procedure

To 1 ml of the amino acid solution taken in a test tube, add few drops of sodium hydroxide (40%) and boil the contents for 5-10 min and cool. Add few drops of lead acetate solution (10%).

(d) Observation

The appearance of a black precipitate of lead sulphide confirms the presence of cysteine or cystine.

Note: The sulphur present in methionine is not released by alkali treatment, thus gives a negative test.

Sodium nitroprusside test

(a) Principle

Under alkaline conditions Sodium nitroprusside reacts with the thiol group of the cysteine to yield an intense purple coloured compound, which fades upon standing.

Note: Cystine gives a positive test only after reduction of the disulphide group to sulphydryl (thiol) group, by chemical agents such as potassium or sodium cyanide.

(b) Reagent

(i) Sodium nitroprusside {[Na_2Fe $(CN)_5$] NO} reagent: sodium nitroprusside (10% w/v) in distilled water (this reagent should be prepared freshly) (ii) liquor ammonia (chilled).

(c) Procedure

To 1 ml of the amino acid solution taken in a test tube, add few drops of sodium nitroprusside reagent and vortex. Add few drops of liquor ammonia and mix the contents.

(d) Observation

An intense purple colour indicates the presence of cysteine. Note that the intensity of the colour fades with time.

Sakaguchi Test

(a) Principle

Under alkaline condition, α-naphthol (1-hydroxy naphth-alene) reacts with a mono-substituted guanidine group of arginine, which upon treatment with sodium hypobromite or hypochlorite, produces a bright red colour. The reaction mechanism is poorly understood.

Note: In addition to arginine, compounds such as α-guanidino butyric acid, monomethylguanidine and octopine, tests positive to this reaction, while guanidine, dimethylguanidine, nitro-arginine, cirtulline, creatinine and creatine do not respond to this reaction.

(b) Reagents

(i) Sodium hydroxide (10% w/v) (ii) α-naphthol reagent 1% w/v in, ethyl alcohol (95%) (iii) Hypobromite reagent (to be freshly prepared). To about 75-80 ml of water add 40 mg of NaOH in a 100 ml measuring cyclinder, Add 10 ml of chilled bromine carefully with constant shaking and cooling. Make final volume to 100 ml by adding water.

Caution: Use chemical safety hood, while adding bromine (corrosive & toxic chemical)

(c) Procedure

To 5ml of chilled amino acid solution taken in a test tube, add 1 ml of sodium hydroxide solution, followed by α-naphthol reagent (1 ml) and vortex. After 2 min, add 10 drops of freshly prepared hypobromite reagent, until the desired colour is achieved.

(d) Observation

Formation of a red colour indicates the presence of arginine.

Isatin test

(a) Principle

Under acidic conditions imino acids such as proline and hydroxy proline condense with isatin to yield a blue coloured product.

(b) Reagent

Isatin reagent: Dissolve isatin (1% w/v) in acetic acid.

(c) Procedure

Apply a drop of imino acid solution on a Whatman No. I filter paper strip (size 25 × 50 mm) and dry the spot by hair dryer or in hot air oven. Apply a drop of isatin reagent on to the dried spot. Repeat the drying procedure for few minutes and observe.

(d) Observation

Appearance of a blue coloured spot on the filter paper, confirms the presence of imino acid.

Note: Other amino acids give pink colour with this reagent.

Millon's test

(a) Principle

The phenolic group of tyrosine reacts with mercuric ions in acidic condition in the presence of sodium nitrite, yielding a red colour complex.

Note: Chlorides, phosphates interfere in the colour development. Tryptophan yields a reddish- brown colour with this reagent.

(b) Reagents

(i) Millon's reagent (15% w/v mercuric sulphate in 6 N sulphuric acid)
(ii) Sodium nitrite solution (5%) – to be freshly prepared.

(c) Procedure

To 1 ml of the amino acid solution in a test tube, add 1 ml of Millon's reagent and vortex. Boil the contents for 3–5 min. A yellow precipitate due to precipitation of protein is obtained as mercury combines with tyrosine of protein. Cool the contents under running tap water and add few drops of sodium nitrite solution. The red colour is developed due to reaction of the precipitate with the nitric acid.

(d) Observation

Appearance of yellow precipitate followed by formation of red colour confirms the presence of tyrosine.

Biuret test

(a) Principle

The peptide bonds of the protein react with cupric copper, under alkaline condition to yield a purple colour.

(b) Reagents

(i) Sodium hydroxide solution (10% w/v), (ii) Cupric sulphate solution (0.15% w/v), (iii) Sodium potassium tartrate solution (0.6% w/v) and (iv) Copper reagent: Mix equal volume of reagent (ii) and (iii) before test.

(c) Procedure

To 2 ml of the protein solution add few drops of alkali and vortex. Add copper reagent drop wise with continuous vortexing till the purple colour appears.

(d) Observation

Formation of a violet-purple colour indicates the presence of protein.

Sulphosalicylic acid test

(a) Principle

The interaction of sulphosalicylic acid with the basic groups of the protein, results in protein aggregation and precipitation.

(b) Reagent

Sulphosalicylic acid reagent: Prepare Sulpho-salicylic acid (5%, w/v) in sodium sulphate solution (20% w/v).

(c) Procedure

To 2 ml of the sample solution add sulphosalicylic acid reagent drop wise with constant mixing.

(d) Observation

Appearance of a turbidity or white precipitate in the sample solution indicates the presence of protein(s).

2.2.2 Titration Curve of Amino Acids

(a) Principle

Amino acids are present as zwitterions at neutral pH and are amphoteric molecules that can be titrated with both acid and alkali.

Table : Summary of the specific groups or amino acids involved in colour reactions

S. No.	Reaction	Specific group or amino acids
1	Biuret reaction	Two peptide linkages
2	Ninhydrin reaction	α-Amino acids
3	Folin–Ciocalteau's test	Phenolic groups (Tyr)
4	Xanthoproteic reaction	Benzene ring of aromatic amino acids (Phe, Tyr, Trp)

S. No.	Reaction	Specific group or amino acids
5	Millon's reaction	Phenolic group (Tyr)
6	Hopkins-cole reaction	Indole group (Trp)
7	Sakaguchi reaction	Guanidino group (Arg)
8	Nitroprusside reaction	Sulfhydryl group (Cys)
9	Sulphur test	Sulfhydryl group (Cys)
10	Sulphosalicylic acid	Basic groups of proteins

$$H_3^+N-\underset{}{\overset{R}{\overset{|}{C}}}H-COOH \rightleftharpoons H_3^+N-\overset{R}{\overset{|}{C}}H-COO^- \rightleftharpoons H_2N-\overset{R}{\overset{|}{C}}H-COO^-$$

Net charge +ve No charge Net charge –ve

Some amino acids have other ionizable groups in their side chains and these can also be titrated.

	Reagents	Volume
1.	HCl (100 mmol/litre)	250 ml
2.	Sodium hydroxide (100 mmol/litre)	250 ml
3.	Amino acids 100 mmol/l glycine, alanine, histidine and lysine, 50 mmol/l glutamic acid.	250 ml
4.	pH meter	
5.	Burette (10 ml)	

(b) Procedure

1. Pipette 10 ml of the amino acid solution into a 100 ml beaker. Standardize the pH meter and determine the pH of the solution. Add the 100 m mol/liter HCl from a burette in small amounts at first and determine the pH after each addition. Continue adding the acid in larger quantities until the pH falls to about 1.3. Now obtain the pH titraction curve.
2. Wash the electrodes in distilled water, restandardize the pH meter, and titrate a further 10 ml with 100 mmol/liter NaOH solution to pH 12.5. Obtain the titration curve for this also.
3. Determine the pK_1 and pK_2 values from titration curves and compare them with the values given in Table.

Table : pK values of some aminoacids

Amino acid	pK_1	pK_2	pK_3 (side chain R)
	-COOH	$-NH_2$	
Glycine	2.34	9.6	–
Alanine	2.35	9.69	–
Serine	2.21	9.15	–
Proline	1.99	10.60	–
Glutamic acid	2.19	9.67	4.3 (carboxyl)
Histidine	1.82	9.17	6.0 (imidazole)
Lysine	2.18	8.95	10.5 (amino)
Cysteine	1.71	10.33	8.33 (sulfhydryl)

Isoelectric point (pI) is calculated as: $\frac{pK_1 + pK_2}{2}$

2.2.3 Protein Assay Methods

Brief summary of spectroscopic methods for Proteins (from zieglerm@u.arizona.edu)

Type	Protein Components	Uses
UV – VIS spectroscopy		
Absorbance	• Peptide bonds (~220 nm) • Aromatic residues [esp.W (280nm), (Y)] • Some ligands & prosthetic groups	• Determine protein concentration • Conformational changes • Ligand binding
Fluorescence	• W [λ_{max} = 280m, $\lambda_{max.em}$=~340nm] • Some ligands & prosthetic groups	• Conformational changes • Ligand binding
Circular Dichroism (CD)	• Secondary structure (180–240nm) (peptide bonds) • Tertiary structure (environment of aromatic R groups) (260–300nm) • Ligands & prosthetic groups	• 2^0 structure (amount & type) (far UV) • Conformational changes (2° & 3° structural changes) • Ligand binding

Type	Protein Components	Uses
NMR	• Nuclei:protons(^{1}H, esp. in aromatic residues & His), but also ^{2}H, ^{13}C, ^{14}N, ^{17}O, ^{31}P, and ^{19}F (natural abundance or isotopically labeled proteins or ligands)	• Complete 3-D structure (only of small proteins) • Conformational changes • Ligand binding including pH titrations (pK$_a$ determination) • Dynamics (motions in structure)

2.2.3.1 λ max for protein and amino acids

(a) Principle

Many compounds have characteristic absorption spectra in UV light and visible region so that identification of these materials in a mixture is possible. Proteins absorb structurally strongly at 280 nm, according to the content of tyrosine and tryptophan. This provides a sensitive and non destructive form of assay. Proteins also absorb in far UV because of peptide bonds.

Absorption at 210 nm: Below 230 nm extinction of protein rises deeply reaching a maximum at 190 nm. This is mainly due to peptide bonds. In practice, it is more convenient to measure the extinction at 210 nm where the specific extinction coefficient is 2.00 for most proteins.

Absorption at 280 nm: Tyrosine and tryptophan absorb at 275 nm and 280 nm respectively so the proteins containing these amino acids will also absorb in this region. Specific extinction coefficient according to how much of these amino acids are present in a particular protein. Pure protein has a ratio of about 1.8 at $\lambda_{280} : \lambda_{260}$ nm.

(b) Requirements

BSA-1 mg/ml, Distilled water, test tubes, foil, 500 ml beaker, measuring cylinder, etc.

(c) Procedure

1. Take 0.1 ml of BSA in a quartz cuvette and add 2 ml of distilled water.

2. Record the absorbance of sample from 230 – 310 nm.
3. Plot the absorbance spectrum of BSA over the entire range (230 – 310 nm) and detect λ max from the graph.

(d) Result

λ_{max} of proteins was found at 280 nm.

(e) Precautions

1. A clean quartz cuvette should be used.
2. Double distilled water should be used.
3. Buffers if used should be of minimal UV absorption at this wavelength.

Note: Values of A_{280} for most proteins lie in the range 0.4–1.5, but extremes include some parvalbumin and related Ca^{+2} binding proteins (0.0) and lysozyme (2.65). The method is rather inaccurate unless the protein is relatively pure and its extinction coefficient is known, and calibrated against dry weight. This technique is relatively insensitive and requires between 0.05 and 2.0 mg of protein. Moreover, it is very sensitive to interference from other compound which also absorb at 280 nm. This is especially true if compounds containing purine and pyrimidine rings (nueclic acids, nucleotides) are present, in such case, the absorbance of the sample is measured at both 280 nm and 260 nm and a correction formula applied.

$$\text{Protein (mg/ml)} = 1.55\ A_{280} - 0.76\ A_{260}$$

If the solvent itself absorbs at 280 nm, an accurate blank must be taken. The very real advantage of this approach, however is that, it is rapid and non-destructive, so that the assayed sample can be used for subsequent processing.

Measurement in the far UV: The peptide bonds in proteins and peptides exhibit absorption maximum at 191–194 nm, but this is more sensitive and less dependent on amino acid composition than measurement at 280 nm.

2.2.3.2 Determination of molar absorbance coefficient (ε) of L-tyrosine

(a) Principle

Molar absorbance coefficient (or molar extinction coefficient) is defined as the absorbance of a one molar solution of pure absorbing material in a 10 mm path length cell under specified conditions of wavelength (λ_{max}) and solvent. The units of ε are M^{-1} cm^{-1}. Molar absorbance coefficient value of a biomolecule can be determined by using the equation A = ε Cl.

Experimentally, the absorbance value of a known concentration of the pure substance can be obtained by recording the absorbance at its λ_{max} using a UV – VIS spectrophotometer.

(b) Reagents

5×10^{-4} M solution of L-tyrosine in 0.1 N NaoH.

Apparatus

UV – VIS spectrophotometer.

(c) Procedure

1. Select the wavelength of UV-VIS spectrophotometer to record the absorbance (λ_{max} = 293.5 nm) of the prepared tyrosine solution.
2. Calibrate the instrument to zero by placing a blank (0.1 N NaOH solution) taken in a 3 ml (1 cm path length) quartz cuvette.
3. Record the absorbance of the tyrosine solution at 293.5 nm.

Calculation: Calculate molar absorbance coefficient as

$$\varepsilon = \text{A} \div 1 \times \text{c} = \frac{\text{A (recorded at } \lambda_{293.5} \text{ nm)}}{(1\,\text{cm})\,(5 \times 10^{-4}\ \text{mol/l})}$$

Here, A = Absorbance

C = Concentration of L-tyrosine

2.2.3.3 Lowry method of protein assay

The Lowry method is reasonably sensitive, detecting down to 10 μg cm^{-3} of protein, and the sensitivity is moderately constant from one protein to other.

(a) Principle

The blue-purple colour developed in Lowry method is a result of (i) reaction of the peptide bonds of the protein with cupric copper under alkaline conditions and (ii) reduction of phosphomolybdic acid by tyrosine and tryptophan (aromatic amino acids) residues of the protein. The colour shows a λ_{max} at 660 nm. The intensity of colour depends on the amount of these aromatic amino acids present.

Protein reference standard: Bovine serum albumin (BSA) is used as a reference standard protein. Weigh accurately 100 mg of BSA and transfer it into a clean and dry 100 ml volumetric flask. Dissolve the protein by adding 5–10 ml of 0.1N sodium hydroxide and make up the volume to 100 ml with distilled water. Dilute the protein solution (1 mg/ml) to give a working standard concentration of 100 µg/ml, (10 X dilution) in a volumetric flask.

(b) Reagents

1. Sodium carbonate (2% w/v) in 0.1 N Sodium hydroxide solution (Reagent A): To prepare 1 L of the reagent, dissolve 20 g of sodium carbonate and 4 g of sodium hydroxide in distilled water and make up the volume to 1 L, using a volumetric flask.
2. Copper sulphate solution (1% w/v) (Reagent B): Dissolve 1 g of copper sulphate in distilled water and make up the volume to 100 ml.
3. Sodium potassium tartrate solution (2% w/v) (Reagent C): Dissolve 2 g of the salt in distilled water and make up the volume to 100 ml.
4. Alkaline copper reagent (Reagent D): To 100 ml of reagent A, add 2 ml of reagent B & C (mix 1 ml, each of B & C in a glass test tube and vortex) and stir the solution and use it for analysis. This reagent should be prepared freshly.
5. Folin's reagent (1 N strength): Check the normality of the commercial Folin's reagent and dilute appropriately, before use.
6. Protein solution (Standard stock solution): Weigh accurately 50 mg of BSA (fraction V) and dissolve in distilled water and make up to 50 ml in standard flask.
7. Working standard: Dilute 10 ml of the stock solution to 50 ml with distilled water in a standard flask so that 1 ml of this solution contains 200 µg of standard protein.

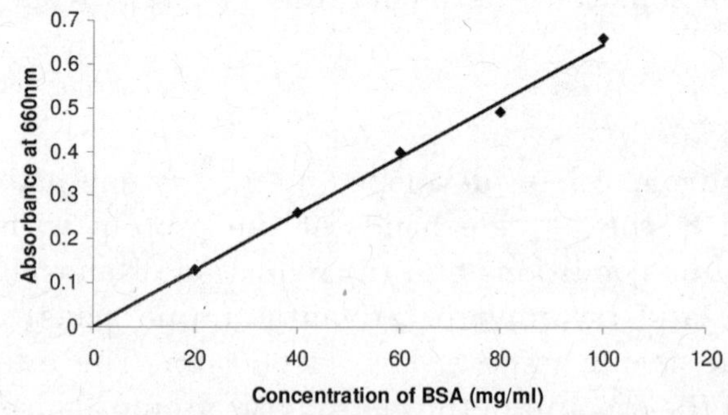

Figure III. 2.1 : Standard graph of Bovine Serum Albumin (BSA) for total protein determination.

(a) Procedure

1. To one ml of the standard protein solution containing 10–100 μg of protein or appropriately diluted unknown protein sample solution, add 4 ml of reagent D and mix the contents.
2. After 10 min of incubation at room temperature, add 0.4 ml of Folin's reagent and vortex the contents immediately.
3. Run a reagent blank with 1 ml of distilled water along with the standards and sample.
4. After 30 min of incubation at room temperature, read the blue colour developed at 660 nm and record the absorbance.
5. Compute the concentration of the protein in the sample from the calibrated curve.

For Bovine serum albumin we typically obtain a linear relationship between absorbance and amount protein over a range of 0.5 to 20 mg protein. The assay has not been reliable for amounts below 0.5 mg, however the actual sensitive range may extend beyond the upper limit.

The Most Highly Cited Paper in Publishing History: Protein Determination by Oliver H. Lowry (http://www.jbc.org)

Oliver Howe Lowry (1910–1996) was born in Chicago, the youngest of five children. Lowry received a bachelor's degree in chemistry in 1932 from Northwestern University and entered the University of Chicago as a graduate student in physiological chemistry.

Lowry's method was based on the "Phenol Reagent" developed by Otto Knut Olof Folin (who was featured in a previous JBC Classic (3)) and Vintila Ciocalteu for use in protein determinations. The Folin-Ciocalteu paper is also reprinted here as a JBC Classic. Lowry determined that the Folin phenol reagent (phosphomolybdic-phosphotungstic acid) would bind readily to copper-treated protein. Over time, the bound reagent was reduced, resulting in a colour change from yellow to blue, which could then be used to determine protein concentration.

Despite the great practical applications of Lowry's method, he did not publish the details immediately. Instead, he passed them on to whoever wanted them. One of the recipients of the method was Earl Sutherland, who complained of being tired of referring to "an unpublished method of Lowry." With Sutherland's prompting, Lowry finally performed a thorough study of the procedure: its limitations and virtues and the results it gave with different proteins and tissues in comparison with other methods. Lowry's first submission to the JBC was returned for drastic shortening, but the paper was eventually accepted in an abbreviated form.

Chemicals and Substances that Interfere with the Lowry Assay

The Lowry method for protein concentration determination is widely used, however can be subjected to interference by many substances and chemicals.

Chemicals that are known to interefere with the Lowry Assay:

- Barbital
- CAPS
- Cesium Chloride
- Citrate
- Cysteine
- Diethanolamine
- Dithiothreitol
- EDTA
- EGTA
- HEPES
- mercaptoethanol
- Phenol
- Polyvinyl pyrrolidone
- Sodium Deoxycholate
- Sodium Salicylate
- Tricine
- TRIS
- Triton X-100

2.2.3.4 Bradford protein assay

The Bradford assay is very fast and uses about the same amount of protein as the Lowry assay. It is fairly accurate and samples that are out of range can be retested within minutes. The Bradford is recommended for general use, especially for determining protein content of cell fractions and assessing protein concentrations for gel electrophoresis.

It is sensitive to about 5 to 200 micrograms protein, depending on the dye quality. In assays using 5 ml colour reagent prepared in lab, the sensitive range is closer to 5 to 100 μg protein.

(a) Principle

The assay is based on the observation that the absorbance maximum for an acidic solution of Coomassie Brilliant Blue G-250 shifts from

465 nm to 595 nm when binding to protein occurs. Both hydrophobic and ionic interactions stabilize the anionic form of the dye, causing a visible colour change. The assay is useful since the extinction coefficient of a dye-albumin complex solution is constant over a 10-fold concentration range.

(b) Reagents

1. Bradford reagent: Dissolve 100 mg Coomassie Brilliant Blue G-250 in 50 ml ethanol (95%); add 100 ml (85% w/v) phosphoric acid. Dilute to 1 litre when the dye has completely dissolved, and filter through Whatman no. 1 paper just before use.
2. (Optional) 1 M NaOH (to be used if samples are not readily soluble in the colour reagent).

(c) Procedure

1. Dilute unknowns if necessary to obtain between 5 and 100 µg protein in at least one assay tube containing 100 µl sample
2. If desired, add an equal volume of 1 M NaOH to each sample and vortex. Add NaOH to standards as well if this option is used.
3. Prepare standards containing a range of 5 to 100 micrograms protein (albumin or gamma globulin are recommended) in 100 µl volume.
4. Add 5 ml Bradford reagent and incubate for 5 min.
5. Measure the absorbance at 595 nm.
6. Prepare a standard curve of absorbance versus micrograms protein and determine amounts from the curve. Determine concentrations of original samples from the amount protein, volume/sample, and dilution factor, if any.

   Note: The Bradford reagent should be a light brown in colour. Filtration may have to be repeated to get rid the reagent of blue components.

Comments: The dye reagent reacts primarily with arginine residues and less so with histidine, lysine, tyrosine, tryptophan, and phenylalanine residues. Obviously, the assay is less accurate for basic or acidic proteins. The Bradford assay is rather sensitive to bovine serum albumin, more so than "average" proteins, by about a factor of two. Immunoglogin G (IgG-gamma globulin) is the preferred protein standard. The addition of 1 M NaOH is to allow the solubilization of membrane proteins and reduce the protein-to-protein variation in colour yield.

2.2.3.5 Bicinchoninic Acid (BCA) Protein Assay

The bicinchoninic acid (BCA) assay is similar to the Lowry method in that it also depends on the conversion of Cu^{+2} to Cu^{+} under alkaline conditions. Thus, this procedure is very applicable to microtiter plate methods. The BCA is used for the same reasons the Lowry is used. Stoscheck (1990) has suggested that the BCA assay will replace the Lowry because it requires a single step, and the colour reagent is stable under alkaline conditions.

The method is more sensible than the Lowry method being able to detect down to 0.5 μg protein cm^{-3}, but perhaps more importantly it is generally more tolerant of the presence of compound that interfere with the Lowry assay.

Both a standard assay for concentrated proteins and a micro-assay for dilute protein solutions are described below.

(a) Principle

BCA serves the purpose of the Folin reagent in the Lowry assay, namely to react with complexes between copper ions and peptide bonds to produce a purple énd product. The advantage of BCA is that the reagent is fairly stable under alkaline conditions, and can be included in the copper solution to allow a one step procedure. A molybdenum/tungsten blue product is produced as with the Lowry.

Method 1: Standard assay for concentrated proteins.

(b) Reagents

1. Reagent A: 1 gm sodium bicinchoninate (BCA), 2 gm sodium carbonate, 0.16 gm sodium tartrate, 0.4 gm NaOH, and 0.95 gm sodium bicarbonate, brought to 100 ml with distilled water. Adjust the pH to 11.25 with 10 M NaOH.
2. Reagent B: 0.4 gm cupric sulphate (5 x hydrated) in 10 ml distilled water.
3. Standard working solution (SWR): Mix 100 volumes reagent A with 2 volumes reagent B.

   Note: The stock solutions are stable. The working solution is stable for 1 week and should be green.

(c) Procedure

Prepare samples containing 0.2 to 50 micrograms protein in microliters.

1. Add 1 ml Standard Working Reagent to each 20 microliters sample and mix. Incubate 30 min. at 60°C.
2. Cool the samples and read at 562 nm. Color will be stable for at least one hour.

Method 2: Micro assay for dilute protein solutions.

(a) Reagents

1. Reagent A: 8 gm sodium carbonate monohydrate, 1.6 gm sodium tartrate, brought to 100 ml with distilled water. Adjust the pH to 11.25 with 10 M NaOH.
2. Reagent B: 4 gm BCA in 100 ml distilled water.
3. Reagent C: 0.4 gm cupric sulfate (5 x hydrated) in 10 ml water.
4. Working solution: Mix 1 volume reagent C with 25 volumes reagent B, then add 26 volumes reagent A to the C/B mixture.

(b) Procedure

1. Prepare samples containing 0.2 to 50 μg protein in 500 μl.
2. Add 500 μl working solution to each 500 μl sample and mix. Incubate 60 min. at 60 degrees C.
3. Cool the samples and read at 562 nm.
4. Determine concentrations of original samples from the amount protein, volume/sample, and dilution factor, if any.

Comments: A longer incubation increases the sensitivity of the assay. The heating can be stopped earlier to prevent the colour from becoming too dark. The assay can be performed at room temperature, but there is greater variability among proteins and the assay is less sensitive.

2.2.3.6 Biuret Protein Assay

The principle of the biuret assay is similar to that of the Lowry, however it involves a single incubation of 20 min. There are very few interfering agents (ammonium salts being one such agent), and Layne (1957) reported fewer deviations than with the Lowry or ultraviolet absorption methods. However, the biuret assay consumes much more material. The biuret is a good general protein assay for batches of material for which yield is not a problem.

2 Biuret + 2 Peptic Chain ---------------> Biuret Complex

Or

Purple coloured peptide-copper co-ordination complex

(a) Principle

Under alkaline conditions substances containing two or more peptide bonds form a purple violet coloured complex with copper salts in the reagent, which shows an absorption maximum at 540 nm.

(b) Reagents

Biuret reagent: Dissolve 1.5 g of cupric sulphate in 250 ml distilled water. Separately, weigh and dissolve 6 g of sodium potassium tartrate in 250 ml of distilled water. Mix the cupric sulphate and tartrate solutions in a 1 litre beaker. To this solution add 300 ml of 10% (w/v) sodium hydroxide solution, with constant stirring. Make up the volume to 1 l with distilled water using a volumetric flask. Discard if a black precipitate forms. Store the biuret reagent in a plastic container.

(c) Procedure

1. Prepare a series of standard solutions using standard bovine serum albumin (1 – 10 mg of protein).
2. To one milliliter of the standard protein solution (bovine serum albumin) add 5 ml of biuret reagent and mix the contents.
3. After 30 min of incubation at room temperature, measure the violet colour developed against the reagent blank at 540 nm.
4. Compute the concentration of the protein in the sample from the calibration curve. While calculating the protein concentration in the unknown sample, the dilution factor has to be taken into account.

Comments: The colour is stable, but all readings should be taken within 10 min. of each other. As with most assays, the Biuret can be scaled down for smaller cuvette sizes, consuming less protein. Proteins with an abnormally high or low percentage of amino acids with aromatic side groups will give high or low readings, respectively.

2.2.4. The Isolation of Casein from Milk

(a) Principle

Most proteins show a minimum solubility at their isoelectric pH. This property of proteins is utilized to isolate the casein. It is done by adjusting the pH of milk to 4-8, its isoelectric point. Casein, the principle phospho-protein of milk is separated from other proteins by isoelectric precipit-ation, *i.e.*, by adjusting the pH of milk to its isoelectric pH (4.8). It is present at a concentration of about 35 g/l.

(b) Reagents

Solvents: Ethanol, diethyl ether. Sodium acetate buffer (0.2 mol^{-1} pH 4.8).

Sample: Cow or Buffalo milk. (1 litre).

Equipment: (i) pH meter (ii) Filtration unit (Buchner funnel, side arm flask, Suction pump) (iii) Top loading balance (iv) Centrifuge.

(c) Procedure

1. Place 100 ml of milk in a 500 ml beaker and warm to 40°C.
2. Centrifuge at 4000 rpm for 20–25 min at room temperature and carefully remove the cream/fat from the surface with the help of a stainless steel spatula.
3. Transfer this skimmed milk into a 500 ml glass beaker. Add all equal volume of distilled water and stir. Note its pH using a pH meter.
4. Drop the pH of the diluted milk to 4.8 by drop wise addition of warm acetate buffer using a burette, with constant stirring. At this pH, casein gets precipitated. Allow the precipitate to settle at room temperature for 30 min.
5. Decant the supernatant carefully and filter the suspension using a filtration unit connected to a suction pump (Buchner funnel fitted Procedure with Whatman No.1 filter paper disc or through a sintered glass funnel).
6. The moist precipitate is washed thrice with 100 -150 ml of distilled water to remove the salts. This is followed by two washes each with 100 ml of ethanol and diethyl ether.
7. Transfer the cake from the funnel onto a clean Petri-plate or a watch glass. Spread material uniformly and allow it to dry at room temperature overnight.
8. Record the weight of dried sample and calculate the percent yield.

Note:

1. Bottle containing diethylether should be kept in chilled condition before use. It is highly inflammable.

2. At pH values above or below the isoelectric point all protein molecules have a net charge of the same sign. Thus they repel each other and prevent the coalescence of single molecules to insoluble aggregates.

2.2.5 Ammonium Sulphate Fractionation of Proteins

Proteins are responsible for many aspects of cell function and structure. To learn more about their precise roles in cellular biology it is often necessary to analyze purified proteins. A protein of interest can be separated from other proteins based upon its unique physical and chemical properties. The first step in protein isolation is to extract the protein from the tissue or cell in a soluble form. In some cases it is desirable to first disrupt the cell and to isolate the appropriate subcellular compartment before solubilizing the protein. Generally, cytosolic proteins are found in the supernatant following cell disruption and centrifugation at 100,000 × g for 30–60 min. Membrane proteins or proteins associated with organelles will require additional solubilization techniques such as detergents or chaotropic agents. Proteins are soluble in aqueous solutions because the charged and polar side-chain groups of the amino acids interact with water. In other words, proteins become hydrated. If the protein-solvent interaction is prevented, proteins will interact with one another and form aggregates that precipitate out of solution. Different proteins will behave differently in terms of interacting with either the solvent or other protein molecules. Therefore, it is possible to separate different proteins based upon different solubility properties. For example, as the salt concentration of a solution is increased the amount of water that is available to interact with proteins is decreased. This will result in the more hydrophobic domains of proteins interacting with hydrophobic domains of other proteins. The interacting proteins will form large masses, or a precipitate, that can be collected by centrifugation. Proteins with a larger proportion of hydrophobic amino acid residues exposed on their surface will be more sensitive to effects of salts than highly charged proteins. Therefore, it is possible to selectively precipitate some proteins under conditions where other proteins remain soluble. One of the most common salts used for this differential precipitation of proteins is $(NH_4)_2SO_4$.

(a) Procedure

A common method for differentially precipitating proteins is by 'salting-out' with $(NH_4)_2SO_4$. As the $(NH_4)_2SO_4$ is added, water molecules interact with the salt ions, thus decreasing the amount of water available to bind to protein. Proteins are differentially precipitated by $(NH_4)_2SO_4$ due to their differences in hydrophobicity. A more hydro-phobic protein will precipitate at a lower salt concentration than a more hydrophilic protein.

1. The $(NH_4)_2SO_4$ is slowly added to a protein solution while gently stirring. The amount to add to precipitate the protein of interest will have to be initially determined empirically. The concentration of $(NH_4)_2SO_4$ needed to precipitate a particular protein is usually expressed as a percent of saturated $(NH_4)_2SO_4$. Either solid or a saturated solution can be added. For large volumes, adding solid is more convenient.
2. Once dissolved, the $(NH_4)_2SO_4$ protein suspension is allowed to slowly stir (usually while kept cold) until equilibrium is reached. This is typically for one hour, however, some proteins require overnight incubations. The precipitated proteins are collected by centrifugation. Typically 10,000 X g for 10 minutes is sufficient to pellet precipitated proteins.
3. The protein pellet is dissolved in the buffer of interest and the remaining $(NH_4)_2SO_4$ is removed by either dialysis or desalting column. Dialysis tubing has defined pore size which allows small molecules to pass through freely, while retaining larger molecules. Multiple buffer changes accelerate the process. $(NH_4)_2SO_4$ precipitation can also be carried out as a two-step procedure. The maximum amount of $(NH_4)_2SO_4$ that does not precipitate the protein of interest is added and this pellet is discarded. $(NH_4)_2SO_4$ is then added to the supernatant to bring the concentration to the minimal percent saturation that completely precipitates the protein of interest. Nomograms for calculating the amounts of $(NH_4)_2SO_4$ needed to achieve the desired percent saturation are available (see Appendix IVB). The pellet is retained and the supernatant discarded at this step.

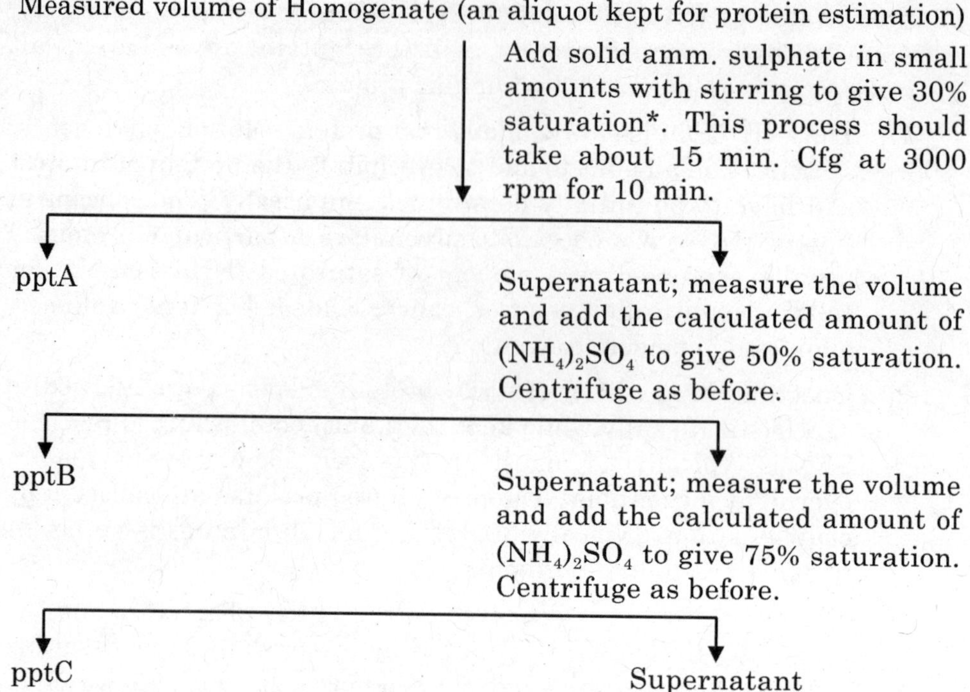

* ***Note:*** *See the Appendix IV B (7) to calculate the amount of* $(NH_4)_2SO_4$ *to be added to give this saturation.*

Another variation of $(NH_4)_2SO_4$ precipitation is 'back extraction'. In this case the total protein is precipitated with 80–90 percent saturated $(NH_4)_2SO_4$ and the pellet is extracted with the appropriate $(NH_4)_2SO_4$ concentration to solubilize the protein of interest. The changes on a protein in solution can also be neutralized by the addition of salts and this also has been used for purification of proteins.

Theoretically any salt can be used but generally ammonium sulphate is preferred for two reasons. One it has a high solubility 840 g/l and secondly, its dissolution in water is exothermic or the solution gets cooled.

Estimate the amount of protein in each fraction and calculate back to per gram tissue weight. Care must be exercised to carry out the protein estimation as ammonium sulphate will interfere with the assay. Therefore, it is preferable to adopt the modified Biuret method after TCA precipitation, alternatively each fraction can be dialysed to remove the salt and then estimation carried out.

Table : Six functional classes of enzymes by the International Union of Biochemists (www.1cro.com/mwking/enzyme-kinetics. html)

S.No.	Classification	Biochemical Properties
1.	Oxidoreductases	Act on many chemical groupings to add or remove hydrogen atoms.
2.	Transferases	Transfer functional groups between donor and acceptor molecules. Kinases are specialized transferases that regulate metabolism by transferring phosphate from ATP to other molecules.
3.	Hydrolases	Add water across a bond, hydrolyzing it.
4.	Lyases	Add water, ammonia or carbon dioxide across double bonds, or remove these elements to produce double bonds.
5.	Isomerases	Carry out many kinds of isomerization: L to D isomerizations, mutase reactions (shifts of chemical groups) and others.
6.	Ligases	Catalyze reactions in which two chemical groups are joined (or ligated) with the use of energy from ATP.

2.3 ENZYMES

2.3.1 Extraction and Purification of Enzyme: General Strategies

Enzymes are biological catalysts responsible for supporting almost all of the chemical reactions that maintain animal homeostasis. Because of their role in maintaining life processes, the assay and pharmacological regulation of enzymes have become key elements in clinical diagnosis and therapeutics. Enzymes are found in all tissues and fluids of the body. Intracellular enzymes catalyze the reactions of metabolic pathways. Plasma membrane enzymes regulate catalysis within cells in response to extracellular signals, and enzymes of the circulatory system are responsible for regulating the clotting of blood. Almost every significant life process is dependent on enzyme activity. Traditionally, enzymes were simply assigned names by the investigator who discovered the enzyme. As knowledge expanded, systems of enzyme classification became more comprehensive and complex. Currently enzymes are grouped into six functional classes by the International Union of Biochemists (I.U.B.) (Table 2.3.1).

I. Extraction of Enzymes

(a) Homogenisation

Fresh tissue is crushed into a paste with an extraction medium (often a buffer) in a mortar and pestle, or in a tissue homogenizer, or in a blender or by ultrasonic vibrations (sonication).

The molarity and pH of the buffer is suitably adjusted (which may vary for different enzymes) to achieve maximum solubility and activity of the enzyme. EDTA (ethylene diamine tetra acetic acid) is often included in the extraction medium to remove heavy metals (which otherwise inhibit enzyme activity), and for disrupting the membranes of cells and cell organelles. Detergents such as Triton-X are also used sometimes to solubilise the membranes.

Many enzyme proteins contain disulfide (S-S) bonds due to the presence of cysteine residues, which are easily broken during enzyme extraction leading to loss of enzyme activity. To overcome this problem thiols are added, such as mercaptoethanol whose sulfhydryl (-SH) group is able to maintain the S-S linkage in enzymes.

If the extract is not homogeneous, the homogenate (extract) is filtered to remove cell debris, fibres etc., otherwise filtration may be avoided. All operations of extraction and purification are generally carried out in cold (0–4°C), since most of the enzymes get inactivated at higher temperatures.

(b) Centrifugation

The enzyme extract is centrifuged to remove cell debris, cell organelles and sometimes other molecular aggregates, leading to partial purification of enzymes.

It also helps in characterization of an enzyme, since, depending upon its mass and shape the enzyme will move through a solution at a definite speed and occupy a characteristic position in the centrifuge tube.

For most cytosolic enzymes, centrifugation at about 30,000 g for 30 minutes is good enough to obtain a fair amount of activity in the supernatant.

However, if the enzyme is located in a specific cell organelle, an extract rich in that organelle is prepared through 'preparative centrifugation'. Centrifugation for different durations at different velocities allows the cell organelles to sediment according to their sizes. All centrifugation operations are conducted in cold (0–4°C).

(c) Precipitation

Enzymes and other proteins are highly charged molecules, and can be precipitated with appropriate charge neutralizing chemicals. Once their charges are broken, they form aggregates and settle down as precipitate.

When, an acid or base is added, the enzyme protein can be brought to its isoelectric pH. At this pH, there is no net charge on enzyme molecules and electrostatic repulsion between them is low so that they tend to aggregate. Therefore, adjusting the pH to the isoelectric point of a protein causes its precipitation.

Acids and bases, however, often inactivate the enzyme, so that their use for precipitation is not recommended in most cases. Instead ammonium sulphate and other salts are used for precipitation in a process called 'salting out'. Salts can change the structure of the solvent, which can lead to large changes in protein conformation by altering the electrostatic interactions between charged groups on the protein surface.

The salt also competes with the protein for solvent molecules and thereby lowers its solvation. In large scale enzyme precipitation, use of many other neutral salts is preferred over ammonium sulphate, which is corrosive and releases NH_3 at higher temperatures.

Some organic solvents like acetone, methanol and ethanol are also used for enzyme precipitation, since water miscible solvents decrease the solubility of proteins, leading to precipitation.

They are cooled upto 40–60°C before their use, and precipitation is carried out at 0°C, because precipitation at room temperature causes denaturation of the enzyme, in most cases. Organic solvents are added drop by drop to avoid local concentration.

Water soluble non-ionic polymers, such as polyethylene glycol, alginate, pectate, carboxymethyl cellulose, polyacrylic and polymeta acrylic acids, etc. also cause enzyme precipitation.

Polyethyleneimine is also widely used as protein precipitant at large scale. They primarily act through the removal of solvent sphere of the enzyme protein.

II. Purification of Enzymes

Enzyme purification involves three steps. These three techniques are described in the following text:

(a) Dialysis, (b) Chromatography (c) Electrophoresis

(a) Dialysis of Enzymes: Dialysis is the process that is used to remove small molecules from enzyme. For this, enzyme precipitate obtained in previous step is dissolved in a small quantity of buffer solution in which the enzyme was originally extracted. The solution is taken in a dialysis bag (may be a cellophane tube) and after sealing securely, the bag is suspended in either distilled water or a buffer of known molarity and ionic composition.

Some other salts or chemicals may have to be added sometimes in the outer solution, to prevent the loss of enzyme activity during dialysis. The dialysis is carried out for a few hours with regular change of the outer solution or distilled water. At large scale enzyme purification 'dialafitration' instead of dialysis is used.

The enzyme solution is filtered .for small molecules through a membrane generally mounted on a fibrous support, by pressure driven operations.

(b) Chromatography for Enzyme Purification: Chromatographic separation of proteins is the most common method of enzyme purification.

Following four types of chromatography are available for this purpose:

(i) adsorption or column chromatography,

(ii) ion exchange chromatography,

(iii) gel filtration chromatography and

(iv) affinity chromatography.

(i) Adsorption Chromatography for Enzyme Purification: In adsorption chromatography, the protein or enzyme solution suspected to contain other proteinaceous impurities is passed through a column of inert material packed in a glass or steel tube. Most commonly used column materials include finely divided solids such as charcoal, silica, alumina, calcium phosphate, hydroxyapatite, etc.

The effluent solution is continuously collected in small fractions of 1.0 to 2.0 l. The protein in each fraction is estimated by measuring the absorption at 280 nm using a UV spectrophotometer. The enzyme is also assayed in each fraction. Various spleen enzymes, such as basic RNAase, acidic RNAase, acidic DNAse, phosphodiesterase, phosphomonoesterase, etc. are often separated from each other using adsorption chromatography. For large scale chromatographic separation of enzymes, the process is accelerated by using motors and other mechanical devices for packing the column, for loading the enzyme on the column and for eluting the enzyme.

(ii) Ion Exchange Chromatography for Enzyme Purification: In ion exchange chromatography, generally a cellulosic ion exchange is taken in the column. The proteins are separated according to their charges.

The resolution is quite high and the technique can facilitate large scale protein purification. This has been successfully employed for insulin purification, plasma fractionation and for purification of many other enzymes.

(iii) Gel Filtration Chromatography for Enzyme Purification: In this chromatography, various proteins are separated on the basis of differences in their molecular sizes. This type of chromatography is also known as molecular exclusion chromatography or molecular sieve chromatography.

The basic arrangement for gel filtration chromatography is similar to that for adsorption chromatography. A column made up of glass or steel is taken and packed with a gel. The most commonly used gel is 'sephadex' which is a cross linked dextran produced by certain strains of bacteria. Several types of sephadex, namely G-10, G-30, G-50, G-100, G-150, G-200, etc. are available, which differ according to their pore sizes.

When a mixture of enzymes or proteins is poured on top of the column, different proteins move downwards according to their molecular sizes and come out from the column in order of decreasing sizes; the, larger molecules are eluted first. The elution volume is logarithmically proportional to the molecular size.

Gel filtration chromatography can also be used for determining the molecular weight of the protein by calibrating the column with proteins of known molecular weights.

(iv) Affinity Chromatography for Enzyme Purification: In this method, enzymes are purified according to their specificity for a particular substrate or cofactor. One component of the mixture containing enzyme binds covalently to the solid support of the column, and the other components percolate down through the column. The basic requirements for affinity chromatography are the same as in adsorption or gel filtration chromatography, but the packing gel must have some component which can bind with one component of the mixture.

Many commercial gels available for affinity chromatography contain functional groups attached to the 'spacer arms' of the gel. The "spacer arm" is a chemical linkage between the functional group and the gel or matrix proper, so that the binding between functional group and the enzyme is kept away from the gel. Thus the steric hindrance will be unlikely to prevent binding of the specific enzyme to the column.

(c) Electrophoesis for Enzyme Purification: Electrophoresis is a technique in which molecules (enzymes, proteins, amino acids, nucleotides and nucleic acids) are separated by differences in their net charge in the presence of an externally applied electric field. This technique is routinely used in enzyme purification and isozymes separation in the laboratories, although it has found only limited application at large scale, since the technique is time consuming and is a bit expensive.

Various types of instrumental approaches have been used to separate and purify charged molecules using electrophoresis. However, the most common method for purifying enzymes is through electrophoresis on polyacrylamide gel.

Gel electrophoresis is also used to separate various isozymes of a given enzyme. Isozymes perform the same catalytic function but differ in their regulatory and some kinetic aspects.

III. Final Step in Processing Enzymes

Most of the commercially available enzyme preparations, purified as above, are concentrated and sterile filtered, after purification. This is done to reduce both, the volume and the microbial contamination of the sample. Often, before storage and transport, the sample is freeze dried with additives, such as sugar substrates and dextrans.

2.3.2. Expression of enzyme activity

Measurement of enzyme activity

Usually, biochemical analysis involves determination of the concentration of an analyte in biological specimen. Unlike these quantitative measurements, the measure of an enzyme molecule is based on their biological activity rather than the measurement of their actual concentration. In determining the enzyme activity, the relationship between the product formed from the substrate and the absorbance due to increase in concentration of the product is a quantitative measure of the velocity or the rate of the reaction as a function of time. Further, the concentration of the product formed can be calculated by using the physical factor, the molar extinction coefficient (ε).

Enzyme activity can be assayed by two methods, namely (i) **equilibrium** or **steady-state** and (ii) **kinetic methods.** In the former, the enzyme is incubated with its substrate for a pre-determined time, at the end of which the enzymatic reaction is stopped and the amount of product formed or substrate consumed is measured. In this type of assay, it is necessary to establish the linear response range, over which the zero order kinetics is followed. If this aspect of the assay is disregarded, the measurements made would lead to an erroneous interpretation.

There are two approaches to assay an enzyme by steady-state method, *viz.*, one-point and two-point assays. One-point assay is based on a single pre-determined time, while the two-point method has two pre-determined time points, representing the linear response of the enzymatic reaction.

In kinetic assay, the reaction rate is continuously monitored at short intervals of time. Hence, any deviation from zero order kinetics can be noted by a non-linear rate of reaction. Kinetic assays are potentially more accurate than steady-state methods of enzymatic analysis.

Assay method: There are two common methods of determining the activity of an enzyme. A **stop time assay** and a **real-time assay**. A stop time assay is just that, start the reaction and stop or read the results at a given time. Second, is for the compound (to be measured) that is stable enough to wait to read and the conditions used to stop the enzyme, *i.e.* acid or base are too harsh to maintain the structure of the readout. For this, there is real time assays: meaning the change in absorbance (also known as optical density – OD) vs. time. From this graph (done on the spectrophotometer) a region is selected that is reasonably linear and the change in OD per min is determined and converted to Units of enzyme activity per ml.

Following points must be kept in mind while performing an enzyme activity

- Absorbance — Most species can only read between 0.01 and 3.0 abs units. At either end of this range there will be too much noise.
- Always run a control assay — This is an assay that does not contain enzyme. It will tell any drift in the baseline absorbance. If there is an appreciable amount of drift, subtract this ΔOD/min from enzyme assay tube. If it is about zero then skip it. The control assay will also tell what the starting absorbance is.
- Proper Rates — This depends on each enzyme. For MDH, a rate of 0.05 to 0.4 ΔOD/min is good enough. If the rate is over too fast (see above) then dilute the enzyme. If one is not certain how much to dilute the enzyme, do a 1:2 or 1:5.
- Run a positive control — Use a sample that has the enzyme. Often this can be from an extract or some purified protein already prepared.
- Temperature — Bring all solutions to room temp before starting assays. The easiest way to do this is mix the next set of tubes while assaying one set. Enzyme should always be on ice before adding to the enzyme cocktail or it will denature.
- Measuring and Pipeting — This is another problem area. Day to day variations or even batch to batch changes in enzyme or substrate solutions will cause a lot of error.

Units of enzyme activity

One international unit (IU): An enzyme unit (IU) is that concen-tration of an enzyme that catalyses the formation of 1 μ mole of product per minute under defined assay conditions. The concentration of enzyme in crude biological preparation is expressed as units/ml (U ml^{-1}).

The Enzyme Commission of IUB, has recommended a new unit namely, Katal. One katal refers to the conversion of one mole of substrate per second. The enzyme activity can be expressed as millikatals, microkatals and nanokatals. Thus one IU of enzyme = 1/60 μ katal = 16.67 n katal.

Specific activity is expressed as the number of enzyme units per milligram protein. It is an index of enzyme purity. During the process of enzyme purification, the value of specific activity increases. A pure enzyme preparation exhibits maximal specific activity.

Total activity of an enzyme is calculated by (Specific activity) × [total protein (mg) in the enzyme preparation].

2.3.2.1 To study the effect of pH and temperature on activity of acid phosphatase (AP)

This assay is based on the method given by Al Omair (2010).

2.3.2.1.1 To determine the activity of acid phosphatase

Acid phosphatases (EC 3.1.3.2) are produced by both prokaryotic and eukaryotic cells and are presumed to convert organic phosphorus into available P_i. Phosphate is an important molecule for cellular growth that is involved in many different biological reactions. The hydrolysis of phosphomonoesters by phosphatases in biological systems is an important process. This process is linked to energy metabolism, metabolic regulation and a wide variety of cellular signal transduction pathways. The role of acid phosphatase in phosphorus metabolism has been extensively studied in prokaryotic and eukaryotic systems. The physiological function of acid phosphatase is to provide inorganic phosphate for the cellular growing.

Several interacellular acid phosphatases have been suggested to have a role in the hydrolysis of interacellular polyphosphates.

Acid phosphatase is found to be localized in the cell walls of *Pisum sativum*, root cap of rice, phloem of *Nicotiana tabacum* and xylem of *Phaseolus vulgaris*. The mechanism which regulates acid phosphatase distribution and activity is unclear, although their abundance may be regulated by the level of phosphate in the environment and their activity is clearly influenced by local polyelectrolytes.

(a) Principle

The enzyme acid phosphatase catalyses the conversion of the substrate, p-nitrophenyl phosphate to p-nitrophenol (a yellow coloured product and inorganic phosphate. The enzyme activity is measured by recording the absorbance due to the formation of p-nitrophenol, at 410 nm. The concentration of p-nitrophenol is calculated by using its molar extinction coefficient value at λ_{max} of 410nm.

Enzyme unit: One enzyme unit hydrolyses 1 μmole of p-nitrophenyl phosphate per minute, at pH 4.8 at 37°C.

Plant Material

- Germinated seeds of *Vigna aconitifolia.*

Seed Germination

Seeds of *Vigna aconitifolia* were sterilized with sodium hypochlorite (0.1%) and soaked overnight in sterile water at room temperature. The wet seeds were germinated on filter papers moistened with distilled water in petri dishes until complete germination. The germinated seeds were sown in plastic pots containing 1 kg of oven dry soil. Then they were placed in a growth chamber maintained at 27/20°C light/dark (12 h) and grown for 10 days. At the end of the experiment, the shoot systems were washed with deionized water and blotted on paper towels.

(b) Reagents

- 100 mM Tris-HCl (pH 5.0)
- 2 mM MgC12
- 2% polyvinylpyrolidone (PVP)
- Sodium citrate buffer, 100 mM, pH 4.8.
- Substrate solution: p-nitrophenyl phosphate, 15 mM dissolved in deionised or double glass distilled water
- Sodium hydroxide (0.2 N).

(c) Preparation of the Crude Extract of AP

Ten gram of collected shoots from 10 days old seedlings of *Vigna aconitifolia* were homogenized in a blender with 30 mL of 100 mM Tris-HCl (pH 5.0) containing 2 mM $MgCl_2$ and polyvinylpyrolidone (PVP) (2%). The homogenate was filtered through four layers of muslin cloth. Centrifuge the filtrate at 10,000g for 10 min. Use the supernatant as enzyme source.

Enzyme assay: Pipette out the following reagents into clean labelled test tubes. Run a blank along with the test sample, in triplicates.

Table : Enzyme assay protocol

Reagents	Blank (ml)	Test (ml)
1. Citrate buffer, 100 mM, pH 4.8	0.5	0.5
2. Substrate solution (p-nitrophenyl phosphate, 15 mM)	0.5	0.5
Mix the contents thoroughly by vortexing.		
3. Enzyme extract	–	0.1
Mix and incubate at 37°C in a water bath for 10 minutes.		
4. Sodium hydroxide solution (0.2 N)	3.9	3.9
Mix the contents		
5. Enzyme extract	0.1	–
Total assay volume	5.0	5.0

Record the absorbance of test and blank at 410 nm, using a spectrophotometer.

Calculation

By using the following equation, the activity, *i.e.*, U ml^{-1} of the enzyme solution:

$$\text{Units/ml enzyme extract} = \frac{(\text{Absorbance of test} - \text{Absorbance of Blank}) \times 5.0}{10 \times 18.3 \times 0.1} \times 10 \text{ df}$$

Where,

5.0 = Total enzyme assay volume in ml

10 = Time of assay in minutes.

18.3 = Millimolar extinction coefficient (ϵ) of p-nitrophenol at 410 nm

0.1 = Volume of enzyme extract.

df = dilution Factor

Note

(i) If the colour developed is too intense, dilute the enzyme extract appropriately and repeat the enzyme assay. Take the dilution factor account while calculating the enzyme activity.

(ii) If the colour development is insufficient for 10 minutes, increase the incubation time to 20 minutes.

Calculate the specific activity of the enzyme by the following equation:

$$\text{Specific activity} = \frac{\text{Units/ml enzyme extract}}{\text{mg protein/ml enzyme extract}}$$

Note: Determine the protein content in the enzyme extract by Lowry method for calculating the specific activity.

2.3.2.1.2 Effect of pH on Acid Phosphatase Activity

The enzyme activity is influenced by various environmental factors including pH. The pH is an important factor in the stability of enzymes. As with activity, for each enzyme there is also a region of pH optimal stability. The enzyme activity is at its peak at a particular pH. The optimum pH value will vary greatly from one enzyme to another as shown in Table 2.3.3.

The pH can influence the enzyme activity in a number of ways. Firstly, it can change the ionization of the enzyme substrate complex. Secondly, it can change the ionization of various groups of the enzyme molecule which may affect the affinity of the enzyme for the substrate. Thirdly, it changes the ionization of the substrate which may affect the binding of the substrate to the enzyme. Fourthly, at extreme pH it can brings about changes in protein structure.

The procedure to find out the optimum pH of an enzyme is given below, with acid phosphatase as an example.

(a) Principle

Enzymes are affected by changes in pH. The most favourable pH value, the point where the enzyme is most active is known as the optimum pH. Extremely high or low pH values generally result in complete loss of activity for most enzymes. Citrate buffer with pH range 3.0–9.0 is used to determine the optimum pH of acid phosphatase.

(b) Procedure

1. Set up a series of test tubes in duplicate each containing 0.5 ml of substrate solution and 0.5 ml of Citrate buffer, 100 mM of the following pH values: 3.0, 3.5, 4.0, 4.5, 5.0, 5.5 and 6.0.
2. Mark one set as 'test' and the other as 'blank'. Proceed to step 3, in order determine the enzyme activity (as described in section 2.3.2.1.1).

3. Draw a graph by plotting the pH values on X-axis and the enzyme activity on Y-axis. Select the pH at maximum enzyme activity as the optimum pH.

Table 2.3.3 : pH optimum for a few enzymes

Enzyme	pH Optimum
Lipase (pancreas)	8.0
Lipase (stomach)	4.0-5.0
Lipase (castor oil)	4.7
Pepsin	1.5-1.6
Trypsin	7.8-8.7
Urease	7.0
Invertase	4.5
Maltase	6.1-6.8
Amylase (pancreas)	6.7-7.0
Amylase (malt)	4.6-5.2
Catalase	7.0

(c) Result

On the study of pH effect on the enzyme activity, it was necessary to fix the other factors affecting the enzyme activity such as enzyme concen-tration, substrate concentration, $MgCl_2$ concentration, as well as the tem-perature under which the reaction mixture was incubated. The results are illustrated graphically in Fig. III.2.2. It appears from the figure that the enzyme activity increased gradually from pH 3.6 to 5.4 which seems to be the optimum value after which the enzyme activity started to decline. Hence, the optimum pH of the acid phosphatase was 5.4.

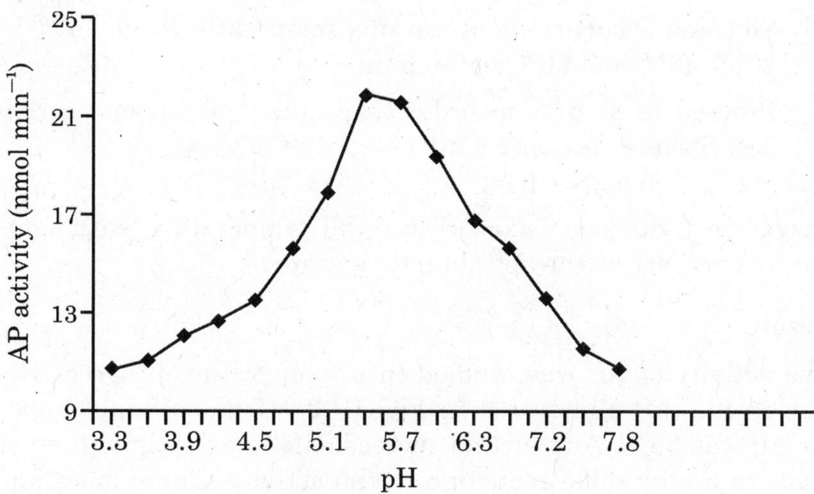

Fig. III.2.2 : Effect of pH on AP activity.

2.3.2.1.3 Effect of temperature on Acid Phosphatase

Like most chemical reactions, the rate of an enzyme-catalyzed reaction increases as the temperature is raised. A ten degree centigrade rise in temperature will increase the activity of most enzymes by 50 to 100 percent. Variations in reaction temperature as small as 1 or 2 degrees may introduce changes of 10 to 20 percent in the results. In the case of enzymatic reactions, this is complicated by the fact that many enzymes are adversely affected by high temperatures. As shown in Figure III.2.3, the reaction rate increases with temperature to a maximum level, then abruptly declines with further increase of temperature. Because most animal enzymes rapidly become denatured at temperatures above 40°C, most enzyme determinations are carried out somewhat below that temperature. Over a period of time, enzymes will be deactivated at even moderate temperatures. Storage of enzymes at 5°C or below is generally the most suitable. Some enzymes lose their activity when frozen.

(a) Principle

The enzyme activity is influenced by temperature and at a particular temperature, the enzyme activity is at its peak. The procedure to find out the optimum temperature of enzyme activity is given below

(b) Procedure

1. Set up a series of test tubes in duplicate each containing 0.5 ml of substrate solution and 0.5 ml of Citrate buffer, 100 mM of the optimum pH. Mark one set as 'test' and the other one as 'blank'.
2. In the 'test' tubes alone add 0.1 ml of the enzyme solution.

3. Mix and incubate at at various temperatures *viz.*, 10°C, 20°C, 30°C, 40°C and 50°C for 10 min.
4. Proceed to step 4, in order determine the enzyme activity (as described in section 2.3.2.1.1).

Draw a graph by plotting temperature values on X-axis and the enzyme activity on Y-axis. Select the temperature with maximum enzyme activity as the optimum temperature.

(c) Result

The activity of AP was studied in a temperature range of 10–80°C. The results are illustrated in Fig. III.2.3. The optimal temperature for AP was 30°C. Any further increase after the optimal temperature leads to a noticeable reduction in the activity. There is an apparent linearity between the enzyme activity and the temperature in the range 10–80°C and the highest enzyme activity is recorded at 30°C, which seems to be the optimum. After 30°C the activity declined gradually. Hence, the optimal temperature of acid phosphatase was 30°C. A higher value was 37°C which was reported for the enzyme from cotton seedlings

This can be further explained that by disrupting the cells using hypotonic shock; the acid phosphatase will leak out of the cells. Then selectively purify the enzyme using a variety of precipitation steps, including a heating step that denatures most protein while leaving acid phosphatase intact. The acid phosphatase at the final stage of purification in this experiment is not pure, but could be purified further by standard biochemical techniques, including ion exchange columns, gel filtration, centrifugation, and so forth.

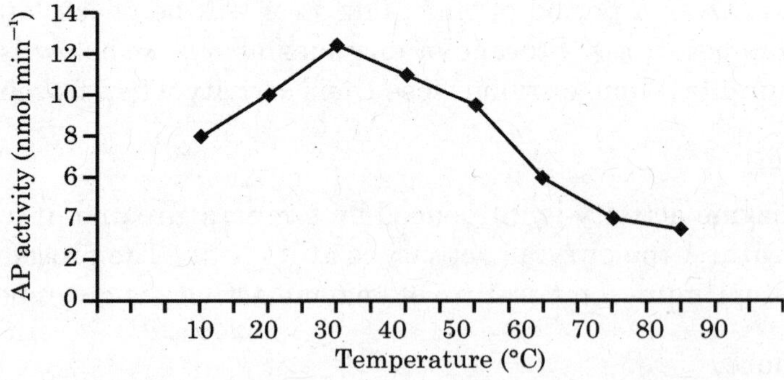

Fig. III. 2.3 : Effect of temperature on AP activity.

2.3.2.2 Purification of acid phosphatase from raw wheat germ

In this experiment, the enzyme acid phosphatase is purified from wheat germ. Acid phosphatase removes phosphate groups from a variety of molecules under slightly acidic conditions. Wheat germ has been found to be a particularly good source of the enzyme.

(a) Materials and Reagents

- Raw Wheat Germ
- Cheesecloth
- 1 M $MnCl_2$
- Saturated $(NH_4)_2$ SO_4 (pH 5.5)
- Cold deionized water
- 0.20 M EDTA (pH 5.7)
- Methanol (cold)
- Dialysis Tubing

(b) Procedure

1. Mix 40 ml of cold water and 10 grams of raw wheat germ. Stir occasionally for 30 minutes on ice.
2. Filter the suspension into a centrifuge bottle through two layers of cheesecloth placed in a funnel. Squeeze the cheesecloth to get most of the liquid out.
3. Centrifuge the filtrate for 10 minutes at 8,000 rpm and 4°C.
4. Decant the supernatant into a graduated cylinder. Mark as supernatant I. Note the volume.
5. Place 0.5 ml sample of supernatant I in a microfuge tube and store it in a –20°C freezer. Later, use it for assays.
6. Put the remainder of supernatant I in a beaker in an ice bath. Stir gently with a magnetic stir bar. Add 2.0 ml of 1M MnCl2 per 100 ml of supernatant I. Stir for 5 to 10 minutes.
7. Centrifuge the mixture for 10 minutes at 8,000 rpm to remove precipitated material.
8. Decant the supernatant into a graduated cylinder. Note the volume of the supernatant II.
9. Save a 0.5 ml sample of supernatant II and store it in the freezer.
10. Put supernatant II into a beaker in an ice bath. Slowly (dropwise) add 50 ml of cold saturated $(NH_4)_2SO_4$ per 100 ml of

supernatant II. (Use a 10 ml pipette and add over a period of 5–10 minutes while stirring the solution gently). The solution is 33 percent saturated.

Note: Add the salt slowly to prevent coprecipitation. Coprecipitation happens when one part of the solution has a higher concentration of salt than another part. The higher part will precipitate proteins. If those proteins are in the middle of an aggregate, they will not redissolve when the solution is thoroughly mixed. The result is that some proteins might precipitate at a lower average salt concentration than they would if the salt is added slowly, giving erroneous results.

11. Stir the solution for an additional 10–15 minutes.
12. Centrifuge for 10 minutes at 8,000 rpm.
13. Decant the supernatant into a graduated cylinder. Note the volume (Supernatant II).
14. Put the supernatant II back into a beaker. Add 62 ml $(NH_4)_2SO_4$ per 100 ml supernatant. Add *slowly* with gentle stirring. The solution is now 60% saturated.
15. Place the beaker in a 70°C water bath. Bring the solution to 60°C while stirring gently, and hold at that temperature for 2 minutes.
16. Plunge the beaker into an ice bath. Stir gently until the solution reaches 6°C to 8°C.
17. Centrifuge for 10 minutes at 8,000 rpm.
18. Decant the supernatant into a graduated cylinder. Record the volume of the supernatant III. SAVE THE PELLET!!
19. Save a 0.5 ml sample of Supernatant III and store it in the freezer.
20. Completely resuspend the pellet obtained from Step 18 in cold water. Use 1/3 the volume of that of Supernatant III or a maximum of 20 ml of water.
21. When an even suspension is obtained, transfer to a 40 ml polypropylene centrifuge tube, and centrifuge for 10 minutes at 8,000 rpm.
22. Decant the supernatant into a graduated cylinder. Note the volume (supernatant IV).
23. Save a 0.5 ml sample of Supernatant IV and store it in the freezer.

24. Store the tubes containing samples of Supernatants I–IV in the freezer.
25. Determine protein concentrations in the supernatants' as

    Final mg/ml Supernatant IV =
26. Unfreeze the sample and centrifuge at 8,000 rpm for 10 minutes.
27. Decant the supernatant and discard. SAVE THE PELLET!
28. Resuspend the pellets in 3 ml (total for all the pellets, not for each pellet) of cold water. Centrifuge at 8,000 rpm for 10 min.
29. While centrifuging, prepare the dialysis tubing. Let the tubing sit in distilled water for 30 or more minutes. Enough water is needed to cover the tubing.
30. Save both the supernatant and the pellet and Resuspend the pellet again in 3 ml cold water to Centrifuge at 8,000 rpm for 10 min.
31. Decant the supernatant. Combine with the supernatants from step 3-6. Record the total volume. Discard the pellet. Save 1.0 ml of Supernatant V for assays.
32. Dialyze Supernatant V against 0.5 litre of 5mM EDTA overnight:
    (a) Prepare a 0.5 L solution of 5 mM EDTA from the 0.2 M EDTA stock solution.
    (b) Wear gloves while handling dialysis tubing.
    (c) Wet a 12 cm piece of 23 mm dialysis tubing with distilled H_2O. It should hydrate for at least 30 minutes.
    (d) Use a dialysis bag clamp to close one end.
    (e) Transfer Supernatant V to the dialysis tubing.
    (f) Put on another clamp at the top of the dialysis tubing.
    (g) Place in a flask or beaker containing the EDTA. Cover the flask or beaker with plastic wrap.
    (h) Either use a magnetic stir bar to stir overnight at 4°C, or let it sit for a week or more without a stir bar at 4°C.

Before dialysis, Supernatant V still has some $(NH_4)_2SO_4$ in it. Acid phosphatase is inhibited by Hg^{++}, Pb^{++}, Fe^{++}, Ag^{+}, Zn^{++}, and Cu^{++}, but stimulated by Mg^{++}. Also, proteases are often stimulated by Mg^{++} and other divalent cations.

Carefully remove the sample from dialysis bag by removing a clip at the end of the dialysis tubing. Do this over a funnel placed on top of a graduated cylinder, or over a beaker. Pour the sample into the cylinder and note the volume of the dialyzed sample.

Characterize the enzymatic activity of the partially purified acid phosphatase fractions according to section 2.3.2.1.1. Construct a purification table for acid phosphatase. For each purification step, the table should include the total protein, total enzymatic activity, specific activity, percent recovery, and fold purification. The total activity is the units of activity if all protein is assayed. The specific activity of each fraction is the total activity/total protein and is a measure of purity of enzyme.

In addition, finally, perform an experiment to determine the Michaelis-Menten constant (Km) of Acid phosphatase (section 2.3.2.3).

Note:

1. *Always keep everything on ice.*
2. *Store 0.5 ml aliquots of the supernatants in microfuge tubes for later analysis.*

2.3.2.3 To determine the Km and Vmax value of acid phosphatase

To determine the Vmax and Km of purified acid phosphatase, the information on V_o is used.

The velocity or rate of reaction can be determined by measuring the decrease in substrate concentration, [S] with time or by measuring the rate of appearance of product, [P] During the early part of the enzymatic reaction, the amount of P formed increases linearly with time. However, in the latter part, the rate of P appearance diminishes, and after certain period the product is no longer formed. The reason for the decline in reaction rate with time includes the depletion of S or the breakdown of the enzyme. First, choose a time point in the enzyme assay that lies within the linear range of the initial velocity. This is critical because enzyme catalytic rates decline with time due to depletion of the substrate, accumulation of the product, and inactivation of the enzyme. Therefore, it is important that the time represents a true initial velocity for the reaction.

The rate measured during this time is the slope of a straight line and is called the initial velocity of the reaction (V_o). The velocity maximum (V_{max}) occurs when all the enzyme molecules get saturated with the substrate. The substrate concentration required to yield half the maximal velocity ($V_{max}/2$) can also be determined and is an important characteristic in describing the enzyme. This constant is known as the *Michaelis-Menten constant (K_m)*. Thus, the K_m reflects the affinity of the enzyme for the S. A higher affinity for the S is reflected in a smaller K_m and vice versa.

(a) Principle

For experimental determination of the K_m and V_{max}, the Michaelis-Menten equation is usually rearranged to a form that is equivalent to the straight-line equation (y = mx + b). One such rearrangement, first described by Lineweaver and Burk, involves taking the reciprocals of each side of Michaelis-Menten equation. If a double reciprocal plot is made with $1/V_o$ on the ordinate and 1/[S] on the abscissa, a straight line relationship is observed, where the slope is K_m / V_{max} and the intercept on the ordinate is $1/V_{max}$.

(b) Procedure

1. Set up a set of acid phosphatase assays containing varying amounts of substrate, but a constant amount of enzyme.
2. Using a stock of 50mM PNPP, prepare at least 0.2 ml each for a set of dilutions of PNPP at the following concentrations.

Table : Set of dilutions of PNPP

Tube	Concentration (mM)	50 mM PNPP	H_2O
1	25		
2	8		
3	4		
4	2		
5	1		
6	0.5		
7	0.25		
8	0		

3. Perform the enzyme assay as given in section 2.3.2.1.1. Read the A_{410} for each tube using the control as the blank.
4. Determine the concentration of nitrophenol produced (m molar) in each reaction using the standard curve in the previous exercise. After incubating the assays for the predetermined time, determine the V_o for each assay in mM/min.
5. Prepare a Michaelis-Menten plot with the above data Fig III.2.4.

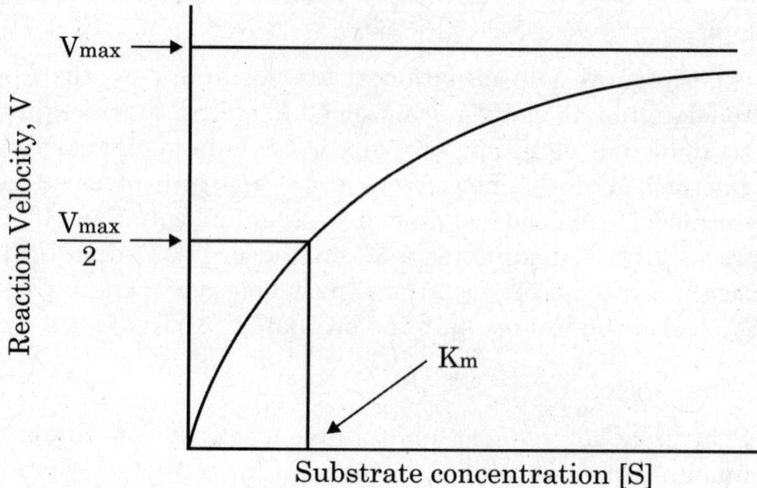

Fig III.2.4 : The effect of [S] on the velocity of an enzyme-catalyzed reaction is described quantitatively by the Michaelis-Menten equation.

6. Plot the inverse of the amount of PNP produced (*i.e.*, 1/ V_o) vs. the inverse of the PNPP concentration (*i.e.*, 1/[S]) for each set of tubes. Express the values in the units of mM and mM/min, respectively. This is a Lineweaver-Burke plot (Fig. III.2.5). The x-intercept is –1/Km, where Km is the Michaelis-Menten constant, the y-intercept is 1/Vmax, and the slope of the line is Km/Vmax.
7. Determine the Km and Vmax from the Lineweaver-Burk plot.

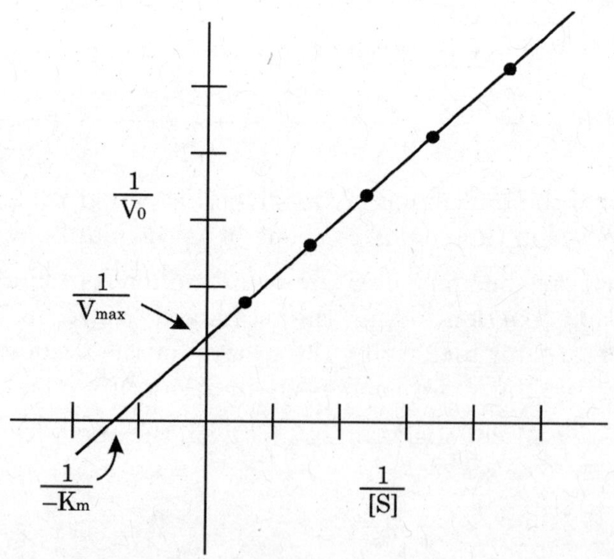

Fig. III. 2.5 : Lineweaver-Burke plot

2.3.3 Extraction and assay of enzymes

2.3.3.1. Malate dehydrogenase (EC 1.1.1.37)

(L-malate: NAD^+ Oxidoreductase)

Malate dehydrogenase (MDH) is one of the enzymes involved in TCA cycle. It catalyses the reversible conversion of oxaloacetic acid to malic acid.

$$\text{Oxaloacetic acid} + \text{NADH} \xleftrightarrow{Mg^{+2}} \text{Malic acid} + \text{NAD}^+$$

This enzyme is also involved in carbon dioxide assimilation in C_4 plants. It is coupled with PEPcase in CO_2 assimilation in the chloroplasts of mesophyll cells. Here it uses NADPH as the coenzyme.

(a) Principle

It is an oxidoreductase involving nicotinamide adenine dinucleotide (NAD), the decrease in absorbance due to the oxidation of NADH is followed.

(b) Reagents

- 50 mM Tris-HCI (pH 8.0)
- 50 mM $MgCl_2$
- 5 mM 2-mercaptoethanol
- 1 mM EDTA
- 5 μ mole Oxaloacetic Acid - 66 mg/50 ml distilled water
- 10 μ mole Magnesium Chloride - 203.5 mg /50 ml distilled water
- 0.1 M Tris-HC1 Buffer (pH 7.8)
- 0.4 μ mole NADH 5.32 mg/10 ml

(c) Procedure

Crude Enzyme Extract

Grind the tissue thoroughly with acid-washed sand in a pre-chilled pestle and mortar in grinding medium (1 ml/g tissue) containing 50 mM Tris-HCI (pH 8.0), 50 mM $MgCl_2$, 5 mM 2-mercaptoethanol and 1 mM EDTA. Pass the homogenate through four layers of cheese cloth, and centrifuge the filtrate at 3000 g for 20 min at 5°C. Use the supernatant as enzyme source.

Enzyme assay: Pipette out the following reagents into clean labelled test tubes. Run a blank along with the test sample, in triplicates.

Table : Enzyme assay protocol

S.No.	Reagents	Test (ml)	Blank (ml)
1.	Oxaloacetic acid	0.5	0.5
2.	Magnesium chloride	0.5	0.5
3.	Tris-HCI buffer, pH 7.8	1.3	1.8
4.	Enzyme extract	0.2	0.2
5.	NADH (Refer step 3)	0.5	–

2. Set the spectrophotometer to zero absorbance at 340 nm without adding NADH in the test against the blank in the reference cuvette.
3. As quickly as possible, add NADH into the test, mix well and record the initial absorbance.
4. Record the absorbance every 30 sec for at least 3 min.

Calculation

1 Unit of enzyme catalyzes the conversion of 1 µmole of substrate to product per minute

To calculate the units in any spectrophotometric based assay, Beer's law is used.

$$A = \varepsilon l C$$

Where, A = absorbance (M^{-1} cm^{-1}), l = path length of the cell (1 cm), c = concentration of the absorbing species (M) and ε = the molar extinction coefficient.

When assaying enzyme activity we use ΔA / min (change in absorbance per time).

So $\Delta A = \varepsilon l (\Delta C)$ – as the concentration of chromophore changes so will the absorbance

$\Delta A/min = \varepsilon l (\Delta C/min)$ – adds in the time factor

$\Delta C / min = (\Delta A/min)/ \varepsilon l$ – rearrange factors

$\Delta C / min = (\Delta A/min)/ (6.2 \times 10^3 \times 1)$

Example of enzymes that use NADH

NADH has an extinction coefficient of 6.2×10^3 $m^{-1}cm^{-1}$

$\Delta C / min = (\Delta A/min \times 0.161 \times 10^{-3})$ M/min – inverse of the denominator

$\Delta C / min = (\Delta A/min \times 0.161 \times 10^{-3})$ (moles/liter)/min – convert M to mole/liter

$\Delta C / min = (\Delta A/min \times 0.161 \times 10^{3})$ (µmoles/liter)/min – convert to µmole

ΔC / min = (ΔA/min × 0.161) (µmoles/ml)/min – convert to ml

This is the Units /ml of enzyme in the assay itself.

Suppose: X ml of enzyme in a final volume of Y ml gives a Δ A / min.

Using a 3.0 ml total assay volume with 0.20 ml of enzyme sample

$$\Delta A/\text{min} \times 0.161 \times (Y/X)$$

Calculate the enzyme activity with decrease in absorbance for one min.

Determine the protein content by the method of Lowry *et al.* in the enzyme extract. Compute the value for mg protein to calculate the specific activity.

This will be different for an assay that uses a different chromophore with a different molar extinction coefficient.

Notes

Stock Solutions: For solutions that need to be made fresh each day make 50 percent more than required.

For all powders stored at –20°C allow to equilibrate at room temperature 10 min before opening as not to let water condense on the material.

2.3.3.2 Peroxidase (E.C. 1.11.1.7)

(Donor: H_2O_2 Oxidoreductase)

Hydrogen peroxide (H_2O_2) is a common end product of oxidative metabolism and, being a strong oxidizing agent, could prove toxic if allowed to accumulate. To prevent this, eukaryotic cells have a membrane-bound organelle, the peroxisome that encloses the enzymes producing peroxides. These peroxides are reduced to water, by peroxidase (POD) that are also present in peroxisomes. These reactions involve a variety of electron donors, including aromatic amines, phenols, and enediols like ascorbic acid.

Peroxidase includes in its widest sense a group of specific enzymes such as NAD-Peroxidase, NADP-Peroxidase, fatty acid peroxidase etc. as well as a group of very non-specific enzymes from different sources which are simply known as POD. POD catalyses the dehydrogenated of a large number of organic compounds, such as phenols, aromatic amines, hydroquinones etc.

POD occurs in animals, higher plants and other organisms. The best studied is horse radish POD.

(a) Principle

Guaiacol is used as substrate for the assay of peroxidase.

$$\text{Guaiacol} + H_2O_2 \xrightarrow{\text{POD}} \text{Oxidized guiacol} + 2H_2O$$

POD activity is a measured by the rate of formation of oxidised guaiacol and can be assayed spectrophotometrically at 436 nm.

(b) Reagents

- Phosphate Buffer 0.1 M (pH 7.0).
- Guaiacol Solution 20 mM
- Dissolve 240 mg guaiacol in water and make up to 100 ml. It can be stored frozen for many months.
- 12.3 mM Hydrogen peroxide solution (0.042%). Dilute 0.14 ml of H_2O_2 (30%) to 100 ml with water. The extinction of this solution should be 0.485 at 240 nm. Prepare freshly.

Enzyme Extraction

Extract 1 g of fresh plant tissue (radish, horseradish root) in 3 ml of 0.1 M phosphate buffer pH 7 by grinding in a pre-cooled mortar and pestle. The homogenate is centrifuged at 18,000 g at 5°C for 15 min. Use the supernatant as enzyme source within 2–4 h. It should be stored on ice till the assay is carried out.

(c) Procedure

1. Pipette out 3 ml buffer solution, 0.05 ml guaiacol solution, 0.1 ml enzyme extract and 0.03 ml hydrogen peroxide solution in a cuvette (should be carried out at 25°C). Mix well.
2. Place the cuvette in the spectrophotometer.
3. Wait until the absorbance has increased by 0.05. Start a stop-watch and note time required in minutes (Δt) to increase the absorbance by 0.1. Most accurate values are obtained when Δt is between 1 and 3 min.

Calculation

The enzyme activity per litre of extract is calculated as below:

The extinction coefficient of guaiacol dehydrogenation product at 436 nm under the conditions specified is 6.39 per micromole, therefore,

$$\text{Enzyme activity units/litre} = \frac{3.18 \times 0.1 \times 100}{6.39 \times 1 \times \Delta t \times 0.1} = \frac{500}{\Delta t}$$

Notes: 1. The enzyme extract has to be diluted appropriately.

2. A dye like o-dianisidine can be used as the electron donor to detect peroxidase because its oxidized product is coloured. The rate of appearance of this coloured pigment (yellow/ orange coloured compound) can be measured at 430 nm.

$$H_2O_2 + \text{Colourless Dye}_{(\text{reduced})} \xrightarrow{\text{peroxidase}} H_2O + \text{Colored Dye}_{(\text{oxidized})}$$

The oxidized O-dianisidine is measured

2.3.3.3 Polyphenol Oxidase (EC 1.14.18.1)

(Monophenol, dihydroxyphenylalanine: Oxygen oxidoreductase)

Browning of the cut surface of some fruits and vegetables is due to the presence of a group of enzymes called polyphenoloxidases. The activities of these enzymes are important with regard to plant defense mechanism against pests and diseases.

These are copper proteins of wide occurrence in nature and catalyse the aerobic oxidation of certain phenolic substrates to quinones which are autoxidized to dark brown pigments. These enzymes are released by the broken cells and they catalyse the reaction between colourless molecules called polyphenols and molecular oxygen. The enzymes are assumed to be single enzymes with broad specificity although there is some evidence for the presence of more than one phenol oxidase in certain tissues. Each individual enzyme tends to catalyze the oxidation of one particular phenol or phenolic compound.

The polyphenol oxidase (PPO) comprises of catechol oxidase and laccase.

The method described here is rapid and sensitive than many other methods available.

(a) Principle

The quinones generated through enzymatic oxidation of 4-methyl-catechol (catechol oxidase) and 1, 4 dihydroxybenzene (laccase) reacts with yellow coloured 2-nitro-5-thiobenzoic acid (TNB) that has an absorption maximum at 412 nm to yield colourless adducts. The decrease in the absorbance of yellow-colour due to enzyme activity is measured.

(b) Reagents

- 0.2 M citrate phosphate buffer (pH 6.0),
- Triton X-100 (1%).
- 500 mg polyamide.

- 2-nitro-5-thiobenzoic acid anion (TNB): Add 30 mg sodium borohydride to a suspension of Ellman's reagent, *i.e.* 5, 5-dithiobis (2-nitrobenzoic acid) (19 mg) in 10 ml water. Within 1 h, the disulphide is quantitatively reduced to the intensely yellow, water-soluble thiol. This solution is stable for at least one week when stored at 4°C.
- Quinine Solutions

  Dissolve 4-methyl-1, 2-benzoquinone in double-distilled water in a 50 ml volumetric flask by bubbling nitrogen gas until the quinone is completely dissolved.

  Similarly, prepare p-benzoquinone solution. Both solutions are stable for 30 min.
- Substrate Solution

  4-methylcatechol (2 mM) for catechol oxidase assay

  Quinol (1, 4 dihydroxybenzene, 2 mM) for laccase assay.

Enzyme Extraction

Prepare acetone powder from the frozen tissue by homogenizing 25 g tissue in two successive 100 ml aliquots of cold acetone. The homogenate was collected by Bűchner filtration through Whatman No. 1 filter paper following both grindings. The homogenate was air-dried until free of acetone odor. The dry acetone powder was weighed and stored in Freezer.

For crude enzyme preparation, mix 100 mg acetone powder with 2.5 ml of 0.2 M citrate phosphate buffer (pH 6.0), 1 ml, of Triton X-100 (1%), 6.5 ml of water and 500 mg polyamide. Shake for 1 h and filter. Use the filtrate as enzyme source.

(c) Procedure

1. Pipette out of 1.4 ml citrate 0.1 M phosphate buffer (pH 6.0), 0.5 ml of TNB and 1 ml of the substrate solution directly into clean and dry cuvette.
2. Initiate the reaction by the addition of 0.1 ml of enzyme preparation
3. Immediately note down the absorbance at 412 nm in a spectrophotometer already set.
4. Record the decrease in absorbance at 30 sec intervals.

Calculation

Note the change in absorbance per minute from the linear part of the curve.

The enzyme units can be calculated according to the following equation:

$$\text{Units in the test} = K \times (\Delta A/\text{min})$$

where, K is 0.272 for catechol oxidase and 0.242 for laccase.

One unit of either catechol oxidase or laccase is defined as the enzyme which transforms 1 μ mol of dihydricphenol to 1 μ mol of quinone per min under the assay conditions. One unit is equivalent to the consumption of 1 μ mol of TNB. The relation between rate and enzyme concentration is linear for all catechol oxidases; however, with laccase it is observed only for low concentrations.

Polyphenol oxidase can also be assayed as given below:

The enzyme extract may be prepared by grinding 5 g leaves with a mortar and pestle in about 20 ml medium containing 50 mM Tris-HCI, pH 7.2, 0.4 M Sorbitol and 10 mM NaCl. Centrifuge the homogenate at 20,000 g for 10 min. and use the supernatant for assay.

Procedure: Add 2.5 ml of 0.1 M phosphate buffer pH 6.5, 0.5 ml of catechol solution (0.01 M) into cuvette and set the spectrophotometer at 495 nm. Now add 0.2 ml, of enzyme extract and start recording the change in absorbance for every 30 seconds up to 5 min.

2.3.3.4 Determination of salivary amylase activity

(a) Principle

This method is based on ability of minimal salivary amylase amount to cleave the certain amount of starch per time unit. The amylase activity in saliva can be expressed in ml of starch solution (0.1%), which is broken down by 1 ml of undiluted saliva during 30 min at body temperature.

(b) Procedure

1. Take 10 tubes and pour 1 ml of water into each tube.
2. Pour 1 ml of saliva into other clean tube and dilute it 10 folds (1 ml saliva + 9 ml water).
3. Add 1 ml of diluted saliva into tube No.1. Shake it carefully.
4. Take 1 ml of solution from tube No. 1 and transfer it into tube No. 2. Shake the content of tube No. 2.
5. Take 1 ml of solution from tube No. 2 and transfer it into tube No. 3. Shake.
6. Take 1 ml from tube No. 3 and transfer it into tube No. 4. Do the same procedure in this way with all subsequent tubes. In this

way a line of dilutions is formed where each preceding solution contains twice more amylase than the subsequent one (1/20; 1/40; 1/80; 1/160 and so on).

7. Add 2 ml of starch (0.1%) solution into each of tubes. Shake the content of each tube.
8. Place the tubes into incubator for 30 min. (Temperature at 37°C)
9. Then take the tubes out and cool immediately in running water.
10. Put a few drops of iodine solution into each tube and mark the colour.
11. The blue colour indicates that starch is not cleaved or incompletely cleaved. Calculate the activity of amylase. The dilution value of the first brown coloured solution is taken for calculation. For example the fifth tube solution is brown coloured. The dilution is 1/320 in this tube. *e.g.* the fifth tube solution is brown coloured. The dilution is 1/320 in this tube.

1ml of saliva would cleave X ml of starch

$$\text{Therefore, } X = \frac{2 \times 1}{\frac{1}{320}} = 640$$

That means:

So, 1 ml of undiluted saliva has cleaved 640 ml of starch solution (0.1%) during 30 min at 37°C.

2.3.3.5 Papain (Papainase, EC 3.4.22.2)

Papain is a cysteine hydrolase that is stable and active under a wide range of conditions. It is very stable even at elevated temperatures. The latex of ***Carica papaya*** is a rich source of four cysteine endopeptidases including papain, chymopapain, glycyl endopeptidase, and caricain. The proteins are synthesized as inactive precursors that become active within two minutes of the plant being wounded and the latex expelled. Papain is a minor constituent, but has been more widely studied because it is more easily purified. It is largely used in brewing and food industries. It is also used in cosmetics and pharmaceutical industries.

(a) Principle

To assay the enzyme an artificial substrate Benzoyl, L-Arginine *p*-nitroanilide (BAPNA) is used. BAPNA is hydrolysed by papain to release *p*-nitroaniline and the colour intensity of the released *p*-nitroaniline is measured at 410 nm. The colour intensity of the released *p*-nitroaniline in the sample may be compared with the standard *p*-nitroaniline. Alternatively, with the molar extinction coefficient of *p*-nitroaniline, the activity of the enzyme is calculated.

(b) Reagents

- 50 mM Tris-HCI buffer (pH 7.5): Dissolve 605 mg Tris in 50 ml of distilled water. Adjust to pH 7.5 with. 0.05 N HCl and make up volume to 100 ml. To the above, add 87.8 mg of cysteine hydrochloride (5 mM) and 74.4 mg of EDTA (2 mM) and dissolve completely.
- Buffered Substrate Solution: Dissolve 43.5 mg BAPNA in 1 ml of dimethyl sulphoxide and make up the volume to 100 ml with above solution (step 1).
- Acetic Acid (30%).
- Enzyme solution: 0.1 mg/ml Papain in distilled water (For crude latex 100 mg/ml will give measurable level of papain).

(c) Procedure

1. Pipette out 0.5 ml (*different aliquots of enzyme can be taken*) of papain solution into a test tube.
2. Make up the volume to 1.0 ml with Tris-HC1 buffer.
3. Add 5 ml of substrate solution.
4. Incubate for 25 min at 25°C.
5. Stop the enzyme reaction by adding 1 ml of acetic acid (30%).
6. Measure the absorbance of the released p-nitroaniline at 410 nm against a control (without the enzyme).

Calculation

Molar extinction coefficient of p-nitroaniline ($E^{1M}_{1\,cm}$)

= 8,800/ 1 mole/cm

$$1 \text{ millimole/L} = \frac{8,800}{1000} = 8.8 \text{ A}$$

$$1\,\mu \text{ mole/L} = \frac{8.8}{1000} = 0.0088 \text{ A}$$

1 μ mole/ml = 8.8

If the absorbance is 8.8, amount of p-nitroaniline = 1 μ mole/ml

If the absorbance of the sample is X,

$$= \frac{1\,\mu \text{ mole/ml}}{8.8} \times X$$

$$= X \times \frac{1\,\mu\text{ mole/ml}}{8.8}$$

This is for 1 ml of the sample for 25 min.

For 7 ml (volume of the assay mixture) of the sample solution

$$= \frac{X\,\mu\text{ mole}}{8.8} \times \frac{7.0}{25\,(\text{min})}$$

The assay mixture contains 0.5 ml of the enzyme.

Activity of the enzyme in 1 ml

$$= \frac{X\,\mu\text{ mole} \times 7.0}{8.8 \times 0.5 \times 25}$$

(μ mole of *p*-nitroaniline released per min)

From this calculate the activity per g sample. For specific activity, determine the protein content of the enzyme solution and express per mg protein.

2.3.4 Immobilization of an Enzyme

To immobilize yeast cells and demonstrate its biological activity by invertase assay.

Enzymes in water-insoluble form show catalytic activity. This activity is retained by coating these enzymes or cells producing these enzymes on a solid surface and their activity is also restricted at that site. This phenomenon of coating enzyme on a solid surface is called immobilization of enzymes.

Immobilization technique was developed in the 1950s. Due to the special characteristics of immobilized enzymes, a considerable amount of work has been carried out on the application of immobilized enzymes on solid catalysts having high substance specificity. Immobilized enzymes are used in various fields, such as chemical processes, analysis, medical treatment, food processing, wine production, affinity chromatography and so forth.

Methods for enzyme immobilization can be classified into three basic categories as follows:

1. Carrier-binding method: the binding of enzymes to water insoluble carriers, such as polysaccharide derivatives, synthetic polymers and porous glass, etc.
2. Cross-linking method: Intermolecular cross linking of enzymes by means of bifunctional or multifunctional reagents, such as glutaraldehyde, bisdiazobenzidine and hexamethylene diisocyanate.

3. Entrapping method: Incorporating enzymes into the lattice of a semipermeable gel on enclosing the enzymes in a semipermeable polymer membrane, such as collagen, gelatin cellulose triacetate polyacrylamide and k-carrageenan, etc.

Sodium alginate, is a linear polysaccharide. It is usually isolated from different brown seaweed and algae, thus the name alginate. Alginic acid can be either water-soluble or water insoluble depending on the type of the salt.

(a) Principle

The process of gelation, which is simply the exchange of calcium ions for sodium ions, is carried out under relatively mild conditions.

$$2\ Na(Alginate) + Ca^{2+} \longrightarrow Ca(Alginate)_2 + 2Na^+$$

Polyvalent cations bind to the polymer whenever there are two neighbouring glucoronic acid residues. Thus, polyvalent cations are responsible for the cross-linking of both polymer molecules and different parts of the same polymer chain.

(b) Requirements

Immobilization of yeast cells

- 1gm dry baker's yeast
- 0.2 M $CaCl_2.\ 6H_2O$: Weigh 0.22 g and dissolve in 100 ml distilled water.

Sterile Sodium Alginate (3%): Dissolve 3 g of sodium alginate in 100 ml distilled water. Sodium alginate solution is best prepared by adding the powder to water with stirring, rather than vice versa, to avoid the formation of clumps. Prolonged stirring may be necessary to achieve com-plete dissolution of sodium alginate. After sodium alginate is completely dissolved, leave the solution undisturbed for 30 min to expel any air bubbles; otherwise, they later entrap and cause the beads to float.

(b) Procedure

1 1 gm baker's yeast was weighed and suspended in 100 ml of sterile sodium alginate solution (3%). One more 250-ml flask containing 100 ml of sterile sodium alginate solution (3%) without addition of baker's yeast is kept as a control.

2 The beads are formed by dripping the polymer solution from a height of approximately 20 cm into an excess (100 ml) of stirred

0.2 M $CaCl_2$ solution using a syringe and needle at room temperature.

3 The beads were kept as it is in chilled calcium chloride solution for 1 hr and then refrigerated overnight.

4. Yeasts are well known to produce invertase enzyme. The production of invertase enzyme was observed by incubating 100 beads of immobilized yeast cells with 0.3 M sucrose solution. 50 ml of 0.3 M sucrose solution was added to 50 ml saline and 100 beads were mixed.

Note:

1. Pressure and the needle gauge can control the bead size. A typical hypodermic needle produces beads of 0.5–2 mm in diameter. Leave the beads in the calcium solution to cure for 0.5–3 hours.
2. The beads can be stored refrigerated up to 6 months period.

2.4 LIPIDS

Introduction

Lipids are non-polar organic and heterogenous biomolecules that are totally or nearly insoluble in water but are quite soluble in non-polar organic solvents like ether, chloroform or benzene. Lipids differ in their structure and perform diverse biological functions. They are the chief storage and transport form of metabolic fuel. They serve as major structural components of the membranes and also form protective coating on many organisms. Some of the vitamins and hormones are lipids.

Classification

There are several different classes of lipids but all of them derive their distinct properties due to the hydrocarbon nature of their structure. Lipids have been classified on the of basis of the structure of their backbone. Based on this, they are divided into two groups: complex lipids and simple lipids. The simple lipids do not contain fatty acids and are called non-saponifiable lipids. They include compounds like terpenes, sterols etc. The complex lipids are esters of fatty acids. In various forms of complex lipids like acylglycerides, phospholipids, sphingolipids and waxes, fatty acids are covalently joined via an ester linkage to a trihydroxy alcohol, glycerol, or its derivative. The complex lipids yield soap on alkaline hydrolysis, they are also called saponifiable lipids.

Triacylglycerols are considered as neutral fats, since there is no electric charge in their molecules. In triacylglycerol molecules, glycerol can be esterified either with molecules of the same fatty acid or with those of different fatty acids. If all three fatty acid groups are the same, the molecule is called a simple triacylglycerol, *e.g.*, tristereoylglycerol (common name: tristearin). Mixed triacylglycerols contain two or three different fatty acids. Triacylglycerols in animals are found primarily in the adipose tissue (body fat), which serves as a depot or store site for lipids. Most natural plant oils and animal fat is composed of mixture of simple and mixed triacylglycerols. If a molecule of triacylglycerol contains only saturated fatty acids, this fat is a solid substance. If unsaturated fatty acids are predominant, this kind of triacylglycerol is oily substance or, simply, oil.

Common biological fatty acids (web.virginia.edu)

Symbol[a]	Common Name	Systematic Name	Structure
12:0	Lauric acid	Dodecanoic acid	$CH_3(CH_2)_{10}COOH$
14:0	Myristic acid	Tetradecanoic acid	$CH_3(CH_2)_{12}COOH$
16:0	Palmitic acid	Hexadecanoic acid	$CH_3(CH_2)_{14}COOH$
18:0	Stearic acid	Octadecanoic acid	$CH_3(CH_2)_{16}COOH$
20:0	Arachidic acid	Eicosanoic acid	$CH_3(CH_2)_{18}COOH$
22:0	Behenic acid	Docosanoic acid	$CH_3(CH_2)_{20}COOH$
24:0	Lignoceric acid	Tetracosanoic acid	$CH_3(CH_2)_{22}COOH$
Unsaturated fatty acids (all double bonds are cis)			
16:1	Palmitoleic acid	9-Hexadecenoic acid	$CH_3(CH_2)_5CH=CH(CH_2)_7COOH$
18:1	Oleic acid	9-Octadecenoic acid	$CH_3(CH_2)_7CH=CH(CH_2)_7COOH$
18:2	Linoleic acid	9,12-Octadeca-dienoic acid	$CH_3(CH_2)_4(CH=CHCH_2)_2(CH_2)_6COOH$
18:3	α-Linolenic acid	9,12,15-Octadecatrienoic acid	$CH_3CH_2(CH=CHCH_2)_3(CH_2)_6COOH$
18:3	γ-Linolenic acid	6,9,12-Octadecatrienoic acid	$CH_3(CH_2)_4(CH=CHCH_2)_3(CH_2)_3COOH$
20:4	Arachidonic acid	5,8,11,14-Eicosatetraenoic acid	$CH_3(CH_2)_4(CH=CHCH_2)_4(CH_2)_2COOH$
20:5	EPA	5,8,11,14,1 7-Eicosapentaenoic acid	$CH_3CH_2(CH=CHCH_2)_5(CH_2)_2COOH$
24:1	Nervonic acid	15-Tetracosenoic acid	$CH_3(CH_2)_7CH=CH(CH_2)_{13}COOH$

[a] Number of carbon atoms: Number of double bonds.

2.4.1 Qualitative Tests of Lipids

Preliminary information about the overall nature of lipids may be obtained from relatively simple qualitative tests. However, complete characterization of a lipid involves detailed analysis of its various chemical constituents. length and degree of unsaturation of fatty acid chains, etc., which is time consuming.

2.4.1.1 Solubility Test for Lipids

(a) Principle

Triacylglycerols having small chain fatty acids are some-what soluble in water, but those containing non-polar long chain fatty acids are insoluble and they form emulsions in water. All triacylglycerols are soluble in diethyl ether, chloroform and benzene. They are slightly soluble in cold methanol, ethanol and acetone, but their solubility increases on warming.

(b) Materials and Reagents

1. Fatty acids (palmitic/oleic acids)
2. Fats and oils (desi ghee, butter, margarine, cod liver oil, phospholipids)
3. Solvents (water, acetone, ethyl alcohol, chloroform, diethyl ether)

(c) Procedure

1. Take small amount of different lipids in various test tubes and add water. Shake well and check their solubility.
2. Note the change, if any, in solubility on warming the above tubes in water bath at 50°C for 5 min.
3. Repeat the solubility test using different solvents.
4. Record the observations regarding solubility of these lipids and conclude about their solubility characteristics.

2.4.1.2 Grease Spot Test

(a) Principle

Oils and fats are greasy in nature therefore they impart their greasiness to the cellulose paper and make it semitransparent.

(b) Reagent

Diethyl ether.

(c) Procedure

Transfer a few drops of oil into a test tube and to this add 2–3 ml of ether. Shake vigoursly and apply a few drops of this solution on a filter paper strip. Let it dry. Appearance of a semitransparent spot on the strip indicates the greasy nature of lipids.

2.4.1.3 Acrolein Test for Glycerol

(a) Principle

When glycerol, either in free form or as an ester of fatty acids, is heated with potassium hydrogen sulphate it gets dehydrated to an unsaturated aldehyde called acrolein. Acrolein can be identified by its characteristic pungent smell.

$$\begin{array}{l} CH_2OH \\ | \\ CHOH + KHSO_4 \\ | \\ CH_2OH \end{array} \xrightarrow{\text{Heat}} \begin{array}{l} CHO \\ | \\ CH + 2H_2O \\ \| \\ CH_2 \end{array}$$

Glycerol Acrolein

(b) Materials and Reagents

1. Lipid samples (butter, oil, glycerol)
2. Anhydrous potassium hydrogen sulphate

(c) Procedure

1. Take approximately 1.5 g potassium hydrogen sulphate in a test tube and add 5 drops of the liquid test sample or an approximately equivalent weight of the test sample if it is solid. Cover the test sample completely by adding more of solid potassium hydrogen sulphate on top of it.
2. Heat the test tube slowly on burner and note the odour of the fumes evolved from the tube.

2.4.1.4 Liebermann-Burchard Test for Cholesterol

(a) Principle

In presence of Lieberman-Burchard reagent, the hydroxyl group (–OH) of cholesterol reacts with the reagents and increases the conjugation of the unsaturation in the adjacent fused ring. Bicholestadiene, the coupled product of dehydrated cholesterols, in the prsesence of acetic anhydride and conc. sulphuric acid, give sulphonated derivatives of different colours.

cholesterol + cholesterol ⟶ Bicholestadiene

(b) *Reagents*

Acetic anhydride, conc. H_2SO_4

(c) Procedure

Transfer 4 ml of the test solution to a clean and dry test tube. Add several drops of acetic anhydride and then 2 drops of conc. H_2SO_4 and mix carefully. The formation of a green or green-blue colour after a few minutes is positive. This colour begins as a purplish, pink colour and progresses through to a light green, then very dark green colour. Since this test uses acetic anhydride and sulpuhric acid as reagents, caution must be exercised so as not to receive severe burns.

2.4.2 Determination of Free Fatty Acids

(a) Principle

Different fat samples may contain varying amount of fatty acids. In addition, the fats often become rancid during storage and this rancidity is caused by chemical or enzymatic hydrolysis of fats into free acids and glycerol. The amount of free fatty acids can be determined volumetrically by titrating the sample with potassium hydroxide (KOH). The acid value is defined as milligrams of KOH required to neutralize the free fatty acids present in one gram of fat or oil. The amount of free fatty acids present or acid value of fat is a useful parameter of keeping quality of fat which gives an indication about the age and extent of its deterioration.

(b) Materials and Reagents

1. Burette.
2. Conical flasks.
3. Test compounds (oil, butter, ghee etc; fresh and the samples that have been stored at room temperature for several days).
4. Standardization of 0.1 N Potassium hydroxide (KOH): Weigh 5.6 g of KOH and dissolve it in distilled water and make the final volume to 1 L. Standardize this solution by titrating a known volume of 0.1 N oxalic acid (630 mg oxalic acid in 100 ml water) using phenolphthalein as an indicator till a permanent pink colour appears. Calculate the actual normality of KOH solution from equation.

$$\text{Normality of KOH} = \frac{\text{Vol. of oxalic acid (ml)} \times \text{Normality of oxalic acid}}{\text{Volume of KOH consumed (ml)}}$$

1. Fat solvent (95% ethanol: ether 1:1, v/v).

2. Phenolphthalein solution (1%) in alcohol (95% w/v).

(c) Procedure

1. Take 5 g of fat sample in a conical flask and add 25 ml of fat solvent to it. Shake well and add a few drops of phenolphthalein solution and again mix the contents thoroughly.
2. Titrate the above solution with 0.1 N KOH until a faint pink colour appears that persists for 20–30 sec.
3. Note the volume of KOH used.
4. Repeat the steps 1 to 3 with a blank (without any fat sample).

Calculations

0.1 N KOH solution used for blank = x ml

0.1 N KOH solution used for sample = y ml

Titre value for sample = (y – x) ml

$$\text{Acid value (mg KOH/g fat)} = \frac{\text{Titre value x Normality of KOH} \times 56.1 \ \text{(Mol. wt. of KOH)}}{\text{Weight of fat (g)}}$$

1ml of 1 N KOH contains 56.1 mg of KOH. Hence a factor of 56.1 is incorporated in the numerator in the above equation to obtain weight of KOH from the volume of 0.1 N KOH solution used during this titration.

2.4.3 Determination of Iodine Number

(a) Principle

The most important analytical determination of an oil/fat is the measurement of its unsaturation. The generally accepted parameter for expressing iodine value. *Iodine value or iodine number is defined as g of iodine absorbed by 100 g of fat.* It is a useful parameter in studying degree of unsaturation or oxidative rancidity of triacylglycerols since higher the unsaturation, greater is the possibility of rancidity.

The reaction is conducted with treatment of a known weight of fat or oil with a known volume of standard solution of iodine monochloride. The unreacted iodine monochloride is determined after converting it to free iodine liberated on addition of excess of KI. The liberated iodine is titrated against 0.1 N sodium thiosulphate solution using starch as an indicator.

$$CH_3(CH_2)_7CH{=}CH(CH_2)_7COOH + ICI$$

Unsaturated fatty acid (Oleic acid) Iodine Monochloride

$$\downarrow$$

$$CH_3(CH_2)_7\underset{\displaystyle I}{CH}-\underset{\displaystyle CI}{CH}(CH_2)_7COOH$$

$$ICI + KI \longrightarrow KCI + I_2$$

$$I_2 + 2Na_2S_2O_3 \longrightarrow 2NaI + Na_2S_4O_6 \text{ (Sodium dithionite)}$$

(b) Materials and Reagents

1. Stoppered bottles.
2. Burette (25 ml).
3 Fat samples: in chloroform (2% w/v).
4. Iodine trichloride solution (Wij's solution): Dissolve 8.5 g of iodine and 7.8 g of iodine trichloride separately in 450 ml acetic acid each. Mix both the solutions and make the volume upto 1 L.
5. Starch indicator (1%): Take 1 g of starch and dissolve it in 100 ml water, boil for a min, cool and centrifuge to get a clear solution.
6. 0.1N sodium thiosulphate (freshly prepared): Dissolve 24.82 g of $Na_2S_2O_3.5H_2O$ in 1 L water. To check its normality, take 20 ml of 0.1 N potassium dichromate, add 10 ml of KI (15%) and then 5 ml of HCI. Dilute to 100 ml with water and titrate with thiosulphate solution till the yellow colour appears. Now add a few drops of starch indicator and continue the titration till the blue colour disappears. Note the volume of thiosulphate solution and calculate its exact normality ($N_1V_1 = N_2V_2$ where N_1 & N_2 is the normality, V_1 is the volume of dichromate solution taken for titration and V_2 is the volume of thiosulphate solution used).
7. Potassium iodide solution (10%): Dissolve 10 g of KI crystals in water and make up the volume to 100 ml.

(c) Procedure

1 Take 10 ml of fat solution into stoppered bottles and add 25 ml of Wij's solution. Shake thoroughly and allow it to stand in dark for 1 h.

2 Similarly prepare a blank in which fat solution is replaced by chloroform.

3 After the reaction time of 1 h in dark, rinse the stopper and neck of the bottle with 50 ml of water and add 10 ml of potassium iodide solution.

4 Titrate the liberated iodine with standard sodium thiosulphate solution till the content of the flask become pale yellow in colour.

5 Add a few drops of starch solution and continue to titrate it further with sodium thiosulphate solution till the blue colour disappears.

Calculations

The difference between the blank and test readings gives the amount of 0.1 N sodium thiosulphate required to react with an equivalent volume of iodine. One L of 0.1 N iodine solution contains 12.7 g of iodine. The iodine number can thus be calculated as follows:

Volume of 0.1 N sodium thiosulfate used for blank = x ml

Volume of 0.1 N sodium thiosulfate used for sample = y ml

$$\text{Iodine number} = \frac{(x-y)\times 12.7}{1000} \times \frac{100}{\text{wt. of sample (g)}}$$

(d) Precautions

The bottles must be shaken thoroughly throughout the titration to ensure that all the iodine is expelled from the chloroform layer.

2.4.4 Determination of Saponification value

(a) Principle

Fats (trigycerides) upon alkaline hydrolysis results in the formation of salts of fatty acids (also called soap) and glycerol. This process is called saponification. The saponification value is defined as mg of KOH required to saponify 1 g of the given fat. This value is useful for a comparative study of the fatty acid chain length in fats/oils.

$$\begin{array}{lcll} H_2C\text{—}OCOR & & & CH_2OH \\ | & & & | \\ HC\text{—}OCOR + KOH & \longrightarrow & 3RCOOK + & CHOH \\ | & & & | \\ H_2C\text{—}OCOR & & & CH_2OH \\ \text{Triacylglycerol} & & \text{Potassium salt of fatty acid} & \text{Glycerol} \end{array}$$

3 molecules of KOH are consumed for saponification of each molecule of triacylglycerol irrespective of chain length of fatty acid. Evidently, each g of a triacylglycerol with shorter chain fatty acids

will contain larger number of molecules of the triacylglycerol and will thus require much more KOH. The saponification value is thus an indication of average molecular weight of the fatty acids in an acylglyceride.

A known quantity of fat or oil is refluxed with a fixed, but an excess of alcoholic KOH. The amount of KOH remaining after hydrolysis is determined by back titrating with standardized 0.5 N HCI and the amount of KOH utilized for saponification can thus be calculated.

(b) Materials and Reagents

1. Burette
2. Reflux condenser
3. Boiling water bath
4. Fat sample
5. Fat solvent: A mixture of ethanol (95%) and ether (1:1, v/v)
6. 0.5 N alcoholic KOH: Prepare 0.5 N solution of KOH by dissolving 28.05g of KOH pellets in 20 ml water and making the volume to 1 L with 95% ethanol.
7. 1% Phenolphthalein solution in 95% alcohol.
8. 0.5 N HCl

(c) Procedure

1. Weigh accurately 1 g of the fat sample in a conical flask and dissolve it in about 3 ml of the fat solvent (Reagent No. 5).
2. Add 25 ml of 0.5 N alcoholic KOH, attach a reflux condenser to it and reflux the contents on boiling water bath for 30 min.
3. Cool it to room temperature and add a few drops of phenolphthalein into the flask.
4. Titrate the contents of the flask with 0.5 N HCl till the pink colour disappears.
5. Similarly, run a blank by refluxing 25 ml of 0.5 N alcoholic KOH without any fat sample.

Calculations

$$\text{Saponification value} = \frac{28.05 \times (\text{titre value of blank} - \text{titre value of sample})}{\text{weight of sample (g)}}$$

1 ml of 0.5 N KOH contains 28.05 mg of KOH.

Precautions

1. Alcohol is highly inflammable. Electric heater should be used during refluxing
2. Alcohol should not get evaporated during saponification. During refluxing, effective cooling of condenser is required.

2.4.5 Estimation of Serum Cholesterol (Total)

Cholesterol is a fat-like substance that is found in all body cells. The liver makes all of the cholesterol the body needs to form cell membranes and to make certain hormones. High blood cholesterol is one of the major risk factors for heart disease. Serum cholesterol varies from 150–240 mg/100 ml in healthy young adults. The level rises with age and may go upto 300 mg/100 ml in the elderly. An increase in serum cholesterol (hypercholesterolaemia) is found in diabetes mellitus, nephritic syndrome, obstructive jaundice, hypothyroidism, xanthomatosis and during ether anaesthesia. An idiopathic hypercholesterolaemia of unknown aetiology occurs in some families. Hypercholesterolaemia, from any cause, predisposes to atherosclerosis. A decrease in serum cholesterol (hypocholesterolaemia) is found in hyperthyroidism, hepatocellular damage, anaemia (except haemorrhagic), acute infections, wasting disease, intestinal obstruction and terminal states of a variety of disease.

Sackett's Method

(a) Principle

Proteins are precipitated and cholesterol extracted in an alcohol-ether mixture. The extract is evaporated and the residue dissolved in chloroform. By Liebermann-Burchard reaction, a green colour is develo-ped and measured colourimetrically.

(b) Reagents

1. Alcohol-ether mixture – Ethyl alcohol (95%) and ether are mixed in a ratio of 3:1.
2. Chloroform-This should be of a high purity and absolutely anhydrous.
3. Acetic anhydride-sulphuric acid mixture-Acetic anhydride and conc. sulphuric acid are mixed in a ration of 20:1 just before use.
4. Stock standard cholesterol solution 200 mg of chemically pure cholesterol is dissolved in and diluted to 100 ml with chloroform.
5. Working standard cholesterol solution – 1 ml of stock standard cholesterol solution is diluted to 25 ml with chloroform. 5 ml of this solution contains 0.4 mg of cholesterol.

(c) Procedure

1. Pipette 12 ml of alcohol-ether mixture in a dry centrifuge tube.
2. Add 0.2 ml of serum slowly. Cork the tube and shake vigorously for one minute.
3. Keep the tube in a horizontal position for half an hour.
4. Centrifuge it at 1,500 r.p.m. for 5 minutes. Pour off the supernatant fluid completely in a small breaker, and evaporate it on a steam-bath or a hot plate. Make up to 5 ml with chloroform. Label the tube 'Unknown'.
5. Label two other test tubes 'Standard' and 'Blank', and pipette respectively 5 ml of working standard cholesterol solution and 5 ml of chloroform to three tubes.
6. Mix and keep in dark for 15 min.
7. Adjust the wavelength of the photometer to 680 nm or put in a red filter. Set the photometer to 100 percent transmittance (or zero obsorbance) with 'Blank' and read 'Unknown' and 'Standard'.

Calculations

$$\text{Serum Cholestrol (mg/100 ml)} = \frac{\text{Absorbance of Unknown}}{\text{Absorbance of Standard}} \times \frac{0.4}{0.2} \times 100$$

$$= \frac{\text{Absorbance of Unknown}}{\text{Absorbance of Standard}} \times 200$$

Recently, an enzymatic method based on the action of cholesterol oxidase on cholesterol has also been developed. Some commercial kits based on the enzymatic method are available in the market, and should be used according to the instructions of the manufacturers.

Determination of cholesterol by enzymatic method

This method is used for auto analysers.

(a) Principle

Cholesterol esters are hydrolysed by cholesterol ester hydrolase to free cholesterol and fatty acids. The free cholesterol produced and pre-existing one are oxidised by cholesterol oxidase to Cholestenone-4-en-3-one and hydrogen peroxide. Peroxidase acts on hydrogen peroxide and liberated oxygen reacts with the chromogen (4-amino phenazone/phenol) to form a red coloured compound which is read at 510 nm (505–530 nm).

Technical contents: Using kit of serum cholesterol.

(b) Procedure

Demonstration and its measurement of serum cholesterol by kit and taking readings. Using a colourimeter or semi-auto analyser.

2.5 NUCLEIC ACIDS

2.5.1 Determination of λ max of DNA

The quality and quantity of isolated nucleic acids can be determined spectrophotometrically.

Nucleic acids have an Amax of 260 nm and proteins have Amax of 280 nm. The A260/A280 ratio is therefore, indicative of the degree of purity of the nucleic acid. A260/A280 ratios of 1.6–1.8 or 1.8–2.0 are usually acceptable for DNA and RNA, respectively. Formulas which take into account protein and other contaminants are also available. Indirect spectrophometric assays for DNA quantification are also available, but rarely used. In some instances fluorometry using fluorescent dyes that bind DNA and/or RNA is used to determine nucleic acid concentrations.

Spectrophotometric Conversions

1 A260 unit of double-stranded DNA = 50 μg/ml

1 A260 unit of single-stranded DNA = 33 μg/ml

1 A260 unit of single-stranded RNA = 40 μg/ml

DNA Molar Conversions

1 μg of 1,000 bp DNA = 1.52 pmole (3.03 pmoles of ends)

1 pmole of 1,000 bp DNA = 0.66 μg

(observed absorbance at λ_{max}) = molar conc. of nucleic acid absorbance at λ_{max} for 1M solution.

(a) Principle

The nucleic acids absorb strongly the wavelengths in the ultra-voilet region of the electromagnetic spectrum due to conjugated double bonds of constituent purines and pyrimidines. They show characteristic maxima at 260 nm which is linearly related with the concentration of DNA or RNA in the solution upto an OD value of 2.

(b) *Requirements*

Nucleic acid (DNA powder) – 0.001 g/100 ml dissolved in distilled water.

(c) Procedure

1. Weigh the required sample of nucleic acid and dissolve in distilled water.
2. Take the sample in cuvette and distilled water is used as blank.
3. Read absorbance at different wavelengths from 210–310 nm at a gap of 10 nm.
4. Plot a graph of absorbance values against wavelength to obtain the lambda max at 260 nm for nucleic acids.

(d) Precautions

1. Weighing of the nucleic acid should be done properly.
2. Use double distilled water only.

2.5.2 Determination of DNA by Diphenylamine Method

Based on the physical or chemical property of the nucleic acids, determination of DNA is possible by a number of methods. A convenient and easy colourimetric method is the quantitative reaction of deoxysugar with diphenylamine reagent.

(a) Principle

The deoxyribose moiety of the purine nucleotides of DNA reacts under extreme acid conditions is depurinated that is quantitatively followed by the dehydration of sugar to hydroxylevulinylaldehyde. This aldehyde condenses in acidic medium, with diphenylamine to produce a deep-blue coloured complex with absorption maximum at 595 nm.

(d) Reagents

1. DNA Standard (250 μg/ml) in citrate buffer saline.
2. Saline citrate: Sodium citrate buffer (150mM, pH 7.0) containing sodium chloride (0.9% w/v)
3. Diphenylamine reagent: Mix 1 g fresh or recrystallized diphenylamine, 100 ml glacial acetic acid with constant stirring, followed by addition of 2.5 ml conc. H_2SO_4.

(c) Procedure

1. To 2 ml of standard solution, containing 50–500 μg of DNA and diluted unknown sample solution, add 4 ml of diphenylamine reagent and vortex.

2. Heat the tubes in a boiling water bath for 10 min. Cool the tubes to room temperature.
3. Read the absorbance of blue solution at 595 nm against the blank.
4. From the standard graph calculate the concentration of DNA dissolved in the saline citrate solution.

Note: Burton has introduced a modified diphenylamine reagent containing acetaldehyde; in this reagent, samples are heated at 30°C for several hours for developing colour.

2.5.3 Determination of RNA by Orcinol Method

Colorimetric procedures suitable for pentose determination have been used for measurement of RNA, and include reactions with orcinol, phloroglucinol, aniline, etc. The reaction of ribose in RNA with orcinol has been widely used.

(a) Principle

The ribose sugar moiety of RNA is converted to furfural under acidic conditions. This derivative in the presence of ferric chloride reacts with orcinol to yield a green coloured complex with an absorption maxima at 665nm. The colour formed largely depends on the concentration of HCl, ferric chloride, orcinol, the time of heating, etc.

(b) *Reagents*

1. Standard RNA (100 μg/ml) in citrate buffer saline, and sample RNA Solutions
2. Orcinol Acid Reagent
3. Add 0.1g of $FeCl_3.6\ H_2O$ in 100 ml of concentration HCl and add 3.5 ml of orcinol (6% w/v).
4. 6% Alcohol Orcinol: Dissolve 6 g orcinol in 100 ml ethanol (95%). Refrigerate in a brown bottle until use. Stable for one month.

(c) Procedure

1. To 3 ml of the standard solution containing 10–150 μg of RNA and appropriately diluted sample solution, add 3ml of orcinol reagent and vortex.
2. Heat all tubes in a boiling water-bath for 30 min.
3. Cool the tubes, and read the absorbance at 665 nm against the blank.

4. Draw a standard curve using A_{600} and the concentration of standard RNA. Calculate the amount in the isolated RNA solution using the graph.

Note: The yield and purity of RNA preparation can be assessed by measuring the absorbance of ultraviolet light by a solution of nucleic acid. A pure RNA solution should give a 260 nm: 280 nm of 2; one unit of A_{260} measured in 1 cm light path length is equivalent to 50 µg/ml.

2.6 CELL ORGANELLES

2.6.1 Isolation of Mitochondria

Most of the problems involved in the isolation of intact mitochondria occur during the initial homogenization as the cell wall of the plant tissues is very rigid and high shearing force may have a deleterious effect on organelle. The integrity and intactness of mitochondrial preparation can be evaluated from degree of respiratory control and P/O ratio, which should be close to the theoretical value of 3. Mitochondria with intact outer membrane show negligible succinate : cytochrome-c reductase activity. Hence these criteria can be employed for judging quality of the mitochondrial preparation.

2.6.1.1 To isolate mitochondria from plant leaves

(a) Principle

The tissue is gently homogenized to disrupt the cells and the mitochondria are pelleted by differential centrifugation. Intact and active mitochondria can be prepared in fairly large amounts by discontinuous density gradient centrifugation on Percoll gradient.

(b) Materials and Reagents

1. Experimental material: Freshly harvested and etiolated tissue.
2. Pestle and mortar.
3. Muslin cloth.
4. Extraction buffer: 0.1 M HEPES-KOH buffer (pH 7.5) containing 0.4 M mannitol, 1 mM EDTA bovine serum albumin (0.1%, w/v) insoluble polyvinylpyrrolidone (PVP) (0.6%, w/v)
5. Suspension buffer: Prepare 20 mM HEPES-KOH buffer (pH 7.5) containing 0.3 M mannitol (0.19c, w/v) BSA.
6. Percoll: Dilute stock solution of Percoll to obtain 62, 45, 28 and Percoll solutions (5% (v/v)) each containing 0.25 M sucrose, 20 mM HEPES-KOH (pH 7.5) and BSA (0.1%, w/v).

(c) Procedure

1. Weigh 50 g of tissue, Cut it into small pieces and homogenize in chilled pestle and mortar in 250 ml of the extraction buffer (perform all the operations at 0-4°C)
2. Filter the homogenate through four layers of muslin cloth and centrifuge at 1,000 × g for 10 min to remove cell debris, chloroplasts and aggregated organelles.
3. Decant the supernatant and centrifuge it at 12000 × g for 10 min to sediment mitochondria.
4. Prepare a discontinuous gradient of Percoll by adding 3 ml of Percoll (60%, v/v) at the bottom of centrifuge tube then 4 ml of Percoll (45%, v/v), 4 ml of Percoll (28% v/v) and finally 4 ml of Percoll (5% v/v).
5. To the pellet obtained, add 1 ml of suspension buffer and disperse it to obtain a homogeneous suspension.
6. Layer the above mitochondrial suspension very gently on top of the gradient and centrifuge at 30,000 × g for 30 min.
7. The mitochondria gather at interface between 45 and Percoll (28%). Remove them using a Pasteur pipette.
8. Dilute the collected mitochondria ten folds with suspending buffer and centrifuge at 10,000 × g for 10 min. This step is carried out to remove Percoll from the preparation. Suspend the sedimented mitochondria in the suspending buffer and recentrifuge at 10,000 × g for 10 min to get rid of residual Percoll.
9. Suspend Percoll-free sedimented mitochondria in 0.5 ml of suspension buffer.

Note: The intactness of outer mitochondrial membrane can be tested by determining the activity of succinate: cytochrome-*c* reductase spectrophotometrically from the increase in absorbance at 551 nm due to reduction of cytochrome-*c*. In a cuvette pipette 0.75 ml of 20 mM phosphate buffer (pH 7.2). 0.1 ml of cytochrome *c* solution, 0.1 ml KCN, 1.7 ml of water and 0.02 ml of mitochondrial preparation. Mix the contents and set the spectrophotometer to zero absorbance at 551 nm. Note any change in absorbance upto 3 min at 15 sec intervals. Start the reaction by adding 0.1 ml of 0.3 M succinate solution and again record the change in absorbance at 15 sec intervals.

2.6.1.2 To Isolate Mitochondria from Rat Liver

(a) Principle

Isolation of mitochondria from rat liver involves centrifugation of liver homogenate directly on a discontinuous gradient of 19, 31, 42 and Percoll (52%, v/v). The mitochondrial band is obtained around the 42%/52% interface and contains nearly 60 percent of mitochondria. This is quite pure though it may contain minor impurities as lysosomes and peroxisomes.

(b) Materials and Reagents

1. Rat liver
2. Homogenization media: 10 mM-HEPES-NaOH (pH 7.4) containing 210 mM mannitol, 60 mM sucrose and 10 mM KCl.
3. Percoll solutions: Percoll diluted in homogenization media to obtain 19, 31, 42 and solutions (52%, v/v).

(c) Procedure

1. Prepare liver homogenate in the homogenization media by standard procedure.
2. Prepare a gradient by pipetting 3 ml of Percoll (52%) at bottom of a 15 ml centrifuge tube followed by layering on top of this 2 ml each of 42% followed by 31% and finally Percoll (19%, v/v).
3. 2 ml of the liver homogenate is layered carefully on surface of the gradient.
4. Centrifuge at 5°C at 37,000 × g for 30 sec with an acceleration time of 95 sec and a braking time of 180 sec.
5. Bands are formed at all the four interfaces along with a pellet. Using Pasteur pipette withdraw the band at the interface of 42%/52% Percoll in which mitochondria originally present in the crude homogenate are recovered.

2.6.2 Isolation of Chloroplasts from Plant Leaves

(a) Principle

Centrifugation separation techniques are based on the behaviour of particles in an applied centrifugal field. The rate of sedimen-tation depends upon the mass, size of particle, sample density and viscosity of medium in which it is sedimenting. The rate of sedimentation is given below:

$$V = \frac{2/9\, r_p^2\, (Pp - Pm)\, w^2\, r}{n}$$

Where, r_p = Radius of particle

Pp = Density of particle

Pm = Density of medium

w = Angular velocity

n = Coefficient of viscosity

r = Radial distance of particle from axis of rotation.

Separation by centrifugation is based on various 'gradients' systems. In different gradient centrifugation, the organelles or cell fraction separation is based upon the centrifugation of different cell fractions at different speeds which courses sedimentation of all cell fractions according to their sizes and densities. This centrifugation at lower speed will lead to the sedimentation of heavier organelles and a gradual increase in centrifu-gation speed will cause sedimentation of higher organelles.

On the contrary a density gradient is carried out at a fixed speed. Here a density gradient is prepared which has zones of increasing density as one moves away from the top to bottom of centrifuge tube. In density gradient centrifugation, the organelles/cell fractions, equilibrate in region of thin matched densities when centrifugation is carried out for a specific duration at a fixed speed. Thus, the cell organelles with higher densities will form lower bands where as those with lesser densities form upper bands.

Choice of solute depends upon the nature of particles to be separated. The gradient material should permit desired type of separation. It should be stable, have negligible osmotic pressure and cause minimum changes in ionic strength, pH or viscosity. Sucrose is widely used as gradient, but is viscous and has high osmotic pressure even at low concentration.

Choice of plant material: Spinach (*Spinach oleracea*) and garden pea (*Pisum sativum*) are the few plants which consistently yield type A chloroplast and are relatively easy available. Preparations of chloroplasts derived by mechanical disruption of leaves of higher plants have little bio-chemical activity (Type A are those which are structurally and functio-nally intact, Type B those in which the outer membrane gets ruptured, but resealed again during isolation and Type C refers to broken chloroplasts).

(b) Requirements

Sucrose solution (30% and 60%) 10 ml each, spinach leaves, homogenization buffer (50 mM tris HCl, 3 mM EDTA, 0.3 M mannitol, 5 mM β- mercaptoethanol (pH = 8 maintained by 0.1 N NaOH).

(c) Procedure

1. Take 5 g of apical leaves, homogenize them in 10 ml of homogenization buffer. Large spinach leaves should be de-ribbeded and cut into 10 mm wide strips.
2 Using gloved hands immediately squeeze the homogenate through four layer of muslin cloth and filter it through eight layers of muslin cloth.
3. Centrifuged the filterate at 4000 rpm for 10 min at 4°C.
4. Decant the supernatant and keeping tubes inverted wipe them insides with tissue, took the pellet and again it was mixed with 5 ml of ice cold buffer and again it was centrifuged.
5. Now again gently the pellet was taken and resuspended with 1 ml of buffer. Then transferred it in 10 ml of 60 percent and 10 ml of 30 percent ice cool sucrose gradient.
6. Centrifuged it at 15,000 for 20 min, at 4°C.

(d) Result

Choloroplast was found lying at the interphase of 60 percent and 30 percent of sucrose gradient, *i.e.*, ring is formed between 60 and 30 percent gradient.

(e) Precautions

1. Buffer should be prepared accurately.
2. Gradient should be prepared carefully.
3. Sample should be loaded carefully along the wall of the centrifuge tube.

Note: In the pelleting and washing procedures it is inadvisable to exceed the centrifugation time and 'g' forces specified, as highly compacted chloroplast pellets are difficult to resuspend without appreciable choloroplast breakage.

2.7 PLANT PIGMENTS AND PHENOLICS

2.7.1 Chlorophylls

There are at least five types of chlorophylls in plants. Chlorophylls 'a' and 'b' occur in higher plants, ferns and mosses. Chlorophylls c, d and e are only found in algae and in certain bacteria. The chlorophylls are readily extracted in organic solvents, such as acetone or ether. Chemically, each chlorophyll molecule contains a porphyrin (tetrapyrole) nucleus with a magnesium atom at the centre and a long-chain hydrocarbon (phytyl) side chain attached through a carboxylic acid group.

(a) Principle

Chlorophylls are extracted in acetone (80%) and the absorption at 663 nm and 645 nm are read in a spectrophotometer. Using the absorption coefficients, the amount of chlorophyll can be calculated.

(b) Materials and Reagent

1. Acetone (80%, prechilled).
2. Fresh spinach leaves : 1 g

(c) Procedure

1. Weigh 1 g of finely cut leaves into a clean mortar and grind the tissue to a fine pulp with the addition of 20 ml of acetone (80%).
2. Centrifuge (5,000 rpm for 5 min) and transfer the supernatant to a 100 ml volumetric flask.
3. Grind the residue with 20 ml of acetone (80%), centrifuge and transfer the supernatant to the same volumetric flask.
4. Repeat this procedure until the residue is colourless. Wash the mortar and pestle thoroughly with acetone (80%). Make up the volume in the volumetric flask to 100 ml with acetone (80%).
5. Read the absorbance of the solution at 645, 663 and 652 nm against the solvent (80% acetone) blank.

Calculation

Calculate the amount of chlorophyll present in the extract mg chlorophyll per g tissue:

$$\text{mg chlorophyll 'a'/g tissue} = 12.7\,(A_{663}) - 2.69\,(A_{645}) \times \frac{V}{100 \times W}$$

$$\text{mg chlorophyll 'b'/g tissue} = 22.9\,(A_{645}) - 4.68\,(A_{663}) \times \frac{V}{100 \times W}$$

$$\text{mg total chlorophyll /g tissue} = 20.2\,(A_{645}) + 8.02\,(A_{663}) \times \frac{V}{100 \times W}$$

where, A = absorbance at specific wavelengths

V = final volume of chlorophyll extract in acetone (80%)

W = fresh weight of tissue extracted.

2.7.2 Carotenes

Carotenoids, the tetraterpenoid (C_{40}) compounds are accessory pig-ments in photosynthetic systems and give characteristic colour to plant parts, particularly flowers and fruits. These exist as hydrocarbons (carot-enes) or oxygenated derivatives. Their levels are altered during physio-logical and pathological conditions.

(a) Principle

On the basis of their solubility, the total carotenoids are extracted and partitioned in organic solvents. The separation of individual compon-ents is carried out by chromatography on activated magnesia.

(b) Reagents

1. Hexane-distilled over KOH
2. Activated Magnesia
3. Diatomaceous Earth-Hyflo Supercel
4. $MgCO_3$
5. Dry Acetone, alcohol-free. To dry treat with anhydrous sodium sulphate and distil over granular zinc (10 mesh).

(c) Procedure

1. Weigh 2–5 g of finely cut sample.
2. Place the weighed sample in a high-speed blender. Add 40 ml acetone, 60 ml hexane and 0.1 g $MgCO_3$. Blend for 5 min.
3. Filter with suction or let residue settle and decant into a separator.
4. Wash the residue twice with 25 ml portions acetone, then with 25 ml hexane and combine the extracts
5. Wash acetone from extract with five 100 ml portions water, transfer upper layer to 100 ml volumetric flask containing 9 ml acetone, and dilute to volume with hexane.

6. To prepare chromatographic column, place small glass wool or cotton plug inside the tube, add loose adsorbent to 15 cm depth. Attach tube to suction flask and apply full vacuum of water pump. Pack activated magnesia-diatomaceous earth (1:1) mixture in this column. The packed column should be about 10 cm high. Place 1 cm layer of anhydrous sodium sulphate above adsorbent.
7. Pour the extract into the column with vacuum continuously applied to the flask.
8. Use 50 ml acetone-hexane (1:9) mixture to develop the chromatogram and wash visible carotenes through adsorbent. Keep the top of the column covered with solvent during entire operation.
9. Collect the entire eluate as carotenes pass rapidly through the column.
10. Transfer the eluate, which has been reduced in volume by the loss of vapour through water-pump, to a 100 ml volumetric flask, dilute to volume with the acetone-hexane mixture.
11. Calibrate the spectrophotometer first with solutions of different concentrations of high purity β-carotene.
12. Determine the absorbance of the solution as soon as possible with a spectrophotometer at 436 nm.
13. Calculate the carotene content (mg/100 g) in the sample material using the calibration curve.

Note:

1. The extract volume gets reduced due to evaporation at room temperature. Volume should be maintained at appropriate stages.
2. Alcohol may be used instead of acetone for extraction. Use 80 ml alcohol and 60 ml hexane in blender: other volumes are same as for acetone extraction.

2.7.3 Lycopene

Lycopene is one of the most important and abundant carotenoids. It has been shown to play a very important role in human nutrition, mainly due to its high antioxidant activity. Lycopene is an open chain unsatu-rated carotenoid containing thirteen double bonds (eleven conjugated and two non-conjugated double bonds). It is chiefly found in tomatoes, apricots, grapefruit, guava and watermelon. Tomatoes and tomato derived products are the richest source of lycopene in the diet. The biological activity of lycopene includes a very important antioxidant activity (protective effects on certain types of cancers have also been described) as well as the induction of cell communication and modulation of hormonal, immune systems and other metabolic pathways. As a result, the determination of lycopene content in food samples is of a relevant importance, providing very important information from the nutritional and physiological point of view.

Chemical structure of lycopene

(a) Principle

One mole of Lycopene when dissolved in one litre petroleum (40–60°C) and measured in a spectrophotometer at 503 nm in 1 cm light path gives an absorbance of 17.2 x 10^4. Therefore, a concentration of 3.1206 µg lycopene/ml gives a unit absorbance. The carotenoids in the sample are extracted in acetone and then taken up in petroleum ether. Lycopene has absorption maxima at 473 nm and 503 nm.

(b) *Reagents*

1. AR grade Acetone
2. AR Petroleum Ether 40-60
3. Sodium Sulphate (5%)
4. Anhydrous Sodium Sulphate

(c) Procedure

1. Take 3–4 tomato fruits (sample) and pulp it well to a smooth consistency in a waring blender. Weigh 5–10 g of this pulp.
2. Extract the pulp repeatedly with acetone using pestle and mortar or a waring blender until the residue is colourless.
3. Pool the acetone extracts and transfer to a separating funnel containing about 20 ml petroleum ether and mix gently.
4. Add about 20 ml of sodium sulphate (5%) solution and shake the separating funnel gently. Most of the colour will be noticed in the upper petroleum ether layer.
5. Separate the two phases and re-extract the lower aqueous phase with additional 20 ml petroleum ether until the aqueous phase is colourless.
6. Pool the petroleum ether extracts and wash once again with a little distilled water.

7. Pour the washed petroleum ether extract containing carotenoids into a brown bottle containing about 10 g anhydrous sodium sulphate. Keep it aside for 30–45 min.
8. Decant the petroleum ether extract into a 100 ml volumetric flask through a funnel containing cotton wool. Wash sodium sulphate slurry with petroleum ether until it is colourless and transfer the washings to the volumetric flask. Make up the volume.
9. Measure the absorbance in a spectrophotometer at 503 nm using petroleum ether as blank.

Note: Volume of petroleum ether might be reduced during processes because of its evaporation. So add 20 ml more of petroleum ether to the separating funnel for clear separation of two layers.

Calculation

$$\text{One unit Absorbance} = 3.1206\ \mu\text{g lycopene/ml}$$

$$\text{mg lycopene in 100 g sample} = \frac{31.206 \times \text{Absorbance}}{\text{Wt of sample (g)}}$$

2.7.4 Phenols

Phenols are the aromatic compounds with hydroxyl groups. They occur in all parts of the plants. Phenols are said to offer resistance to diseases and pests in plants. Phenols include an array of compounds like tannins, flavonols etc. Total phenol estimation can be carried out with the Folin-Ciocalteau reagent. The Folin-Ciocalteau method has the advantage of a fairly equivalent response to different phenols with the disadvantage of responding to sulfur dioxide and sugar.

(a) Principle

Folin-Ciocalteau (FC) colourimetry is based on a chemical reduction of the reagent, a mixture of tungsten and molybdenum oxides. Phenols react with phosphomolybdic acid in Folin-Ciocalteau reagent in alkaline medium and produce blue coloured complex (molybdenum blue) with a maximum at 650 nm. The intensity of light absorption at that wavelength is proportional to the concentration of phenols.

Colour development is slow, but can be accelerated by warming the sample. With excessive heating, however, subsequent colour loss is quite rapid, and timing the colourimetric measurement becomes difficult to reproduce. The reagent is commercially available, but can be prepared (Singleton and Rossi, 1965).

(b) *Reagents*

1. Ethanol (80%)
2. Folin-Ciocalteau reagent stored in the dark and discarded if reagent becomes visibly green.
3. Na_2CO_3 (20%) solution: Dissolve 200 g anhydrous sodium carbonate in 800 ml water and bring to a boil. After cooling, add a few crystals of sodium carbonate and let sit 24 hr at room temperature. Filter through Whatman no. 1 filter paper and add water to 1 liter. Store indefinitely at room temperature.
4. Standard (1 mg/ml Catechol in water)
5. Working standard: Dilute standard 10 times.

(c) Procedure

1. Weigh exactly 1.0 g of the sample and grind it with a pestle and mortar in 10 X volume of ethanol (80%).
2. Centrifuge the homogenate at 10,000 rpm for 20 min. Save the supernatant. Re-extract the residue with five times the volume of ethanol (80%), centrifuge and pool the supernatants.
3. Evaporate the supernatant to dryness.
4. Dissolve the residue in 5 ml of distilled water.
5. Pipette out different aliquots (0.2–2 ml) into test tubes.
6. Make up the volume in each tube to 3 ml with water.
7. Add 0.5 ml of Folin-Ciocalteau reagent. Shake well.
8. After 3 min, add 2 ml of Na_2CO_3 (20%) solution to each tube.
9. Mix thoroughly, place the tubes in a boiling water for exactly one min, cool and measure the absorbance at 650 nm against a reagent blank.
10. Prepare a standard curve using different concentrations of catechol.

Calculation

From the standard curve find out the concentration of phenols in the test sample and express as mg phenols/100 g material.

Interferences

Since the colour formation of the Folin-Ciocalteau reaction is based on chemical reduction of the reagent, this reaction is general enough to allow for interference from a number of sources. The Lowry method for protein analysis is based on the reaction of the Folin-Ciocalteau reagent with tyrosine phenolic groups, so samples with high protein levels will not be suitable for phenolic analysis by Folin-Ciocalteau. In addition, it should be noted that phenolics will interfere with protein analysis by the Lowry method.

2.8 VITAMINS

2.8.1 Determination of Thiamine

(a) Principle

In order to free the thiamine from the natural ester and protein bonds, the material to be examined is digested with sulphuric acid and subsequently treated with a phosphatase preparation. In order to exclude foreign fluorescing material the thiamine is separated from the extract by adsorption. The thiochrome resulting from oxidation with potassium ferricyanide in alkaline solution is extracted with isobutyl alcohol. The intensity of the blue fluorescence of the isobutyl alcohol extract is compared with that of the standard solution. The intensity of fluorescence is measured photoelectrically and fluorescence value of a blank test is subtracted from that of the sample extract.

(b) *Reagents*

1. 0.1 N Sulphuric acid
2. Potassium ferricyanide (0.03%): Weigh 30 mg of potassium ferricyanide and dissolve in 100 ml of distilled water.
3. NaOH (15%)
4. Isobutyl alcohol.
5. Sodium sulphate (anhydrous).
6. Stock thiamine solution: Dissolve 5 mg of thiamine hydrochloride in 100 ml of water.
7. Working thiamine solution: Take 2 ml of the above solution and dilute it to 100 ml with water. This solution contains 1μg thiamine hydrochloride per ml.

(c) Procedure

1. Grind 2 g of sample (cereal) in a pestle and mortar. Add to it about 20 ml of 0.1 N H_2SO_4. Centrifuge at 5000 rpm for 20 min. Decant the supernatant and extract the residue again with 10 ml of the acid. Pool the supernatant fractions into 100 ml separating funnel. Make the final volume to the 50 ml with 0.1 N H_2SO_4.

2. To 10 ml of thiamine extract add 3 ml of the potassium ferricyanide reagent followed by 3 ml of NaOH (15%) to make the solution alkaline. Shake the contents for 10 min in shaker.
3. Add 10 ml of iso-butanol and shake the mixture thoroughly. Allow it to stand till the two phases separate out completely.
4. Discard the lower aqueous layer and using Pasteur pipette remove traces of water in the upper organic layer by adding a small amount of solid anhydrous sodium sulphate.
5. Take appropriate aliquots of standard solution containing 0 to 5µg thiamine hydrochloride and proceed as in Steps 2 to 4.
6. Set fluorimeter by selecting suitable primary filter of 366 nm and a secondary filter of 435 nm.
7. Measure the fluorescence intensity of the standards and samples.
8. From the reference curve, determine the amount of thiamine in sample and express the results in terms of mg of thiamine/100 g of sample.

2.8.2. Estimation of Ascorbic Acid in Lemon Juice

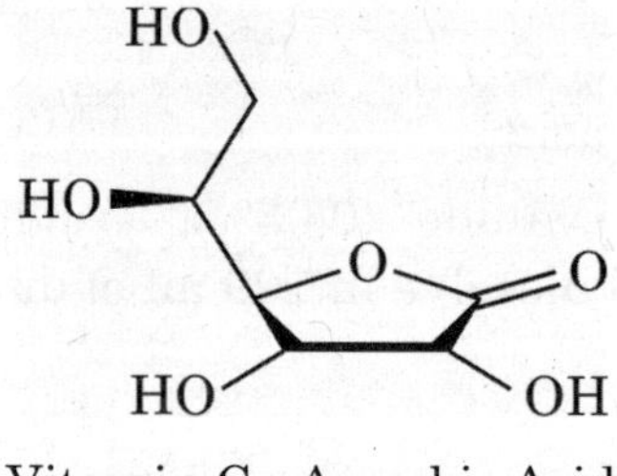

HO

HO

O O

HO OH

Vitamin C - Ascorbic Acid

(a) Principle

Vitamin C (L-ascorbic acid) is stable in an acidic solution however, it gets oxidized to its dehydro form by air at alkaline pH. Ascorbic acid reduces the 2,6-Dichlorophenol indophenol dye to a colourless leuco-base. The ascorbic acid gets oxidised to dehydro-ascorbic acid. Though the dye is blue coloured compound, the end point is the appearance of pink colour. the dye is pink coloured in acidic medium. Oxalic acid is used as the titrating medium.The redox reaction occurring during the titration is

L-Ascorbic acid + Sodium salt of 2,6 dichlorophenol indophenol ⟶ Dehydroascorbic acid + Leuco form of the dye

(b) *Reagents*

1. 2,6-Dichlorophenol indophenol solution: Weigh 42 mg of sodium bicarbonate into a small amount of distilled water. Dissolve 52 mg of of the dye. Make up the final volume to 200 ml.
2. Oxalic acid (4%).
3. Stock vitamin C solution: Dissolve 100 mg of vitamin C in 100 ml of oxalic acid (4% w/v).
4. Working standard: dilute 10 ml stock solution to 100ml with oxalic acid (4%). The concentration of working standard is 100μg/ml.

(c) Procedure

1. Squeeze fresh lemons to obtain lemon juice. Strain the juice through muslin cloth and immediately dilute (10 folds) with oxalic acid (4%).
2. Pippete out 5 ml of the working standard solution into a 100 ml conical flask.
3. Add 10 ml of oxalic acid (4%) and titrate against the dye solution till the appearance of a light pink colour. Note down the volume of the dye used.
4. Similarly, titrate 5 ml of test solution (diluted lemon juice), add 10 ml of oxalic acid (4%) and titrate against the dye solution and record the volume of dye used.
5. Calculate the amount of ascorbic acid in mg/100 g of sample

(d) Precautions

1. The total volume of the extract must be measured. The measurement must be as accurate as possible.
2. The extract should be split into aliquots before titration. However, the measurement of aliquot volume must be accurate; the use of pipettes is recommended.
3. Determination of the endpoint is subjective and may result in variations in the determined concentrations between groups.

2.8.3 Measurement of Riboflavin (Vitamin B_2) in Human Urine

Riboflavin is used in veterinary and medical practices for supplementation of animal feeds and as natural colouring agent in food products. It is eliminated in urine and with an average diet, the daily losses amount to 12 percent of the intake. A fall in the level of riboflavin (Vit. B_2)

excretion occurs before deficiency Symptoms of Vit. B_2 are noticed. Under protein deficiency the urinary output of Vit. B_2 increases. Riboflavin deficiency can be assessed by measurement of urinary excretion of riboflavin. Urine is collected in a dark container using glacial acetic acid as preservative. The interfering impurities are removed by oxidation with potassium permanganate. This is followed by removing excess of potassium permanganate by reacting it with hydrogen peroxide.

(a) Principle

Riboflavin is extracted in to a butanol-pyridine mixture and fluorescence intensity is measured in a fluorometer using excitation light of 450 nm and recording emission at 545 nm. The riboflavin is measured in terms of the difference in fluorescence before and after chemical reduc-tion. The intensity of fluorescence is proportional to the concentration of riboflavin in dilute solutions.

(b) *Reagents*

1. Glacial Acetic acid
2. Potassium permanganate (5% w/v) Prepare fresh every time
3. Hydrogen peroxide (3%)
4. Sodium sulphate (anhydrous, solid)
5. Butanol-pyridine mixture (92:8)
6. Riboflavin standard: Dissolve 25 mg of riboflavin in 500 ml distilled water, add 1.2 ml of glacial acetic acid and make volume to I L. Dilute 5 ml of this solution to 50 ml with distilled water to get working solution containing 2.5 μg riboflavin/ml protect from light.

(c) Procedure

1. Take 2 ml urine sample in a test tube and add 0.5 ml glacial acetic acid. Then add to this 0.5 ml of potassium perman-ganate solution (5%), shake well and leave for one min.
2. Add 0.5 ml of H_2O_2 (3%) solution until the purple colour of excess potassium permanganate disappears completely.
3. Now add 10 ml of butanol-pyridine mixture and shake for 30 sec. Keep the tubes undisturbed for 10 min and then discard the aqueous layer.
4. Add 1.5 g sodium sulphate anhydrous with small dropping funnel to remove the traces of moisture. Shake and keep in dark for 10–15 min.

5. Read fluorescence intensity at 450 nm excitation light and 545 nm emission light.
6. Prepare a blank and standards as described above and record their fluorescence.

Calculations

Let the fluorescence intensity of sample be S_1, and fluorescence intensity of standard containing x μg riboflavin be S_2 and volume of urine taken for estimation be V ml

$$\text{Riboflavin content (mg/ml)} = \frac{x}{S_2} \times \frac{S_1}{V}$$

(d) Precautions

1. As far as possible, carry out all the steps in a diffused light to prevent destruction of riboflavin.
2. Sometimes a fine precipitate of $KMnO_4$ may be produced. This should be removed by centrifugation.

2.8.4 Estimation of Vitamin A

(a) Principle

Vitamin A is a fat-soluble vitamin. Vitamin A – palmitate with trichloroacetic acid (TCA) produces blue colour which is measured at 620 nm.

(b) *Reagent*

1. Trichloroacetic acid solution – Dissolve 15 g trichloroacetic acid in 25 ml of chloroform and store in dark
2. Saponification mixture: 2 N Potassium hydroxide in alcohol (90%)
3. Petroleum ether
4. Sodium sulphate (anhydrous)
5. Standard vitamin A solution: 1.5 mg vitamin A palmitate dissolved in 10 ml chloroform

(c) Procedure

1. Extraction of Vitamin A

   Grind 1–5 g of the sample material to a fine paste and add 1.0 ml of Saponification mixture. Reflux the tubes gently for 20 min at 60°C. Cool the tubes at room temperature; add 20 ml distilled water and mix well. Extract vitamin A with 10 ml portions of petroleum ether in a separating funnel. Repeat this. Pool the extract (organic layer) and add sodium sulphate (anhydrous) to remove the moisture. Keep it for 30–60 min. Evaporate 5 ml aliquot of the ether extract to dryness at 60°C. Add 1.0 ml of chloroform to the dried residue.

2. To a series of clean, dry test tubes, pipette out aliquots of the standard in the concentration range 1.5–7.5 µg.
3. Make up the volume in each test tube to 1.0 ml with chloroform.
4. Rapidly add 2.0 ml of TCA solution from a fast delivery pipette, mixing the contents of the tube.
5. Immediately record the absorbance at 620 nm in a spectrophotometer.
6. Determine the absorbance of the sample tubes in similar manner.
7. Construct a standard graph and calculate the amount of vitamin A/g tissue of the samples.

Table : A summary for extraction of different vitamins from samples

Vitamin	Procedure
Thiamine	The sample is steamed for 30 min in 0.1 N H_2SO_4, followed by enzymatic hydrolysis with papain and takadiastase at pH 4.5 for 18–24 h at 37°C.
Riboflavin	Hydrolyse the sample by autoclaving at 15 psi for 15 min in presence of 0.1 N HCl and treat the autoclaved sample with mixture of enzyme (papain and takadiastase) at pH 4.5 to 4.7 at 37–45°C for 24 h. This procedure can be employed for extraction of almost all acid stable vitamins.
Nicotinic acid	Suspend the powdered sample in 0.1 N H_2SO_4 and autoclave at 15 psi for 30 min.
Pantothenic acid	Hydrolyse the finely ground experimental material with calf int-estinal phosphatase and pigeon/chick liver enzyme preparation in 0.02 M $NaHCO_3$ buffer (pH 7.2) and incubate at 37°C for 4 h.
Vitamin B6	Autoclave the sample in presence of 0.5 M HCl at 15 psi for 4 h.
Vitamin B12	Suspend the sample in extraction solution (pH 4.5) consisting of 13 g anhydrous $NaHPO_4$, 1.2 g citric acid and 1.0 g anhydrous Na metabisulphite per 100 ml and autoclave at 15 psi for 10 min.

Vitamin	Procedure
Biotin	Hydrolyse by autoclaving with 6 NH_2SO_4 at 15 psi for 2 h.
Folic acid	Liberation of folic acid from conjugated form is achieved enzym-atically using one of the following conjugases: Conjugates from hog kidneys (acetone dried powder) at pH 4.5 for 48 h at 37°C or conjugases from rat liver/chicken pancreas at pH 7.0 for 48 h at 37°C.
Inositol	Hydrolyse with HCl (18%) by refluxing for 6 h.
Choline	Autoclave in H_2SO_4 (3%) at 15 psi for 2 h.

2.9 BIOCHEMICAL SEPARATION TECHNIQUES

Separation, identification and purity are three main criteria in the analysis of biomolecules for assessing their properties, activity and structure. The routinely used biochemical separation techniques in laboratory are based on various characteristics, such as solubility, density, viscosity, charge, adsorption, biological specificity, etc. There are many biochemical separation techniques, but the chromatographic and electrophoretic techniques are widely used in laboratories.

2.9.1 Paper Chromatography

(a) Principle

Paper chromatography is an example of partition chromatography. Cellulose contains a large amount of bound water even when extensively dried. In paper chromatography, cellulose in the form of filter paper sheets provides the support medium where water is adsorbed between the Cellulose fibres and forms a stationary hydrophilic phase. The distrib-ution of solute occurs between the immobilized water (the stationary phase) and the developing partially miscible organic solvent (the mobile phase). A solute mixture introduced into such a system which can separate its components, provided that they possess significantly different partition coefficients between the mobile phase and water-saturated cell-ulose. When the partition favours the aqueous phase, the solute will tends to remain near the point of application. Conversely, when the mobile organic phase is favoured, the solute will move with the solvent flow.

The solute mixture to be analyzed is placed at a marked spot on a strip or sheet of paper and allowed to dry. This spot defines the origin. The paper chromatogram is developed by allowing the solvent to flow along the sheet in a suitable chromatography chamber. Capillarity draws the solvent through the paper, dissolves the sample as it passes the origin, and moves the components in the direction of flow. Since the solubility of different components in the mobile phase varies, they get separated from each other. The solvent front is marked and, after drying the paper, the positions of the compounds present in the mixture are detected by spraying the paper with indicating reagents. The different compounds are identified on the basic of their Rf values. The Rf (Relative to front) value is the ratio of the distance moved by a compound to that moved by the solvent fronts:

$$\text{Rf} = \frac{\text{The distance moved by the compoud from origin}}{\text{The distance moved by the solvent from origin}}$$

This value is more or less constant for a particular compound in a particular solvent system and paper under a given set of physical conditions of solute concentration, temperature and pH.

(b) Apparatus

Chromatography chambers: The major prerequisite of a suitable chamber for paper chromatography is that it should be air-tight and capable of maintaining a saturated atmosphere with respect to the developing solvents. Suitable chambers may be constructed from plywood. The overall dimensions are about 30" x 30", with the width determined by the number of troughs to be employed. The end walls should be provided with windows. A 2" ledge near the top of each end wall serves as a support for the solvent troughs. A rubber tubing gasket around the top of the box will make a satisfactory closure against which the lid is held tightly in position by means of window latches or suitcase clamps.

Unless it is possible to reserve a separate chamber for each solvent system, chambers should be constructed from materials impervious to the components of the system. Bare wood will absorb a variety of solvents; paraffin-impregnated wood will absorb n-butanol. The vapours subsequently given off by the absorbed substances may disturb the phase relationship of other solvent systems if these are used in the contaminated chambers. Various difficulties, such as streaking and variable rates of flow may result as a consequence. For utilizing a chamber for more than one solvent system it would be desirable to choose one with an impervious, cleanable interior such as glass, formica, or stainless steel. It should be noted that the stainless steel is subject to corrosion by HCl. Formica when properly bonded to wood with a phenolic resin glue makes a very satisfactory lining resistant to all the solvents in common use. For convenience the cabinet should be equipped with leveling devices.

Trough Assembly: Glass troughs are preferred owing to their chemical inertness. Stainless steel troughs are not so versatile as glass because it precludes the use of corrosive HCl either in the solvent or in the atmosphere of the chamber. Some stainless steels, owing to the catalytic action of heavy metals, cause excessive oxidation of phenol-containing solvents. The resulting oxidation products which remain on the paper after drying interfere with the flow of the second solvent, particularly in the region of the phenol front. Retardation of flow in this region produces an uneven front during development in the second solvent. Troughs are usually 24" long.

Trough supports, antisiphoning rods, and weights for holding the papers in the trough in descending chromatography are available commercially.

Drying Chambers: In between solvents (in two-dimensional chromatography) and at the end of the run, the paper must be dried to ensure adequate removal of the solvents. For small-scale sporadic work, a conventional fume hood may prove adequate. For a routine use of paper chromatography a specially designed drying chamber with forced draft ventillation may be used. The drying cabinet should be designed to accommodate the antisiphoning rods. Completed chromatograms may be clipped to the rods and, with the latter serving as handles, transferred to the drying chamber where the ends of the rods fit into notched cleats. The cleats are affixed to opposite sides of the box 2" to 3" below the rim. At least 2" should be allowed between notches. The inside width of the chamber will thus be slightly in excess of the length of the antisiphoning rod. Its length will depend on the capacity required. The overall depth should be at least 29" to provide vertical room for standard chromatographic papers.

Drying of chromatographs with the aid of heat should be avoided, as elevated temperatures, besides having possible destructive effects on the experimental compounds, accelerate the oxidation of phenol and thereby cause discolouration of the paper.

Auxiliary Equipment

(i) Micropipettes, syringe/microsyringes or a glass capillary for spotting samples onto the paper.

(ii) Hair dryer used to evaporate the solvent while spotting samples onto the paper.

(iii) A sprayer used for spraying indicating reagents on the devel-oped chromatograms. The sprayer should be capable of delivering a uniform, finely dispersed spray.

Heat source for colour development: The colour development with many compounds occurs only when the chromatogram after spraying is heated for a reasonable length of time. For this, an oven large enough to accommodate that sheet is needed. Usually 5 to 10 minutes heating at 100 to 110°C in the oven is sufficient for colour development, however, it is necessary to observe the course of the reaction so, that the paper may be removed when maximum spot intensity with minimum background colouration has been obtained. Better control can be exercised if the paper is heated by holding it before a radiant heat source. An infrared lamp is a good source for paper strips, and a radiating electric space heater gives good results with large sheets.

Papers: Papers are available which have different running characteristics, for example, slow, medium and fast. Others have been acid washed to remove traces of impurities which may affect certain analyses.

In one-dimensional chromatography, the paper should always be developed in its machined direction which is normally marked on the packet. For adsorption and normal phase partition paper chromatography, the paper in suitable form is commercially available. Paper for reverse phase chromatography must be prepared immediately before use.

Whatman No. 1 is the paper most frequently used for analytical purposes. For separating large quantities of material Whatman No. 3 MM is employed, however the resolution is inferior to Whatman No. 1. For a rapid separation, Whatman Nos. 4 and 5 are convenient, although the spots are well defined.

The handling of the chromatography paper with bare hands should normally be avoided. Amino acids transferred to the paper from the sweat on the fingers may cause spurious spots to appear. Similarly, the oil left behind as a result of handling may have a waterproof effect and thus produce localized areas where the distribution equilibria between mobile and stationary phases will be disturbed. These difficulties can be overcome by wearing clean surgical gloves while handling the paper.

Preparation of the sample: Samples containing excess salts should be desalted before chromatography since the presence of salts results in a poor chromatogram with spreading of spots and changes in their Rf values. It can also interfere with the chemical reactions used to detect the compounds being separated. Macromolecules such as proteins are also removed prior to chromatography by ultrafiltration or gel filtration.

Sample application: The amount of sample applied is critical. There must be enough sample so that the developed spots can be detected, but overloading will lead to "tailing" and lack of resolution. The proper sample size is determined by trial and error. It is usually recommended that two or three spots of different concentrations can be applied for each sample tested. The sample (10–50 µl) is applied onto the paper with a micropipette, syringe, microsyringe, or a glass capillary in small volumes at a time. A small drop of sample is spotted onto a marked spot on a line drawn with a pencil 5 cm from the edge of the paper and allowed to dry; then the spotting process is repeated by superimposing more drops on the original spot. The area of the sample spot should be as small as possible.

Solvent system: The choice of a solvent system is largely empirical and will depend on the mixture investigated. If the compounds move close to the solvent front in solvent A, then they are too soluble, while if they are crowded around the origin in solvent B then they are not sufficiently soluble. A suitable solvent for separation would, therefore, be an appropriate mixture of A and B, so that the Rf values of the compounds of the mixture are spread across the length of the paper. The pH may also be important in a particular separation, and many solvents contain acetic acid or ammonia to create a strongly acidic or basic environment. A wide selection of solvent systems is available in the biochemical literature.

Development of paper Chromatogram: Paper chromatograms can be developed by either ascending or descending solvent flow in chambers that must be air tight and saturated with solvent vapours. The experimental apparatus for each is shown in Figure III.2.6.

In the ascending chromatography the paper is supported on a frame with the bottom edge in contact with a trough containing the developing solvent. The sample spots should be in a position just above the surface of the solvent so that, as the solvent rises vertically up the paper by capillary action, separation of the sample is achieved. In the descending chromatography, the end of the paper near which sample spots are located is held in a trough at the top of the tank and the rest of the paper allowed hanging vertically. The solvent is then added to the trough. As the solvent moves downwards under gravity sample components get separated. Although ascending chromatography is often preferred because of the simplicity of the set-up, descending chromatography has two advantages: (1) it is faster because gravity assists the flow, (2) it can offer better resolution for compounds with small Rf values (which require long runs) because the solvent can be allowed to run off the paper. Rf values cannot be determined under these conditions, but it is useful for quantitative separations.

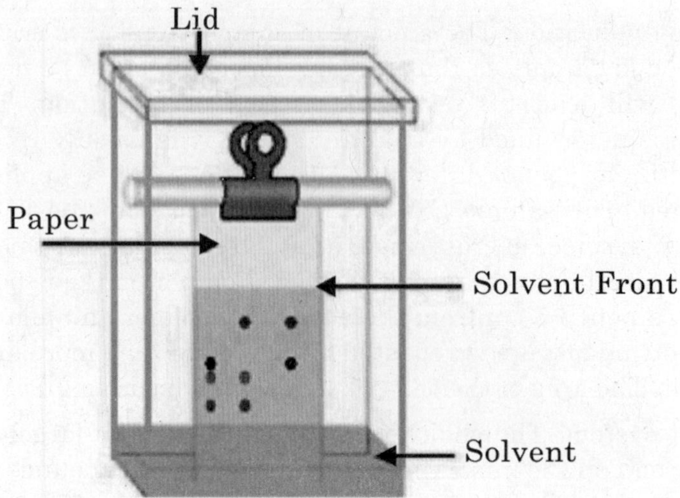

Fig. III. 2.6 : Chromatography jar (http://en.wikipedia.org/)

Detection and identification of spots: Spots in paper chromatograms can be detected by their colours, by their fluorescence, by the chemical reactions that take place after the paper has been sprayed with various reagents, or by radioactivity. Identification is usually based on comparison with standards of known Rf or by elution. Elution is accomplished by cutting out the spot and soaking the paper in the appropriate solvent; this can often be done in a quantitative manner.

2.9.1.1 Separation of Amino Acids by one Dimensional Paper Chromatography

(a) Principle

Amino acids whose Rf values differ significantly can be separated conveniently by one dimensional paper chromatography. Using this technique the various amino acids can also be quantitated, Amino acids on the paper chromatogram are visualized by their reaction with ninhydrin.

(b) Reagents and materials

1. Solvent system: Butanol-acetic acid-water 4:1:5 (v/v); Mix the solvents thoroughly and separate the two layers.
2. Spray reagents Ninhydrin (0.25% in 95% alcohol).
3. Amino acids (2 mg/ml).
4. Chromatography chamber.
5. Whatman filter paper No. 1.
6. Air oven.

7. Sprayer.
8. Dryer.
9. Micropipette or Microcapillary.

(c) Procedure

Saturate the chamber with the lower or upper layer of the solvent. Then cut a piece of filter paper depending on the size of the chromatographic chamber and the number of samples to be spotted. Draw a pencil line 5 cm from one end and mark the position of spots with 1" distance between any two spots. Spotting should not be done within 1" distance from both the end walls (side edges of the paper) Spot the mixture and the standard amino acids (each 20 μl) at the marked points. Care should be taken not to handle the paper with bare hands. After applying the sample transfer the paper to the saturated chromatography chamber and fix it according to ascending or descending chromatography. Place the upper layer of the solvent in the trough at the bottom (ascending chromatography) or at the top (descending chromatography). The paper should not touch the sides of the chamber. Develop the chromatogram till the solvent reaches just few centimeters to the upper edge (in ascending) or the lower edge (in descending). Remove the paper and mark the solvent front, Dry thoroughly in air (or an air oven) and spray with the spray reagent. Warm the paper by placing in an oven at 105°C 10 minutes to make the spots visible. Mark the spots with a pencil and calculate the Rf value for each amino acid. Find out the unknown amino acids by comparing their Rf values with those of standard amino acids.

$$\text{Rf} = \frac{\text{The distance moved by the compoud from origin}}{\text{The distance moved by the solvent from origin}}$$

For quantitative estimation of individual amino acids, cut the spots in a sprayed chromatogram and extract the coloured complex with 5 ml methanol and determine the colour intensity at 570 nm. A more accurate way is to develop two paper chromatograms under identical conditions, spotted with the same amount of the sample. Spray one of the papers with ninhydrin and identify the spots by using chromatograms with known amino acids. On the other paper, mark positions corresponding to these spots. Cut each spot and elute it with 5 ml of ethanol (80%) taken in a test tube. Determine the concentration of the amino acids after developing colour with ninhydrin as described in the estimation of amino acids.

2.9.2 Separation of Plant Pigments by Thin Layer Chromatography (TLC)

(a) Principle

Like all chromatographic methods, TLC takes advantage of the different affinity of the analyte with the mobile and stationary phases to achieve separation of complex mixtures of organic molecules. Molecules get separated between the hydrated stationary phase and non-polar mobile phase. Hydrophilic analytes have more affinity to the polar stationary matrix, while less hydrophilic molecules tend to have more affinity towards mobile phase, resulting in its faster movement and separation. This technique is rapid as compared to paper chromatog-raphy. The separated analytes are identified by comparing their Rf values to that of reference standards.

The Rf value depends upon the (i) solvent system, (ii) degree of saturation of the mobile phase in the chromatographic chamber, (iii) particle size of the adsorbent, (iv) type of adsorbent, (v) temperature and humidity. The R_f value for an analyte is constant for a given set of experimental conditions.

Commonly used stationary matrix for TLC include, silica gel-G, alumina. Normally, glass support is used for coating stationary matrixes. TLC requires only a few ng of sample for a successful analysis and can be accomplished in a matter of minutes.

(b) Material

1. Glass plates (20 × 20 cm or 20 × 10 cm thickness 3 mm).
2. Chromatographic glass jar with lid.

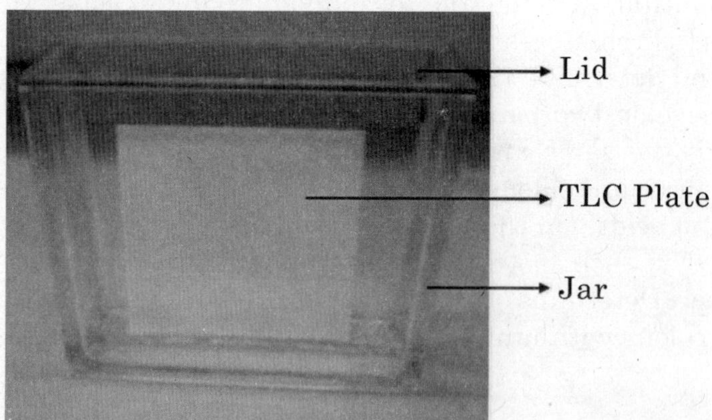

Fig. III.2.7 : TLC apparatus
(from "A Manual For Biochemistry Protocols" by MR Wenk & A Z Fernandis)

3. Spreader.
4. Capillary tubes (5 μl/10 μl capacity) or micro-syringe (10 μl).
5. Silica gel-G. TLC grade.

Sample: Extract the sample following the procedures indicated for each group of compounds Solvent system: Petroleum ether: Acetone (7:3 v/v).

(b) Procedure

1. On a balance weigh out 0.5 gram of fresh spinach and combine with 0.5 gram of anhydrous magnesium sulfate and 1.0 gram of sand. Transfer these materials to a mortar and using a pestle grind the mixture until a fine dry powder is obtained (grind the mixture really well). The anhydrous magnesium sulfate will remove the water from the leaves.
2. Transfer the powder (2.0 grams total) to a small test tube and combine with 2.0 ml of acetone. Stopper the test tube and shake vigorously for approximately one minute. Make sure that the solid and solvent are well mixed.
3. Allow this mixture to stand for 10 minutes, and then using a pipette carefully transfer the solvent above the solid into a small microcentrifuge tube. Use care not to transfer any of the solid material. The solvent extract should be green. Cap the micro-centrifuge tube to minimize solvent evaporation.
4. Obtain a TLC chamber (a 400 ml glass beaker covered with parafilm or aluminum foil) and add developing solvent. The solvent should completely cover the bottom of the chamber to a depth of approximately 0.5 cm. Keep the chamber covered so that evaporation doesn't change the composition of the solvent.
5. Obtain a TLC plate (a silica gel coated plastic sheet) which has been precut (5.5 × 9.0 cm) and make four dots equal distance apart with a pencil on the coated side approximately 1.0 cm from the bottom of the strip. The dots should be parallel with the bottom of the strip.
6. Fill a capillary tube (TLC applicator) by placing it in the leaf extract (it will fill by capillary action) and keep the finger off the end of the capillary tube. Apply the extract to the centre of the first dot on the TLC plate by quickly touching the end of the TLC applicator to the plate. Allow to dry. Repeat several times to make a concentrated dot of extract. Be sure to let dry between applications. In a similar manner, place an additional dot on

each side of the first dot to make a short line of extract. Try to make the spots as small as possible, but dark enough to see the colour clearly.

7. Carefully place the TLC plate in the TLC chamber. The TLC plate should sit on the bottom of the chamber and be in contact with the solvent (solvent surface must be below the extract dots). Cover the TLC chamber.
8. Allow the TLC plate to develop (separation of pigments) for approximately 10 minutes. As the solvent moves up the TLC plate you should see the different coloured pigments separating.
9. Remove the TLC plate from the chamber when the solvent front is approximately 1.0 cm from the top of the TLC plate. With a pencil, mark the level of the solvent front (highest level the solvent moves up the TLC plate) (the solvent evaporates and disappears quickly). Also measure the pigment distances quickly as some pigments (especially the beta-carotene) may fade over time.
10. The rate at which a pigment moves up the plate is reported as an R_f value which is defined as the ratio of the distance moved by the spot to the distance moved by the solvent. Determine the R_f values for each of the pigments observed using the formula provided below.

$$R_f = \frac{\text{distance moved by solute (pigment)}}{\text{distance moved by solvent}}$$

For the purpose of identification, following are the reference Rf values of the pigments in the petroleum ether: acetone solvent system.

Table III. 2.9.1 : Rf values of plant pigments

Pigment	R_f Value
Chlorophyll a	0.68
Chlorophyll b	0.54
Chlorophyll c	0.03
β-carotene	0.94
Fucoxanthin	0.51
Lutein	0.43
Violaxanthin	0.22

Note: The separated pigment spots on the TLC plate should be scored immediately, as the pigment undergoes destruction/bleaching in the presence of light.

2.9.3 Separation of Blue-dextran and Potassium Dichromate by Gel–Filtration Chromatography

(a) Principle

In gel filtration chromatography (also known as size exclusion, gel permeation, or molecular sieve chromatography), molecules are separated based on size. As with SDS PAGE (sodium dodecyl sulfate polyacrylamide gel electrophoresis), a gel filtration chromatography column can be calibrated by using molecules of known size; the gel filtration column can then be used to determine the size of an unknown protein. In contrast to SDS PAGE, which measures the molecular weight of denatured protein, gel filtration measures the size of the native protein. Gel filtration columns are made of porous beads packed into a column. The beads can be polymers of dextran (Sephadex), agarose (Sepharose and Superose), agarose cross-linked to dextran (Superdex), polyacrylamide (Sephacryl), or related compounds. Different types of beads have somewhat different physical properties that may make them more appropriate for different proteins. As a solution containing molecules of varying sizes passes through the column, the molecules distribute between the inside and outside of the pores depending on their size. Molecules too big for the pores are totally excluded, and elute from the column first. Smaller molecules fit in the pores, and therefore elute later. The elution volume for a molecule is thus inversely related to the size of the molecule. Gel filtration beads are characterized by their composition. They are also characterized by their exclusion limit, which is the size of the smallest molecule incapable of entering the beads. Beads with a variety of compositions and pore sizes are available. For example, a Sephadex G-100 bead is made of a dextran polymer and should be able to separate molecules of up to 100 kDa; molecules bigger than 100 kDa will be completely excluded and will run at the void volume. The "void volume" is the elution volume of a molecule so large that it is totally unable to enter any of the pores. The void volume therefore represents the minimum possible elution volume for any molecule. Another useful parameter for a gel filtration column is "column volume", which is the volume of an empty cylinder the size of the column being used. This represents the maximum elution volume for a very small molecule. The protein sample should be applied to the column in the smallest reasonable volume to ensure good separation. The molecules do not form tight associations with the column, and if loaded in a large volume, will elute in a large volume with little separation between large and small molecules. The maximum volume that can be applied to a column varies somewhat with bead type (*e.g.*, about 2 to 3% of the bed volume for Sephadex, and about 0.5% of the bed volume for Superdex).

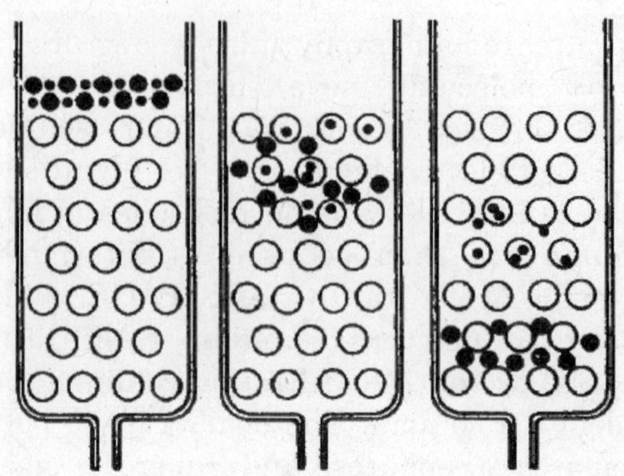

Fig: III. 2.8 : Gel Filtration Chromatography

Materials: Glass column with a tapered end [10 cm x 1.5 cm (dia)] or 10 ml plastic disposable syringe, Sephadex G-10, Blue-dextran, potassium dichromate, siliconized glass wool or glass wool, screw clip, silicone tube, burette stand with clamp.

Sample: Prepare a mixture of blue-dextran (MW 2000 kDa-high molecular weight polysaccharide) and potassium dichromate (MW 294 Da) in distilled water (concentration – 2 mg of each in 1 ml of distilled water).

(b) Procedure

1. Take a 150 ml glass beaker, add 8–10 g of Sephadex G-10 into it. Add 50 ml of distilled water and stir the contents. Allow the gel to swell for about 4 hours. Decant the solution to remove the fines. Wash the gel suspension with distilled water. Pour the gel into a column, plugged with glass wool at the bottom end, so as to hold the gel matrix from running out of the column. Allow the gel to settle under gravity. Avoid the formation of air bubbles while packing the gel matrix. Equilibrate the column by passing 20–25 ml distilled water.
2. Apply 0.3 ml sample solution onto the top surface, without disturbing the gel matrix. After the sample enters the gel matrix, elute the sample components with distilled water (flow rate, 20 ml/h). Collect 1 ml fractions in graduated plastic eppendorf tubes. Blue dextran does not enter the gel pores, hence is excluded out earlier. Potassium dichromate enters the gel pores and is retarded by the gel matrix thus it elutes later.

3. The separation in the gel column can be monitored visually, as they are coloured. Record the absorbance of the fractions collected (at 630 nm for blue dextran and 340 nm for potassium dichromate).
4. Plot an elution profile graph between Fraction number (X-axis) and Absorbance (Y-axis).

2.9.4 To Separate Proteins by SDS – PAGE

(a) Principle of Electrophoresis

Any charged ion or molecule migrates when placed in an electric field. The rate of migration of a compound depends on its net charge, size, shape and the applied current. This can be represented by the following equation:

$$v = \frac{E.q}{f}$$

where, v = velocity of migration of the molecule

E = electric field in volts/cm.

q = the net electrical charge on the molecule.

f = the frictional coefficient which is a function of the mass and shape of the molecule.

The movement of a charged molecule in an electric field is often expressed in terms of electrophoretic mobility (μ) which is defined as the velocity per unit of electric field.

$$\mu = \frac{v}{E} = \frac{E.q}{f.E} \qquad \text{Since } \mu = \frac{E.q}{f} = \frac{q}{f}$$

For molecules with similar conformation, f varies with size, but not with the shape. Thus electrophoretic mobility (μ) of a molecule is directly proportional to the charge density (charge/mass ratio). Molecules with different charge/mass ratio migrate under the electric field at different rates and hence get separated. This is the underlying basic principle for all the electrophoretic techniques.

Polyacrylamide Gel Electrophoresis

Electrophoresis in polyacrylamide gel is the most widely used technique for analysis and characterization of proteins and nucleic acids. Use of polyacrylamide gels has several advantages, such as it is chemically inert, gives superior resolution, is amenable to preparation of gels with wide range of pore sizes and that the gels are stable over a wide range of pH, temperature and ionic strength.

Polyacrylamide gel is prepared by polymerization of acrylamide monomers into long chains and cross-linking through bifunctional reagents such as N, N'-methylene bisacrylamide. Polymerization is initiated by ammonium persulfate (APS) or riboflavin. N, N, N', N'-tetramethy-lethylene diamine (TEMED) catalyses formation of free radical from persulfate which in turn initiate polymerization. In the case of riboflavin-TEMED system, light causes photodecomposition of riboflavin and production of necessary free radicals. Oxygen inhibits polymerization. Gel solutions have, therefore, to be degassed prior to the polymerization step.

The pore size of polyacrylamide gels can be varied by changing total concentration of acrylamide and bis-acrylamide monomers in a fixed volume of gelation solution. Gels ranging from 3 to 30 percent acrylamide concentration can be made and these can be used for separation of molecules of size upto 1×10^5 daltons. A gel with low percentage has larger pore size and is suitable for separation of high molecular weight compounds and a high percentage gel has smaller pore size and is used for separation of relatively low molecular weight compounds. Alternatively, gels of different pore sizes may be prepared by varying the relative concentrations of acrylamide and bis-acrylamide. Gradient gels with linear gradients of increasing acrylamide concentration give better resolution and are quite commonly used. The pore size created in the gel are inversely proportional to the concentration of acrylamide. A 10 percent polyacrylamide gel will have a larger pore size than a 15 percent gel. Proteins migrate faster through a gel with high pore size. Ideally, the percentage gel selected will place the target protein at the centre of the gel. T gel (10%) is suitable in the range 15 – 200 kd.

Polymerization

The rate of polymerization depends on concentration of APS and it is inhibited by oxygen. After electrophoresis proteins are commonly visualised by staining with commassie brilliant blue dye that is commonly used stain as SDS in gel interfers with amido black.

In stacking gel mobility order is: $Cl^- >$ Protein > glycinate. In separating gel mobility order is $Cl^- >$ glycinate > protein. Average charge on glycinate ion at pH 8.5 is about –0.2.

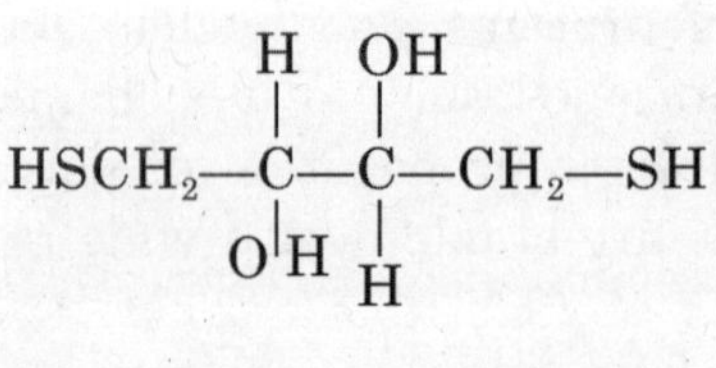

H OH

$HSCH_2—C—C—CH_2—SH$

OH H

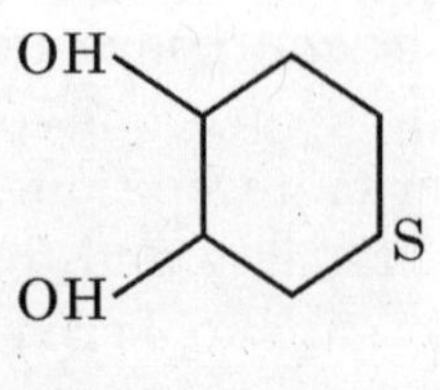

OH

S

OH

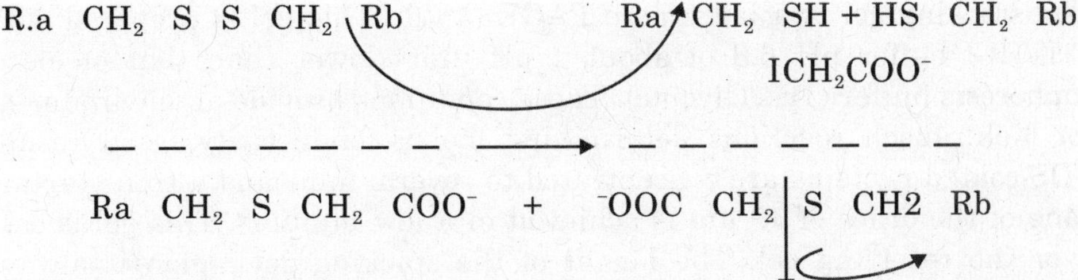

Fig. III.2.9 : Reduction and carboxymethylation of proteins containing disulphide bridges (From Protein analysis and purification: benchtop techniques, by I.M. Rosenberg).

A B C

S-S

+ SDS

+ β-ME (β-Mercaptoethanol)

+ heat

C

negatively charged SDS molecules

A B

SDS–PAGE

⊖

B

C

A

⊕

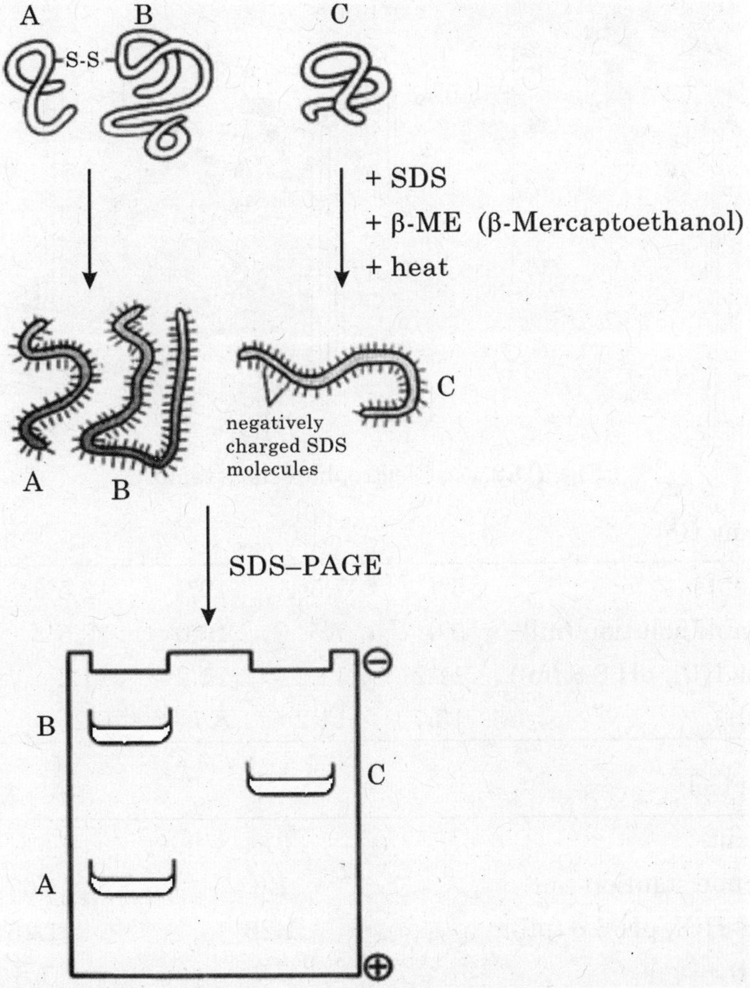

Fig. III. 2.10. Reduction and carboxymethylation of proteins containing disulphide bridges (From Protein analysis and purification: benchtop techniques, by I.M. Rosenberg).

Stacking gel

The stacking gel is a large pore PAGE (4%T). This gel is prepared with Tris/HCl buffer pH 6.8 of about 2 pH units lower than that of electrophoresis buffer (Tris/Glycine). These conditions provide an environment for Kohlrausch reactions determining molar conductivity, as a result, SDS-coated proteins are concentrated to several folds and a thin starting zone of the order of 19 μm is achieved in a few minutes. This gel is cast over the resolving gel. The height of the stacking gel region is always maintained more than double the height and the volume of the sample to be applied. This is based on isotachophoresis. Stacking gel is used to compress applied components to thin bands before it enters resolving gel.

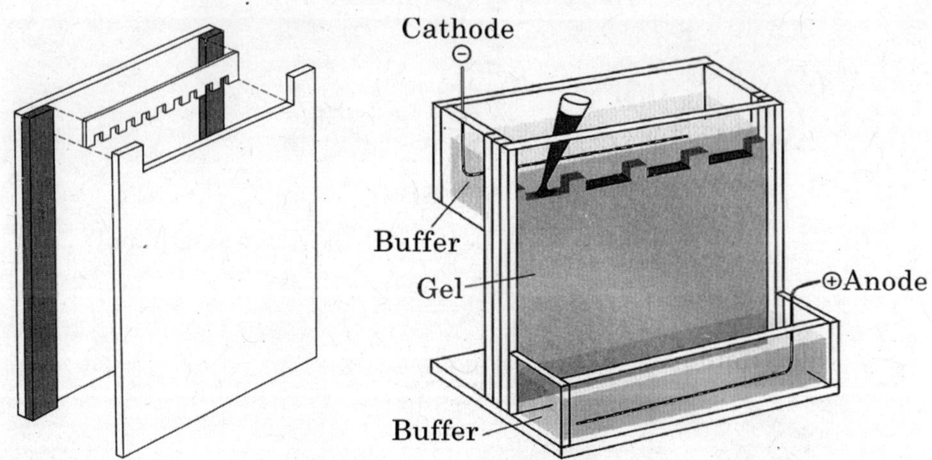

Fig. III.2.11 : Electrophoresis assembly

Separating Gel

Ingredients	5%	7.5%	10%	12.5%	15%
Acrylamide solution (ml)	5.0	7.5	10.0	12.5	15.0
1 M Tris HCl, pH 8.8 (ml)	11.2	11.2	11.2	11.2	11.2
H_2O (ml)	13.7	11.2	8.7	6.2	3.7

Stacking Gel

Ingredients	3%	5%
Acrylamide solution (ml)	1.0	1.67
1 M Tris HCl, pH 6.8 (ml)	1.25	1.25
H_2O (ml)	7.7	7.8

(b) Requirements

Apparatus

It consists of a power pack and an electrophoresis unit. The power pack supplies a stabilized direct current at controlled or required voltage and current output. The electrophoresis unit contains the electrodes, buffer reservoirs and gel casting assembly.

Reagents used

1. Stock acrylamide solution (30%)
   a. Acrylamide – 30 % (30/29.2 gm in 100 ml)
   b. Bis-acrylamide – 0.8 (0.8 gm in 100 ml)
2. Separating or resolving gel buffer (pH 8.8)
   Tris HCl – 1.875 M (22.69 gm in 100 ml)
3. Stacking gel buffer (pH 6.8)
   Tris HCL – 0.6 M (7.26 gm in 100 ml)
4. Polymerising agents:
   a. TEMED – Fresh from refrigerator
   b. APS (10%), w/v (stock solution) always fresh.
5. Running / Electrode buffer (pH 8.2 to 8.4)
   a. 0.05 M tris (11 gm)
   b. 0.019 M glycine (28.8 gm)
   c. 0.1% SDS (2 gm)
   d. Water – 2 litre
6. Sample buffer (5x concentration)
   a. 0.6 M Tris HCl buffer (pH 6.8) (0.5 ml)
   b. SDS - 0.5 gm
   c. Sucrose or glycerol - 5 gm
   d. β - mercaptoethanol - 0.25 gm
   e. Bromophenol blue - 0.5% w/v in H_2O
   Total volume of sample buffer - 10 ml
7. 10% SDS solution :
   1 or 2 gm in 20 ml (stored at room temperature)
8. Protein stain solution
   a. Commassie brilliant blue dye - 0.1 gm

b. Methanol - 40 ml

c. Acetic acid - 10 ml

d. Water - 50 ml

9. Destaining solution :

a. Methanol - 40 ml

b. Acetic acid - 10 ml

c. Water - 50 ml

(c) Procedure

1. The glassware and spacers are thoroughly cleaned and dried, and assembled properly.
2. The assembly is held together in upright position with the help of clamps.
3. While petroleum jelly or agar/agarose (2%) is applied around spaces of spacers and hold them in upright place and make assembly leakage proof.
4. The sufficient volume of separating gel mix about 30 ml (for chamber of about 18 x 9 x 0.1 dimensions) is prepared by mixing following :

   gel (12%) is used:

   a. Distilled water = 13.92 ml.

   b. Stock acrylamide (30%) = 15.96 ml.

   c. Tris 1.875 M (pH = 8.8) = 9.6 ml

   d. SDS (10%) = 0.1 ml

   e. APS (10%) = 0.1 ml

   f. TEMED = 20μl

5. After mixing separating gel solution is poured carefully in chamber between glass plate.
6. A mixture of n-butanol and distilled water (1:1) is overloaded on top of gel and leave it to set for 30 to 60 min (water saturated butanol)
7. Stacking gel (4%) is prepared by mixing

   Stacking gel (4%) = 10 ml

   a. Distilled water = 7.5 ml

   b. Acrylamide (30%) = 1.35 ml

   c. Tris (pH = 6.8) = 1 ml

d. SDS (10%) = 0.1 ml

e. APS (10%) = 50 μl

f. TEMED = 10 μl

8. The upper aqueous layer of water saturated n-butanol is removed from top of gel and stacking solution is poured into chamber as polymerised separating gel.
9. Comb is placed in stacking gel and is allowed to set for 30 – 60 min.
10. After polymerisation the stacking gel the non-polymerised gel is removed with the help of water.
11. Comb is removed carefully without disturbing shapes.
12. Prepare samples for electrophoresis, following suitable extraction procedure. Adjust the protein concentration in each sample using the sample buffer and water in a way that the same amount of protein is present per unit volume. Heat the solution in boiling water for 2–3 min to ensure complete interaction between protein and SDS.
13. Cool the sample solution and take up the required volume in a microsyringe and carefully inject it into a sample well through the electrode buffer. Marking the position of wells on the glass plate with marker pen.
14. Turn on the current to 10–15 mA for initial 10–15 min until the sample travel through the stacking gel. Then continue to run at 30 mA until bromophenol blue reaches the bottom of the gel (about 3 hrs). However, the gel may run at a high current (60–70 mA) for short period (1 mm) with proper cooling.
15. After the run is complete carefully remove the gel from the plate and immerse in staining solution for at least 3 hrs. or overnight with uniform shaking.
16. Transfer the gel to a suitable container with atleast 200–300 ml destaining solution and shake gently and continuously. Dye that is bound to protein is thus removed.
17. Change in destainer frequently particularly during initial periods until the background of the gel becomes colourless.
18. The protein fractioned into band is seen coloured. The gel can now be photographed or stored in polyethylene bags or dried in vacuum for permanent record.

SDS-polyacrylamide gel electrophoresis is rapid, sensitive and capable of a high degree of resolution. As little as 0.1 μg of a protein gives a distinct band when stained with Coomassie blue, and even less (0.02 μg) can be detected with a silver stain. Proteins that differ in mass by about 2 percent can usually be distinguished.

Precautions

1. All chemicals and distilled water should be of high quality.
2. Old solution results in poor resolution of proteins.
3. The solution should be filtered before use.
4. Acrylamide as a monomer is highly neurotoxic, handle with extreme care.
5. Degassing of gel mixture can be done for easy polymerization.

2.9.5 Silver Staining of Proteins

The fractionated polypeptides on gels are visualized by staining generally either with coomassie brilliant blue R 250 or amido black 10B dye. The above dyes can detect a band containing as little as 0.1 μg of polypeptide. In many occasions, the available protein for electrophoresis is so small or some proteins occur in minute amounts, the detection becomes extremely difficult with these dyes. Under such circumstances a higher sensitive detection system is required. Silver staining is a very useful method in this regard with about 10–100 fold greater sensitivity over dye staining. It is comparable in sensitivity to autoradiography of labelled polypeptides. The stain is useful for protein concentrations ranging from approximately 1 ng to 1 mg. If the staining does not work, gels can be destained and restained again. It should be noted that while this method of staining is more sensitive and much faster than colloidal coomassie staining, the recovery of peptides from an in-gel digest is decreased.

There are different methods described by different workers for silver staining. These methods are based upon binding silver ions to proteins and reducing the Ag^{2+} to metallic silver. The silver interacts with primary amines or sulfhydryls. In the diamine method, silver diamine complexes are stabilized with ammonium ions and the free amine groups of proteins. The silver bound to protein is then reduced to metallic silver with formaldehyde. In the non-diamine method, $AgNO_3$ reacts with protein under acidic conditions, followed by reduction to metallic silver with formaldehyde under alkaline conditions. Photodevelopment relies on light (photons) to reduce silver ions to metallic silver. Some silver stain protocols result in certain types of proteins being stained particular colours, thus aiding in the identification of proteins.

The method given below is very simple and rapid.

(a) Principle

The amino acids particularly aromatic amino acids in the protein reduce silver nitrate and form complexes with metallic silver of yellowish- brown to brown colour.

(b) Reagents

1. Silver nitrate
2. HPLC-water or Milli-Q water.
3. HPLC-grade methanol
4. Acetic acid
5. Sodium Thiosulphate
6. Sodium Carbonate
7. Formaldehyde (37%)
8. Disodium, Ethylene Diamine Tetra-Acetic Acid (Na_2-EDTA)
9. Potassium Ferricyanide (for destaining)

1. Gel staining protocol

Prepare all solutions and reagents just prior to use

Step	Reagent	Final volume =100ml	Operation	Time
A	Methanol (50%) acetic acid (10%)		Fix	30 min
B	Methanol (5%)		Incubate	15 min
	Milli-Q water		Wash 3 times	3 × 5 min
C	$Na_2S_2O_3.5H_2O$ (Sol C)	0.2 g/ml Cold and fresh	incubate	120 sec
	Milli-Q water		Wash 3 times	3 × 30 sec
D	Silver nitrate ($AgNO_3$)	0.2g/100 ml Cold and fresh	Incubate	25 min
	Milli-Q water		Wash 3 times	3 × 60 sec
E	Sodium Carbonate Formaldehyde (37%) solution $Na_2S_2O_3.5H_2O$ (Sol C)	3g/100 ml 50µl/100 ml 2 ml/100 ml	Develop	10 min. Maximum
F	Na_2-EDTA	14 g/l	Stop Develop	10 min
	Milli-Q water		Wash	

2. Destaining protocol

   Dissolve 0.4g potassium ferricyanide ($K_3Fe(CN)_6$) in 200 ml sodium thiosulphate solution (0.2 g/l, step c above).

Destain

   a. Destain until no bands are visible; the gel may be slightly yellow.
   b. Wash gel 4–5 times for 15 min with milli-Q H_2O until gel is transparent and has no background colour.
   c. Starting with step c from the gel staining protocol, restain gel.

3. Gel Storage

   Gel can be stored in acetic acid solution (5%) at 4°C for several weeks prior to in-gel digestion. Alternatively, gel pieces can be stored dry.

(c) Precautions

   1. Use Analytical grade reagents. All glassware should be thoroughly cleaned.
   2. Wear gloves while working, particularly when handling the gel; otherwise fingerprints will be a problem.
   3. The gel becomes very fragile during the treatment; exercise adequate care.
   4. The method is very rapid and easy. The proteins can be fixed in the gel, if necessary, by immersing it in glutaraldehyde (10%) for 30–60 min and washed in water prior to washing the gel in washing solution mentioned above.
   5. Continuous gentle shaking of the gel in every step is important.
   6. The colour formation is inorganic in reaction and hence proceeds very rapidly. Stop the reaction once the required intensity of bands is obtained by pouring stopper solution and with shaking.

2.9.6 Separation of Proteins by Isoelectric Focusing

Principle

Isoelectric focusing is an electrophoretic method in which proteins are separated on the basis of their pI. It makes use of the property of proteins that their net charges are determined by the pH of their local environments. Proteins carry positive, negative, or zero net electrical charge, depending on the pH of their surroundings. Proteins are positively charged in solutions at pH values below their pI and negatively charged above their isoelectric points. Thus, at pH values below the pI of a particular protein, it will migrate toward the cathode during electrophoresis. At pH values above its pI, a protein will move toward the anode. A protein at its isoelectric point will not move in an electric field. Proteins show considerable variation in isoelectric points, but p*I* values usually fall in the range of pH 3–12 with a great many having p*I*s between pH 4 and pH 7. When a protein is placed in a medium with a linear pH gradient and subjected to an electric field, it will initially move toward the electrode with the opposite charge. During migration through the pH gradient, the protein will either pick up or lose protons. As it does, its net charge and mobility will decrease and the protein will slow down. Eventually, the protein will arrive at the point in the pH gradient equaling its p*I*. There, being uncharged, it will stop migrating. In this way, proteins condense, or focus, into sharp bands in the pH gradient at their individual, characteristic pI values.

Focusing is a steady-state mechanism with regard to pH. Proteins approach their respective pI values at differing rates, but remain relatively fixed at those pH values for extended periods. This type of motion is in contrast to conventional electrophoresis in which proteins continue to move through the medium until the electric field is removed. Moreover, in IEF, proteins migrate to their steady-state positions from anywhere in the system. Thus, the sample application point is arbitrary. In fact, the sample can be initially distributed throughout the entire separation system.

1. Establishing pH gradients

Stable, linear pH gradients are the keys to successful IEF. Establishment of such gradients is accomplished in two ways with two different types of molecules, carrier ampholytes and acrylamido buffers.

Carrier ampholytes (*amphoteric* electrolytes) are mixtures of molecules containing multiple aliphatic amino and carboxylate groups. They are small (about 300–1000 Da in size) multicharged organic buffer molecules with closely spaced p*I* values and high conductivity. Ampholytes are included directly in IEF gels. In electric fields, carrier ampholytes partition into smooth pH gradients that increase linearly from the anode to the cathode. The slope of a pH gradient is determined by the pH interval covered by the carrier ampholyte mixture and the distance between the electrodes. The use of carrier ampholytes is the most common and simplest means for forming pH gradients.

Acrylamido buffers are derivatives of acrylamide containing both reactive double bonds and buffering groups. Their general structure is CH_2=CH-CO-NH-R, where R contains either a carboxyl [–COOH] or a tertiary amino group [*e.g.*, –$N(CH_3)_2$]. They are covalently incorporated into polyacrylamide gels at the time of casting. The key acrylamido buffers have pK values at pH 1, 3.6, 4.6, 6.2, 7.0, 8.5, 9.3, 10.3, and >12. They can be used to cast just about any conceivable pH gradients. In any given gradient, some of the acrylamido compounds act as buffers, while others serve as titrants. Since the buffering compounds are fixed in place in the separation medium, the gels are called "immobilized pH gradients", or IPGs. IPGs offer the advantage of gradient stability over extended runs. They are, however, more cumbersome and expensive to cast than carrier ampholyte gels. IPGs are commercially available in sheet form in a few pH ranges. A greater variety of pH ranges are available in IPGs that have been cut into strips for the IEF first dimension of 2-D PAGE.

IEF is a high-resolution technique that can routinely resolve proteins differing in p*I* by less than 0.05 pH unit. Antibodies, antigens, and enzymes usually retain their activities during IEF. The proper choice of ampholyte or IPG range is very important to the success of a fractionation. Ideally, the pH range covered by an IEF gel should be centered on the p*I* of the proteins of interest. This ensures that the proteins of interest focus in the linear part of the gradient with many extraneous proteins excluded from the separation zone.

With carrier ampholytes, concentrations of about 2 percent (w/v) are best. Ampholyte concentrations below 1 percent (w/v) often result in unstable pH gradients. At concentrations above 3 percent (w/v) ampholytes are difficult to remove from gels and can interfere with protein staining.

2. Gels for isoelectric focusing

As an analytical tool, IEF is carried out in large-pore polyacrylamide gels (5%T, 3%C) which serves mainly as anticonvective matrices. Polyacrylamide IEF gels are polymerized with an initiator system including riboflavin for photopolymerization. Photochemical initiation of polymerization with a combination of the three compounds riboflavin, ammonium persulfate, and TEMED, results in more complete polymerization of IEF gels than does chemical polymerization in gels containing low-pH ampholytes. Suitable initiator concentrations are: 0.015% ammonium persulfate, 0.05% TEMED, and 5 μg/ml riboflavin-5'-phosphate. Photochemical polymerization is allowed to continue for 2 hr, with the second hour under direct lighting from a nearby fluorescent lamp.

The most common configuration for analytical IEF is the horizontal polyacrylamide slab gel. Gels are cast with one exposed face on glass plates or specially treated plastic sheets. They are placed on cooling platforms and run with the exposed face upward. Electrolyte strips, saturated with 0.1–1 M phosphoric acid at the anode and 0.1–1 M sodium hydroxide at the cathode, are placed directly on the exposed surface of the IEF gel. Electrodes of platinum wire maintain contact between the electrical power supply and the electrolyte strips. In another possible configuration, the gel and its backing plate are inverted and suspended between two carbon rod electrodes without the use of electrolyte strips. IPG strips for 2-D PAGE are often run with the gel facing down in dedicated IEF cells.

Casting gels for isoelectric focusing

(a) Equipment and reagents

Electro-foucussing apparatus: A vertical slab gel electrophoresis apparatus can be used.

1. Gel casting apparatus for a horizontal electrophoresis cell; e.g., the GE MultiPhor.
2. Fluorescent lamp.
3. T (30%), acrylamide (3% C) stock solution: dissolve 29.1 g of acryla-mide and 0.9 g of bisacrylamide in 72.5 ml of water. The final volume will be 100 ml.
4. glycerol (50%): mix 50 g of glycerol with 63 g of water (100 ml final volume).
5. riboflavin-5'-phosphate (FMN) (0.1%): dissolve 50 mg of FMN in 50 ml of water.
6. Carrier ampholytes (40%, (w/v) concentration). with a pH range spanning the pIs of the proteins of interests.
7. APS (10%): APS solutions should be made fresh daily, but may perform satisfactorily for up to a week. Store at room temperature.
8. TEMED.

(b) Procedure

The formulation given here is for 12 ml of gel solution containing T (5%), (3%C) acrylamide, carrier ampholytes (2%), glycerol (5%). The volume needed depends on the casting apparatus that is used; adjust volumes accordingly. This recipe is sufficient for casting one gel of 100 × 125 × 0.8 mm (10 ml) or four 100 × 125 × 0.2 mm (10 ml total). Assemble the casting apparatus according to the manufacturer's instructions. The use of gel support film for polyacrylamide is highly recommended.

1. Dissolve the following components completely:

a.	T (30%), C (3%) acrylamide stock	2.0 ml
b.	Carrier ampholytes (40%)	0.6 ml
c.	Glycerol (50%)	1.2 ml
d.	Water	8.2 ml
e.	FMN (0.1%)	60 μl
f.	APS (10%)	18 μl
g.	TEMED	4 μl

2. Transfer the gel solution to the casting apparatus with a pipette and bulb,
3. Position a fluorescent lamp about 3–4 cm from the gel solution and illuminate the solution for about one hour.
4. Open the gel cassette or lift the gel from the casting tray to expose the face of the gel. Place the gel with the open face upward and illuminate it with the fluorescent lamp for an additional 30 min.
5. The gel may be used immediately or it can be covered with plastic wrap and stored at 4°C for several days. Best results are sometimes obtained when IEF gels are left overnight at 4°C before use.

Good visualization of individual bands generally requires a minimum of 0.5 gm each with dye staining or 50 ng each with silver staining. One of the simplest methods for applying samples to thin polyacrylamide gels is to place filter paper strips impregnated with sample directly on the gel surface. Up to 25 μl of sample solution can be conveniently applied after absorption into 1-cm squares of filter paper. A convenient size for applicator papers is 0.2 × 1 cm, holding 5 μl of sample solution. Alternatively, 1 to 2 μl samples can be placed directly on the surface of the gel. In most cases, IPG strips (which are provided in dehydrated form) are rehydrated in sample-containing solution prior to electrophoresis. Rehydration loading allows higher protein loads to be applied to gels than do other methods. It is particularly popular because of its simplicity.

There are no fixed rules regarding the positioning of the sample on the gel. In general, samples should not be applied to areas where they are expected to focus. To protect the proteins from exposure to extreme pH, the samples should not be applied closer than 1 cm from either electrode. Forming the pH gradient before sample application also limits the exposure of proteins to pH extremes.

Isoelectric focusing

(a) Equipment and reagents

1. Flat-bed electrophoresis cell; *e.g.*, the GE Multiphor
2. Power supply 0–500 volts.
3. Refrigerated water circulator if required.
4. 1N NaOH catholyte if required: dissolve 4 g of NaOH in 100 ml of water.
5. 1 N H_3PO_4 anolyte if required: dissolve 2.3 ml of H_3PO_4 (85%) (14.6 M, 44 N) in 97.7 ml of water.
6. Electrode strips, if required: For electrode strips, cut thick filter paper about 7 mm wide and about 4 mm shorter than the gel.
7. Sample application strips: For sample application strips, cut thin filter paper to 0.2×1 cm, one strip per sample.

(b) Procedure

1. Set up the IEF cell as recommended by the manufacturer. This includes connecting a water circulator, if used, and preparing the cooling platform and electrode strips, if necessary.
2. Place sample application strips on a glass plate and pipette 5 µl of a protein sample to each strip. Place the application strips 1 cm from the anode end of the gel.
3. Position the gel in the IEF cell and make electrode contact as specified for the particular cell.
4. Close the the electrophoresis cell and connect the leads to the power supply; the red lead is the anode and the black lead is the cathode.
5. Set the running conditions as recommended by the manufacturer of the electrophoresis cell.
6. When the focusing is complete, disconnect the power supply, remove the electrodes and lift the gel carefully.
7. Place the gel in the fixing solution for 15–20 min. Transfer the gel to the staining solution for 30 min, and finally to the destaining solution for about 20 min. until the protein bands are clearly visible. This should be done at room temperature.
8. To preserve the gel, immerse it in glycerol (10%) for 30–60 min. Soak a cellophane sheet in the same solution for a few min. and wrap it around the gel. Keep it at 50°C
9. The gel may be photographed.

Chapter - 3

C. MOLECULAR BIOLOGY

INTRODUCTION

Chromosomes consist of nucleoproteins, *i.e.,* nucleic acids and proteins, and one of them should constitute the genetic material. The genetic material consists of nucleic acids and not proteins, proved by transformation experiments conducted by F. Grifith in 1928. Later, in 1944 Avery, Macleod and Macarthy repeated Grifith's experiments to identify the transforming substance responsible for converting non-virulent into virulent strains of Pneumococcus (*Diplococcus pneumonia*) by their experiments in mice. Their experiments confirmed that DNA and not proteins is the genetic material and the phenomenon was called **transformation.**

Although double stranded RNA is not uncommon, but RNA is generally found as single stranded. DNA is always genetic in nature, but only RNA is rarely genetic in nature as found in plant viruses. Experiments conducted by H. Fraenkel and Conrat with tobacco mosaic virus proved that RNA can also be the genetic material. Mostly RNA is non-genetic in nature and found in cell as tRNA, mRNA and rRNA.

Structure of DNA

Nucleic acid is a long chain polymer (polynucleotides) composed of monomeric units called nucleotides. Each nucleotide consists of a nucleoside and a phosphate group, and each nucleoside in turn consists of a sugar molecule and a base. Bases are of two types, purines and pyrimidines.

Purines – Adenine and Guanine

Pyrimidines – Thymine and Cytosine are present in DNA, whereas Uracil in RNA

Sugars are deoxyribose in DNA and ribose in RNA

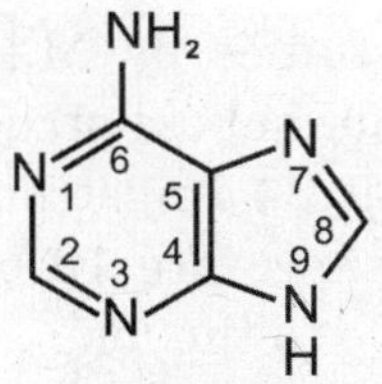

Adenine
(6-Aminopurine)

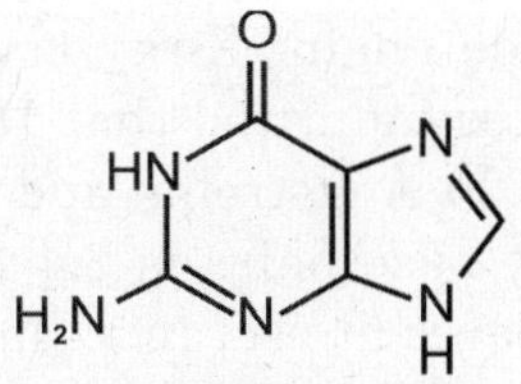

Guanine
(2-Amino-6-oxopurine)

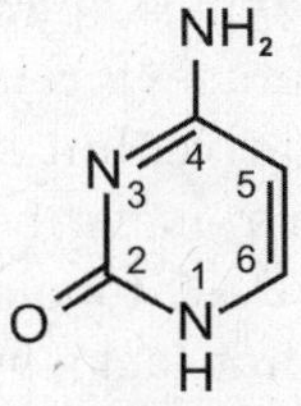

Cytosine
(2-Oxo-4-aminopyrimidine)

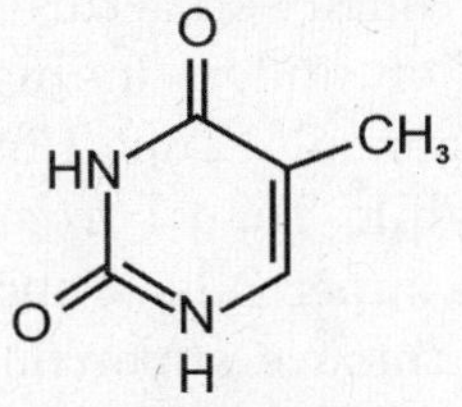

Thymine
(2,4-Dioxo-5-methylpyrimidine)

In a nucleoside carbon No. 1 in sugar is linked with nitrogen at position No. 3 in pyrimidines at position No. 9 in purines. The phosphate molecule is linked with sugar molecule at carbon No. 5 or at carbon No. 3 and the nucleotides are called as 5'p3'OH nucleotide and 3'p5'OH nucleotide.

Table III.1 : Types of Nucleic acids.

S.N.	Base	Nucleoside	Nucleotide	Abbreviation	
				RNA	DNA
1.	Adenine	adenosine	adenylic acid	AMP	dAMP
2.	Guanine	guanosine	guanylic acid	GMP	dGMP
3.	Cytocine	cytidine	cytidylic acid	CMP	dCMP
4.	Thymine	thymidine	thymidylic acid		dTMP
5.	Uracil	uridine	uridylic acid	UMP	

Polynucleotide – A number of nucleotide monomer units may give rise to a polynucleotide chain through the formation of phosphodiester bonds. A polynucleotide chain has a direction of 3' to 5' or from 5' to 3'. A phosphate group is present at the 5' end and a hydroxyl (OH) group at the 3' end.

The structure of DNA was given by J.D. Watson and H.F.C. Crick in 1953 according to which DNA has a double helical structure with two polynucleotide chains connected by hydrogen bonds and they run in opposite directions. The number of adenine molecules is always equal to the number of thymine molecules. Similarly, the number of cytosine molecules is always equal to the number of guanine molecules. Further, although A=T and G≡C, neither there is any restriction on the ratio (A+T/(C+G), nor is there any restriction on the sequence of bases in a polynucleotide chain. X-ray crystallographic studies by M.H.F. Wilkins and associates indicated that DNA has a double helical structure with a diameter of 20 Angstroms and a pitch of about 34 Angstroms. Distance between two base pairs is 3.4 Angstrom and there are 10 base pairs in each turn. Bases are set in a plane at right angle to the long axis.

The double helix structure of DNA represents a linear DNA molecule, but DNA *in vivo* often has a closed structure without free ends and exists as a supercoiled structure. Topoisomerase enzymes participate in overwinding or underwinding of DNA. Topoisomerase I relax DNA by removing supercoils. Topo I from E. coli act on negative, but not positive, supercoils, while Topo I from eukaryotic cells can remove both positive and negative supercoils. Topoisomerase II change DNA topology and can introduce or remove supercoils. A DNA molecule that lacks supercoiling, whether closed or open is said to be relaxed. The supercoils may be negative (as found *in vivo*) when they are in a direction opposite to the cloakwise turn of right handed DNA. The positive supercoils does not occur in nature, but can be created *in vitro*

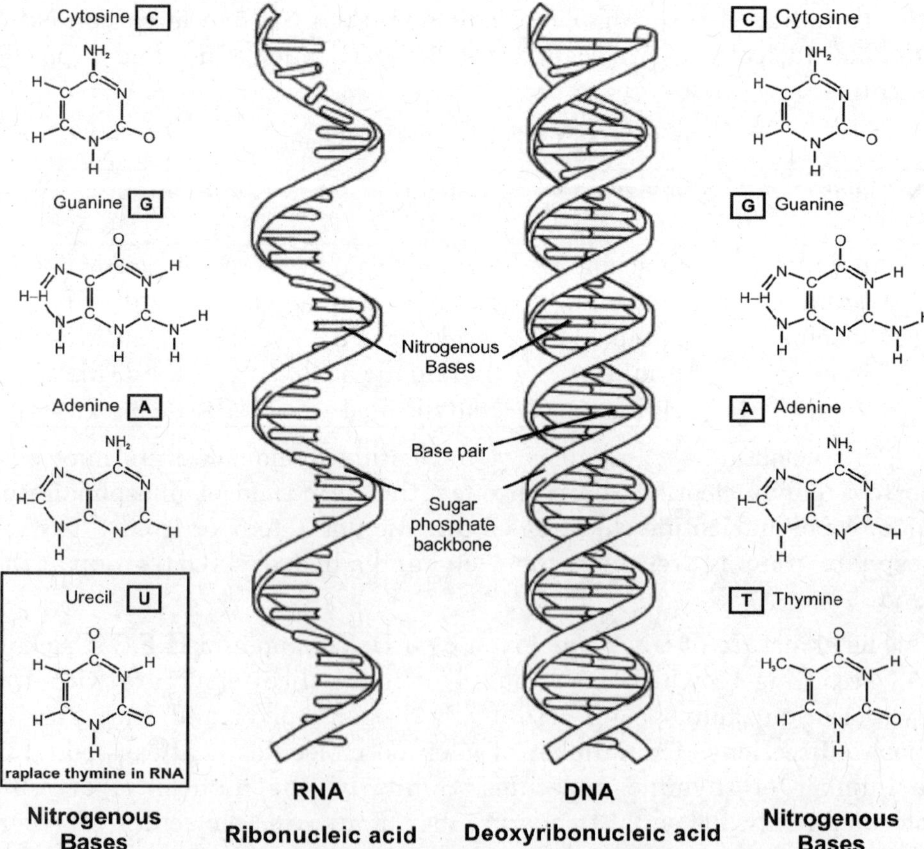

Fig. 3.1 : Structure of nucleic acids DNA and RNA.

Alternative forms of DNA

The helical coiling of DNA structure as given by Watson and Crick is assumed to be right handed and this form of DNA has been called the B form of DNA. However, DNA may exist in other alternative forms which differ in the number of residues per turn (n) or in the spacing of residues along the helical axis (h). The other alternative forms of DNA are A, B, C and Z. Apart from these, the D and E forms are also rarely found as extreme variants with only 8 and $7\frac{1}{2}$ base pairs per turn, respectively.

Table III.2 : Properties of different types of DNA.

S.N.	Properties	A DNA	B DNA	Z DNA
1.	Handedness	Right	Right	Left
2	Helical diameter (Angstrom)	25.5	19	18
3.	Base pairs /turn	10.9	10	12
4.	Helix rise/base pair (Angstrom)	2.56	3.38	5.71

3.1 ISOLATION OF NUCLEIC ACID

A genetic engineer will require either total DNA, plasmid DNA, chromosomal DNA or phage DNA. When selecting for a procedure for isolation of DNA, factors which should be considered are the size of the plasmid, the simplicity and reproducibility of the procedure, the bacterial genus to be lysed, the gentleness of the procedure (for plasmids), the DNA yield and the subsequent use of the plasmid DNA (*i.e.*, for restriction endonuclease analyses, transformation, etc.).

For the rapid isolation of plasmid DNA, one should aim for a few manipulations as possible. Screening bacterial colonies for plasmid content, or using plasmid DNA for certain transformation and cloning procedures does not require the preparation of pure plasmid DNA. Pure DNA may be required for producing restriction maps, in certain cloning procedures, etc. For this, a series of "options" may be included to the basic plasmid isolation technique. For example, a RNase might be used to eliminate RNA, a phenol extraction step might be added to remove proteins, or the addition of the nuclease inhibitor, diethyl pyrocarbonate might be used to eliminate endogenous nucleases present in some genera like *Clostridium*.

3.1.1 Isolation of DNA

Isolation of DNA can be of two types, *i.e.*, **plasmid DNA** and **genomic DNA**.

Plasmid DNA protocols can be divided into 4 steps:

1. Culture of Bacterial cells.
2. Harvesting of DNA.
3. Isolation of plasmids.
4. Purification of DNA.

For isolation of genomic DNA from bacteria, the cell extract is prepared by cell lysis brought about by exposure of cells to chemical agents that affect the integrity of the cell barriers. Weakening of the cell wall is brought about by lysozyme, EDTA or a combination of both. Lysozyme is an enzyme present in egg white, tears and saliva, and digests the polymeric compounds that give rigidity to the cell wall. EDTA removes magnesium ions which are essential for maintaining the overall structure of the cell envelope and also inhibit cellular enzymes. A detergent such as SDS may also be incorporated which aids the process of lysis by removing the lipid molecules present in the cell membrane. In *E. coli* and related species, cell lysis is carried out in presence of sucrose which prevents the cells from bursting straight away and instead sphaeroplasts are formed. Cell lysis is then carried out using a non-ionic detergent, *e.g.*, Triton X-100 as ionic detergents cause chromosomal breakage. The insoluble cell debris is removed by centrifugation, the cell extract deproteinized and RNA removed by ribonucleases. First the total DNA has to be isolated and then the plasmid DNA removed at some step by centrifugation, using size or conformation as the basic difference between the two types of DNA, *i.e.*, genomic and plasmid DNA.

Note: *The preparation of reagents and solutions required in the experiments of molecular biology have been given in Appendix IVC of this book.*

3.1.1.1 Rapid Method For Isolation of Plasmid DNA

Plasmids, the extrachromosomal DNA are autonomously replicating body, have become almost indispensable in the construction of useful recombinant micro-organisms. The foreign gene for particular protein is cloned in the plasmid which contains a number of unique restriction sites, a strong selection marker and exists in high copy number.

pUC 19/ HB 101 is a high copy number plasmid which is designed to allow simple and rapid preparation of cloned recombinant DNA constr-ucts. This involves: isolation of the plasmid, digestion of desired insert to the vector, transformation of the construct into a suitable recipient strain and finally checking for the presence of the recombinant construct.

Amplification of plasmid DNA

If plasmid replication does not require new protein synthesis (*i.e.*, not inhibited by chloramphenicol), the amplification of plasmid DNA may be achieved through the addition, at midlog phase, of chloramphenicol powder (150 μg/mL), followed by further incubation of 3 to 5 hours. Two plasmids that may be amplified are pBR322 and ColE1.

Method 1: Alkaline detergent method

This is a rapid, reproducible, and gentle procedure for the recovery of bacterial plasmids and has the further advantage that large plasmids are readily visualized. This method has been used successfully with a variety of bacterial genera, including enteric species as well as species of *Staphylococcus, Pseudomonas, Rhizobium, Agrobacterium, Clostridium,* etc.

Procedure

1. Grow cells in enriched medium. Aerated cultures should be grown to late exponential phase, while in case of nonaerated cultures overnight incubation should be done.
2. Pellet the cells by centrifugation for 10 minutes, at 25°C. Remove the supernatant and resuspend the cells in 1 ml of TE buffer. Transfer the cells to a 1.5 ml microfuge tube and pellet by centrifugation for 2 minutes at 4°C in a microfuge.
3. Remove the supernatant and resuspend the cells in 25 μl of TE buffer and vortex vigorously.
4. Add 400 μL of freshly prepared lysis buffer (TE buffer at pH 12.45 containing 1.5% SDS).
5. Mix the suspension by inverting the microfuge tube. Allow the lysis mixture to stand either for 30 minutes at room temperature or for 60 minutes at 56°C. Mix the tubes by inversion at least once during the incubation period.
6. Neutralize the mixture by adding 150 μl of 3M sodium acetate, pH 4.8 (adjust the pH of 3M sodium acetate with acetic acid). Gently mix several times by inversion. Lowering the pH causes chromosomal DNA to form an insoluble complex that, together with protein-SDS complexes, is precipitated by the high salt concentration.
7. Place the suspension on ice bath for 60 minutes to complete precipitation of high molecular weight complexes.
8. Centrifuge for 15 minutes in a microfuge at full speed. Recover the supernatant using a polyethylene Pasteur pipette and transfer to a clean microfuge tube.

Phenol/Ether Extraction (Optional)

Phenol-chloroform extraction is a liquid-liquid extraction technique in biochemistry. It is widely used in molecular biology for isolating DNA, RNA and protein. Equal volumes of a phenol:chloroform mixture and an aqueous sample are mixed, forming a biphasic mixture.

(a) Add 0.5 ml of sodium acetate (1M) saturated phenol to the neutralized mixture (preparation of saturated phenol given in appendix).

(b) Invert the tube several times in order to mix the two phases.

(c) Centrifuge the tubes in a microfuge for 10 minutes, at full speed, at 4°C.

(d) Remove the upper aqueous phase to another tube using a polyethylene Pasteur pipette.

(e) Add an equal volume of reagent-grade ether to the aqueous phase and mix by inversion.

(f) Allow the phases to separate and then discard the upper ether layer. The DNA remains in the lower layer.

(g) Repeat 2–3 times.

(h) Remove residual ether by evaporation under an air stream (since trace amounts of phenol may interfere with the activity of certain restriction endonucleases and other enzymes, ether is used to remove residual phenol from the aqueous phase).

9. Precipitate the DNA by adding 2 volumes of cold ethanol to the supernatant. Mix by inversion and store at -20°C for 2 hours.
10. Pellet the DNA by centrifuging the tubes at full speed in a microfuge for 10 minutes at 4°C.
11. Pour off the ethanol, invert the tube and air dry.
12. Add 20 to 100 µl of the TE buffer and store at 4°C till further use.

Method 2: Boiling method

This procedure is an adaptation of the method developed by Holmes and Quigley (1981). It is used for screening large number of colonies for plasmid content. Plasmids ranging in size from 3.2 to 39.2 kilobases (kb) are easily visualized by this procedure and has been used for isolating plasmids in *Neisseria* species and *Escherichia coli*.

Procedure

1. Suspend either a single colony (4 mm) or a loopful of cells from solid medium in 500 µl of STET buffer in a 1.5 ml microfuge tube and vortex.
2. Add 100 µl of freshly prepared lysozyme (10 mg/ml), mix by vortexing and allow to stand for 10 minutes at room temperature.
3. Place the suspension in a boiling water bath for 60 seconds (boiling temperatures denature nucleases that may be present in the extract. Longer periods of boiling may result in the appearance of chromosomal DNA in final preparations due to the fragmentation of the long strands).
4. Centrifuge the suspension in a microfuge at maximum speed (*e.g.*, 15,600 ×g) for 15 minutes (after centrifugation, a fluffy, whitish pellet generally indicates good lysis. The pellet comprises chromosomal DNA and denatured proteins).
5. Transfer the supernatant to another microfuge tube.
6. Precipitate the DNA by adding either 2 volumes of cold ethanol or an equal volume of cold isopropanol. Store at -20°C for 2 hours or at -70°C for 10–20 minutes.

   DNA may be precipitated overnight at -20°C if desired. The solution may be cloudy or slightly frozen, but this will not affect the final result.
7. Pellet the precipitated DNA by centrifuging for 15 minutes at maximum speed.
8. Decant alcohol. Air dry the pellet for about 10 minutes or until all of the alcohol has evaporated.
9. Resuspend the pellet in 50–100 µl of TE buffer.

Method 3: Alkaline Lysis method

A. Growth of *E. coli* strains containing plasmid

Plasmids such as pUC and pBR are always isolated from liquid cultures containing appropriate antibiotic because they carry an anti-biotic marker gene. Use of antibiotic is to enable cells to retain the plasmid.

Procedure

1. Take 10 ml LB medium in 100 ml flask. Add ampicillin upto 50 µg/ ml final concentration aseptically. Transfer a single colony of *E. coli* strain using a sterile loop. Grow the bacterial cells in an incubator shaker for 15–16 hours at 37°C, 200 rpm speed.

2. Transfer 1 ml grown culture of *E. coli* in 100 ml fresh LB medium. Take out an aliquot of 1 ml in a sterile tube and put the rest on shaker for growth.
3. Read Absorbance of the sample at 600 nm on a spectrophotometer.
4. Continue the experiment by taking out aliquots of 1 ml at regular intervals of 30 minutes till O.D. is reached to 2–2.5.

B. Isolation of plasmid DNA

Sodium dodecyl sulphate (SDS) is a detergent and extracts lipids from cell wall and cell membrane of the bacterial cell, and thus its osmotic balance is disturbed and cells are lysed. Sodium acetate of pH 4.8 is added to neutralise the pH so that the plasmid DNA remains intact.

Procedure

1. Transfer a single colony from given strain of *E. coli* in 10 ml LB containing appropriate antibiotic.
2. Incubate at 37°C, 200 rpm for overnight.
3. Inoculate 1 litre LB with 10 ml overnight culture. Incubate at 37°C, 200 rpm for overnight.
4. Harvest the bacterial cells at 4000 rpm for 15 minutes at 4°C. Discard the supernatant.
5. Thoroughly resuspend the cell pellet in 20 ml of lysis buffer. Add 1 ml lysozyme solution. Incubate at 37°C for 5–10 min.
6. Add 40 ml of freshly prepared lytic mixture.
7. Mix the contents thoroughly by gently inverting the tube several times. Incubate on ice for 10 min.
8. Add 30 ml of ice-cold 3 M sodium acetate (pH 4.8) buffer. Cap the tube tightly and shake the mixture several times. Incubate on ice for 20 min.
9. Centrifuge the bacterial lysate at 8000 rpm for 15 min. at 4°C.
10. Transfer the supernatant in another tube and add 400 μl of RNase buffer. Incubate at 37°C for 1 hour.
11. Add equal volume of isopropanol. Incubate at 4°C for 30 min.
12. Centrifuge at 10,000 rpm for 15 min. at 4°C.
13 Discard the supernatant carefully. Wash the pellet with ethanol (70%).
14. Dry the pellet and dissolve in 5 ml of TE buffer.

C. Purification by phenol / chloroform extraction

Phenol denatures proteins and also removes RNA. Chloroform extracts lipids and also phenol from nucleic acid preparations. Isoamy-lalcohol stops frothing and helps in separation of layers. Phenol is distilled at 160°C to remove oxidation products, such as quinones, that cause breakdown of phosphodiester bonds.

Procedure

1. Take DNA dissolved in TE buffer and add equal volume of phenol-chloroformisoamyl alcohol.
2. Cap the tube tightly and vortex.
3. Centrifuge at 10,000 rpm for 5 min at 4°C.
4. Transfer the aqueous phase in a fresh tube and repeat the process.
5. Add equal volume of chloroform-isoamylalcohol to aqueous phase taken in fresh tube. Vortex and Centrifuge.
6. Remove the phenolic/chloroform phase.
7. Add proteinase K to a final concentration of 50 mg/ml.
8. Incubate at 37°C for 45 min.
9. Add one fifth volume of 3 M sodium acetate (pH 4.8)
10. Add double volume of absolute ethanol. Keep at -20°C for overnight. Centrifuge at 10,000 rpm for 15 min at 4°C.
11. Discard the supernatant. Wash the pellet with ethanol (70%).
12. Dry the pellet and resuspend in 2 ml TE buffer.

3.1.1.2 Isolation *E. coli* genomic DNA

Procedure

1. Transfer a single colony from given strain *of E. coli* in 10 ml LB media. Add ampicillin to a final concentration of 100 μg/ml. Incubate at 37°C (200 rpm) for 15–16 hours.
2. Incubate 1 litre LB with 10 ml grown culture. Incubate at 37°C (200 rpm) for 15–16 hours.
3. Harvest the bacterial cells at 4000 rpm for 15 minutes at 4°C. Discard the supernatant.
4. Thoroughly resuspend the cell pellet with 20 ml of lysis buffer. Add 1 ml of lysozyme solution. Incubate at 37°C for 5–10 min.
5. Add 40 ml of lytic mixture (1% SDS).

6. Mix the contents throughly by gently inverting the tube several times.
7. Add 30 ml of ice cold 3M Sodium acetate (pH 4.8) buffer. Shake the mixture.
8. Incubate on ice for 20 min.
9. Centrifuge the bacterial lysate at 8000 rpm for 15 min at 4°C.
10. Add 400 μl of RNase buffer. Incubate at 37°C for 1 hour.
11. Extract with phenol-chlorofrom-isoamyl alcohol twice and chloroform-isoamyl alcohol once as done in plasmid DNA isolation.
12. Add proteinase K to a final concentration of 50 μg/ml. Incubate at 37°C for 45 min. Add double volume of absolute ethanol and keep at -20°C for overnight.
13. Centrifuge at 10,000 rpm for 15 min. at 4°C.
14. Discard the supernatant. Wash the pellet with ethanol (70%). Dry and resuspend in 2 ml TE buffer.

3.1.1.3 Isolation of bacteriophage λ DNA

For phage DNA isolation the phage particles can be obtained in large numbers from the extracellular medium of an infected bacterial culture. Such a culture can be centrifuged, the bacteria pelleted, leaving the phage particles in suspension and their DNA extracted by a single deproteinization step for removing the phage capsid. The bacterial cells are killed by chloroform treatment and phage particles are separated by centrifugation. The crucial step is however, to grow an infected culture in such a way that the extracellular phage titre is high. Large culture volumes are needed if sufficient quantities of phage DNA are required.

Preparation of Plating Bacteria

1. Take 50 ml of autoclaved LB media supplemented with maltose (0.2%) in a sterile 250-ml flask and inoculate with a single bacterial colony. Grow the culture overnight at 37°C with slow agitation (250 rpm in incubator shaker).
2. Centrifuge the cells at 4000g for 10 minutes at room temperature.
3. Discard supernatant and resuspend the cells by vortexing in a 0.01 M $MgSO_4$ (~20 ml) to an approx. density (OD_{600} = 2, *i.e.*, ~1.6×10^9 cells/ml).
4. Store the cell suspension at 4°C till further use.

Plating Bacteriophage λ

1. Prepare tenfold serial dilution of bacteriophage stocks in SM.
2. Dispense 0.1 ml of each dilution to be assayed into a sterile tube.
3. Add 0.1 ml of plating bacteria to each tube. Mix by shaking or vortexing. Incubate for 20 minutes at 37°C to allow the bacteriophage particles to adsorb to the bacteria.
4. Add 3 ml of molten (47°C) agar or agarose (0.7%) to the first tube, vortex it gently, and immediately pour the entire contents to a petri-plate containing 30–35 ml of hardened bottom agar medium (LB).
5. Allow the top agar/agarose to harden at room temp. Invert the plates and incubate at 37°C. Plaques begin to appear after about 7 hours of incubation and should be counted or picked after 12–16 hours.

Picking Bacteriophage λ Plaques

1. Place 1 ml of SM in a polypropylene tube. Add 1 drop of chloroform.
2. Using a micropipette, stab through the chosen plaque into the hard agar beneath. Apply mild suction so that the plaque, together with the underlying agar is drawn into the pipette.
3. Place the fragments of agar in the SM/chloroform prepared in step 1 and stand at room temperature for 1–2 hours to allow the bactoriophage particles to diffuse out of the agar.

Preparation of stocks of bacteriophage λ from single plaques

To achieve maximum yield, an inoculum of 10^5 pfu is usually sufficient to produce confluent lysis of a bacterial lawn.

1. Mix 10^5 pfu of bacteriophage with 0.1 ml of plating bacteria. Incubate for 20 minutes at 37°C.
2. Add 3 ml of molten top agar/agarose (0.7%), mix, and pour immediately onto a plate containing 30–35 ml of hardened agar.
3. Incubate the plate for 6–8 hours at 37°C. At the time of harvesting, the plaques should touch one another.
4. Add 5 ml of SM and store at 4°C for several hours with intermittent, gentle shaking.
5. Harvest the SM using a pasteur pipette and place in a polypropylene tube.
6. Add 0.1 ml of chloroform, vortex briefly, and centrifuge at 4000g for 10 minutes at 4°C.

7. Recover the supernatant, add 1 drop of chloroform, and store at 4°C. The titer of the stock should be approximately 10^{10} to 10^{11} pfu/ml.

Purification of Bacteriophage λ

The bacteriophage DNA is purified by adding pancreatic DNAse 1 and RNAse. To this solid NaCl is added to a final concentration of 1M. (NaCl promotes dissociation of phage particles from bacterial debris and is required for precipitation of phage particles from PEG, added to a final concentration of 10% w/v). The precipitated phage particles are obtained by centrifugation at 11,000g for 10 min at 4°C.

Extraction of bacteriophage λ DNA

1. Transfer the phage suspension to another centrifuge tube. Add EDTA from a 0.5 m stock solution (pH 8.0) to a final concentration of 20mm.
2. Add pronase to a final concentration of 0.5 mg/ml, or add proteinase K to a final concentration of 50 µg/ml.
3. Add SDS from a stock solution (10% w/v) in water to a final concentration of 0.5 percent. Mix by inverting the tube several times.
4. Incubate for 1 hour at 37°C (pronase) or 56°C (proteinase K).
5. Cool the digestion mixture to room temperature. Add an equal volume of phenol equilibrated with 50 mm Tris (pH 8.0). Mix by inverting the tube many times until a complete emulsion has formed. Separate the phases by centrifugation at 3000g for 5 minutes at room temperature.
6. Extract the aqueous phase once with a 50:50 mixture of equilibrated phenol and chloroform.
7. Recover the aqueous phase and extract once with an equal volume of chloroform.
8. Add sodium acetate from a 3 M stock solution (pH 7.0) to a final concentration of 0.3 M. Mix well.
9. Add 2 volumes of ethanol and mix. Store the solution at room temperature for 30 minutes. The phage DNA can be seen as a threadlike precipitate, which is removed from the solution on the outside of a pasteur pipette.
10. Transfer the DNA to a microfuge tube containing 1 ml of ethanol (70%).
11. Recover the DNA by centrifugation at 12,000g for 2 minutes at 4°C in a microfuge.

12. Discard the supernatant carefully, and allow the pellet of DNA to dry at room temperature.
13. Redissolve the DNA in an appropriate volume of TE (pH 7.6).

Precautions

1. Air bubbles should be avoided while preparing plates for plaque formation.
2. Gently swirl the plate to obtain an even distribution of bacteria and top agar/agarose for plaque formation.
3. Sterilization conditions should be maintained at all steps of the experiment.
4. Reagents and solutions containing viruses should be discarded properly after autoclaving.

3.1.1.4 Extraction of plant DNA using CTAB

DNA extraction from plant tissue can vary depending on the material used. Essentially any mechanical means of breaking down the cell wall and membranes to allow access to nuclear material, without its degradation is required. For this, usually an initial grinding stage with liquid nitrogen is employed to break down cell wall material and allow access to DNA, while harmful cellular enzymes and chemicals remain inactivated. After grinding the tissue, it is resuspended in a suitable buffer, such as Cetyltrimethylammonium bromide (CTAB). In order to purify DNA, insoluble particulates are removed by centrifugation, while soluble proteins and other material are separated through mixing with chloroform and centrifugation. DNA must then be precipitated from the aqueous phase and washed thoroughly to remove contaminating salts. The purified DNA is then resuspended and stored in TE buffer.

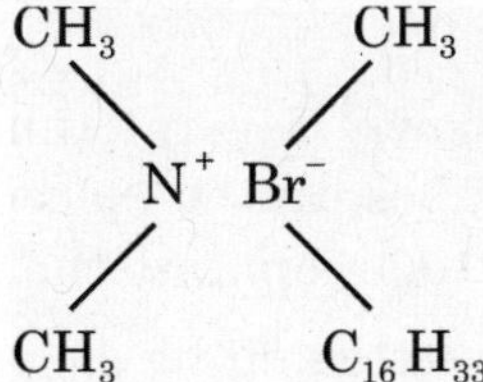

Structure of CTAB $[(C_{16}H_{33})N(CH_3)_3Br]$

CTAB is a cationic detergent that has the useful property of precipitating nucleic acids and acidic polysaccharides from solutions of low ionic strength. Under these conditions, proteins and neutral polysaccharides remain in solution. In solutions of high ionic strength, CTAB forms complexes with proteins and all, but the most acidic polysaccharides, but will not precipitate nucleic acids. CTAB is therefore, particularly useful for purification of DNAs from organisms that produce large quantities of polysaccharides, *e.g.,* plants. For preparation of genomic DNAs, the detergent is added to bacterial or cell lysates that have been adjusted to high ionic strength (>0.7 M NaCl). After removing the CTAB/polysaccharide/protein complexes by sequential extraction with chloroform and phenol, the genomic DNA is recovered from the supernatant by precipitation with isopropanol or ethanol.

Most commercial preparations of CTAB are mixtures of trimethylammonium bromides with varying lengths of aliphatic tails and about 80 percent of a typical preparation consists of the cetyl form (M = 364.48). Because CTAB precipitates in the cold, solutions containing the detergent should be stored at temperatures >15°C.

Procedure

1. Grind 200 mg of plant tissue to a fine paste in approximately 500 µl of CTAB buffer.
2. Transfer CTAB/plant extract mixture to a microfuge tube.
3. Incubate the CTAB/plant extract mixture for about 15 min at 55°C in water bath.
4. Centrifuge the CTAB/plant extract mixture at 12000 g for 5 min to precipitate cell debris. Transfer the supernatant to another microfuge tube.
5. Add 250 µl of Chloroform: Isoamyl alcohol (24:1) and mix the solution by inversion. After mixing, spin the tubes at 13000 rpm for 1 min.
6. Transfer the upper aqueous phase only (contains the DNA) to a clean microfuge tube.
7. Add 50 µl of 7.5 M Ammonium acetate followed by 500 µl of ice cold absolute ethanol.
8. Invert the tubes slowly several times to precipitate the DNA. Alternatively the tubes can be placed for 1 hr at -20°C after the addition of ethanol to precipitate the DNA.
9. The DNA can be pipetted off by slowly rotating/spinning a tip in the cold solution. The precipitated DNA sticks to the pipette and is visible as a clear thick precipitate.
10. To wash the DNA, transfer the precipitate into a microfuge tube containing 500 µl of ice cold ethanol (70%) and slowly invert the tube.
11. Repeat this step (alternatively the precipitate can be isolated by spinning the tube at 13000 rpm for a minute to form a pellet.

Remove the supernatant and wash the DNA pellet by adding two changes of ice cold ethanol (70%)).

12. Centrifuge the DNA at 13000 rpm for 1 min. Remove all the supernatant and allow the DNA pellet to dry.
13. Resuspend the DNA in sterile DNase free water (approx. 50–400 µl H_2O; the amount of water needed to dissolve the DNA can vary, depending on the amount isolated). RNaseA (10 µg/ml) can be added to the water prior to dissolving the DNA to remove any RNA in the preparation (10 µl RNaseA in 10ml H_2O).
14. After resuspension, the DNA is incubated at 65°C for 20 min to destroy any DNases that may be present and store at 4°C.
15. The isolated DNA can be quantified and purity can be checked either spectrophotometrically or by agarose gel electrophoresis of the DNA.

Precautions

1. For collecting the precipitated DNA, a tip should be gently rotated in the solution.
2. For transfer of upper aqueous phase, care should be taken so that the precipitated proteins do not get transferred

3.1.1.5 Isolation of mammalian DNA

Mammalian DNA may be isolated either from cell monolayers, cell suspension cultures or from mammalian tissue. If the DNA is to be isolated from cell monolayers then the cells will have to be scraped from the monolayers and suspended in TBS and then cell suspension taken for DNA isolation. In case of cells growing in suspension culture the DNA isolation can be proceeded directly after centrifugation to collect the cells, washing and then proceed for DNA isolation. For isolation of DNA from animal tissue, the tissue has to be first crushed in liquid nitrogen and then extracted in extraction buffer. Since animal cells are devoid of cell wall mere lysis of the cell membrane is sufficient to obtain the cell extract.

3.1.1.5.1 Isolation of genomic DNA from mammalian tissue

Procedure

1. Inoculate the bacterial culture in 10 ml of medium and incubate overnight with shaking at appropriate temperature.
2. Centrifuge at 5000 rpm for 15 min. and resuspend the pellet in 400 µl of TE buffer. Alternatively resuspend 1–3 loopful of cells from a plate in 400 µl of TE buffer.

3. Add 15 ml of freshly prepared lysozyme (10mg/ml).
4. Vortex briefly and incubate at 37°C for 30 minutes.
5. Add 5 μl of Proteinase K (10 mg/ml) and 20 μl of SDS (20%).
6. Mix and incubate at 55°C for 30 minutes.
7. Extract DNA once by adding equal volumes of phenol: chloroform : isoamyl alcohol (25 : 24 : 1). Mix contents and spin for 15 min at 10 000 rpm.
8. Remove the aqueous layer and extract once with chloroform: isoamyl alcohol (24 : 1). Centrifuge for 15 minutes at 10 000 rpm for 15 minutes.
9. Repeat the above procedure until there is no white precipitate visible at the interphase.
10. Add 0.1 volume of 3M sodium acetate, pH 5.2 and add 0.7 volume of isopropanol. Mix gently and spool out DNA with a sterile tip of centrifuge to pellet DNA. If DNA is not visible, centrifuge at 10 000 rpm for 20 minutes.
11. Wash pellet with 200 μl of chilled ethanol (70%).
12. Air dry the pellet and resuspend in 50 μl of TE buffer.
13. Add 5μl of RNase (10 mg/ml) and incubate at 37°C for 20 min.
14. Quantify or check the quality of DNA on agarose gel (1%).
15. Store at 4°C till further use.

Precautions

1. Tissue or blood samples may contain infectious material, *e.g.*, viruses, therefore, such samples should be handled appropriately by wearing gloves.
2. Use mechanical pippeting devices.
3. Work in a biosafety cabinet.
4. All waste material should be disinfected before disposal.

3.1.1.5.2 Isolation of genomic DNA from blood

Principle

The basic principle of extraction is that all other components of the chromatin are removed leaving behind the DNA. The proteins are digested by the enzyme Proteinase-K in the presence of SDS, which act as catalyst and also help in lysis of cells (WBC, bacterial cells). Sequestration of magnesium ions by EDTA serves to inhibit nuclease activity; phenolchloroform is used to remove the protein from the aqueous phase. Chloroform eliminates any traces of phenol, as phenol can cause phosphodiester breakage. Also the pH of phenol should be maintained above 7.8 as all eukaryotic RNA with poly-A tails dissolve in alkaline phenol, but in the acid range, the DNA will portion into the organic phase. The DNA is finally precipitated with alcohol or isopropanol from a salt solution of moderate concentration of monovalent cations. An average of 400–500 g of DNA can be extracted from 100 ml of blood.

This method of genomic DNA isolation from blood samples by phenol: chloroform extraction method is based on the method described by Sam brook and Russell (2001) with slight modification.

Procedure

For DNA isolation from blood, fresh blood should be collected in tubes containing anticoagulants such as acid citrate dextrose. For fresh blood centrifuge and discard the supernatant plasma, whereas for frozen blood, thaw the blood in a water bath at room temperature, dilute with PBS, centrifuge and proceed with the pellet containing lysed cells.

1. Take the blood samples, stored at -20°C and thaw at room temperature.
2. Thawed blood samples are centrifuged at 3000–3500 rpm for 15–20 min at room temperature.
3. Discard the reddish tinged supernatant containing plasma and lysed RBC by pipetting.
4. Mix the remaining cells containing RBC and WBC with more than two volumes of chilled RBC lysis buffer (1X), mixed end to end, incubated in ice for 10min and centrifuge at 3000–3500 rpm for 15–20 min at room temperature.
5. Again discard the reddish tinged supernatant containing plasma and lysed RBC by simple inversion of the tubes or by pipetting.
6. Repeat the step 4 and 5 till the WBC pellet appears nearly white in colour.
7. Add 3 ml of extraction buffer (at the rate 3 ml/10 ml of blood), tap to disperse the WBC pellet in extraction buffer (it was insured that the pellet is completely suspended so that the cells are accessible to SDS and Proteinase K).
8. Add 200 µl of 10 percent SDS (at the rate 200 µl/10 ml of blood) and mix the tubes gently by inverting. The content of the tubes will appear viscous indicating lysis of WBC.

9. Add 25 μl of Proteinase K (at the rate 25 μl of 25 mg Proteinase K/ml of triple distilled water for 10 ml blood) in two pulses *i.e.*, half the requirement was added to the tube in first pulse, mix gently end to end and then keep in water bath at 50°C. After 3–4 hours add the second pulse of remaining amount of Proteinase K and incubate tubes were 50°C overnight.
10. Add equal amount of equilibrated phenol (pH>7.8), and mix by inverting gently for 15 min till a light coffee coloured uniform solution, without any balls of phenol is formed and centrifuge at 3000–3500 rpm for 15–20 min.
11. Transfer the upper aqueous phase containing DNA to fresh 15 ml, clean, sterile and autoclaved propylene tubes using a 1 ml tip whose tip has been cut to widen the bore.
12. Repeat extractions (as in step 10–11) once with equal volume of phenol:chloroform:Isoamyl alcohol (25:24:1) and with chloroform: isoamyl alcohol (24:1).
13. To the final aqueous phase obtained, add 3M sodium acetate (at the rate 100 μl/ml (pH 5.2) of aqueous phase and mix gently.
14. Add two volumes of absolute ethanol (chilled) or isopropanol (at room temperature), mix by inversion and keep at room temperature to allow precipitation of DNA.
15. Transfer DNA along with 500 μl of isopropanol to fresh centrifuge tubes and centrifuge at 10000 rpm for 10 min.
16. Discard the supernatant by gentle inversion or aspirate.
17. Wash the DNA pellet twice in 500 μl of ethanol (70%).
18. Finally air dry the DNA pellet in the incubator after inverting on blotting paper so that last traces of ethanol are evaporated.
19. Add about 200 μl of TE buffer and put in water bath at 60°C for 2 hours to inactivate DNase and other enzymes.
20. Store the DNA samples in microcentrifuge tube at 4°C for a week so that DNA is dissolved and subsequently store at -20°C, indefinitely.

Precautions

1. Maintain sterilized conditions.
2. Care should be taken while transferring the upper aquous layer containing DNA after phenol treatment.
3. Do not over dry the final pellet of DNA.

3.1.1.6 Isolation of chloroplast DNA

The only cell organelles having their own DNA (extranuclear DNA) are chloroplast and mitochondria. The genomes of chloroplast and mitochondria are closed circular DNA molecules containing genes coding for many enzymes necessary for the function of the organelles. Isolation of chloroplast DNA is difficult compared to nuclear DNA isolation because apart from the disruption of the cell membrane and cell wall, the chloroplast membrane has to be disrupted to release the chloroplast DNA.

The present procedure given by Triboush *et al.* (1998) is easy to perform, inexpensive and requires short-time. The protocol includes organelle isolation, DNase treatment, lysis, deproteinisation and a final DNA purification with sodium dodecyl sulphate and potassium acetate. Same buffers can be used for chloroplast and mitochondrial DNA isolation. The organelle DNA yield is 5–0 μg/gram of tissue. Further, this procedure does not require CsCl/sucrose/percol gradient ultracentrifugation and gives restrictable chloroplast DNA free from mitochondrial DNA.

All the steps are to be carried out at 4°C unless otherwise stated.

Chloroplasts Isolation

1. Prepare STE buffer, add BSA and β-mercaptoethanol before use.
2. Green seedlings (5 gm), kept in dark for 48 hrs (to reduce the starch level) are harvested, thoroughly rinsed, cooled to 0°C and ground in a cooled mortar with 20 ml of STE buffer.
3. Filter the homogenate twice through four layers of muslin cloth.
4. Centrifuge at 1000 rpm (200xg) for 20 min.
5. Discard the nuclei pellet and centrifuge the supernatant at 4000 rpm (3700× g) for 20 min.
6. Resuspend the resulting chloroplast pellet in same buffer using a soft paint brush and centrifuge at 4000 rpm (3700x g) for 20 min.

DNase Treatment

1. Re-suspend the organelle pellet in 0.2ml of ST buffer.
2. Prepare DNAse by dissolving 125μg in 0.2 ml of ST buffer with 0.02 M Mg-acetate and add to above suspension and adjust the final volume to 0.5 ml (final DNase concentration is 250μg/ml).
3. Incubate at 37°C for 20 min with periodic mixing.
4. Add EDTA-Na_2 to a final concentration of 0.2 M to stop the reaction.

5. Wash the organelle in 25–50 ml of NETF buffer and collect the pellet by centrifugation.
6. Lyse immediately or keep the pellet at -20°C.

Lysis and Purification

1. Resuspend the pellet in 600µl of TEN buffer at 0°C and add 30 ml of SDS (20%) by drops, shaking the mixture constantly.
2. Warm the tubes at 60°C in water bath for full lysis of the organelles.
3. Remove proteins by adding an equal volume of equilibriated phenol and shake the mixture for 5 min.
4. Centrifuge at 12,000 rpm for 10 min.
5. Remove the upper aqueous phase.
6. Add phenol:chloroform (1:1) and centrifuge as above.
7. Add chloroform:isoamyl alcohol (24:1) and centrifuge as above.
8. Precipitate DNA with 1/10 vol. of 5M ammonium acetate and 1 vol. of isopropanol at -20°C for 2–3 hrs or overnight.
9. Centrifuge for 3–5 min. at 12,000 rpm.
10. Wash the DNA pellet with ethanol (70%) and then with pure ethanol to remove residual salt.
11. Dry the pellet at room temp and re-dissolve in TE buffer.
12. Add 150 µl of TEN buffer after the DNA is completely dissolved.
13. Incubate the DNA solution at 65°C for 15 min. with 10 µl of SDS (20%).
14. Add 50 µl of 5 M potassium acetate. Shake mixture at 65°C till the precipitate dissolves.
15. Incubate at -20°C for 30 min and centrifuge at 12,000 rpm for 10 min.
16. Collect supernatant carefully in another tube and pour it into 10 µl of 5M ammonium acetate and 100 µl of isopropanol. Mix and keep at -20°C for 30 min. to precipitate the DNA.
17. Centrifuge at 12,000 rpm for 10 min to pellet the DNA.
18. Wash the pellet with 70 percent and then with 96 percent ethanol, dry at room temp. and dissolve in TE buffer.

The isolated DNA can be electrophoresed as given under section 3.1.1.8. or used for restriction enzyme digestion.

3.1.1.7 Isolation of mitochondrial DNA

Mitochondrial DNA was discovered in the 1960s by Margit M.K. Nass and Sylvan Nass by electron microscopy as DNase-sensitive thread inside mitochondria. In humans, 100–10,000 copies of mtDNA are usually present per cell. In mammals, each double-stranded circular mtDNA molecule consists of 15,000–17,000 base pairs. The two strands of mtDNA are differentiated by their nucleotide content with the guanine rich strand referred to as the heavy strand, and the cytosine rich strand referred to as the light strand. The heavy strand encodes 28 genes and the light strand encodes 9 genes for a total of 37 genes. Of the 37 genes, 13 are for proteins (polypeptides), 22 are for transfer RNA (tRNA) and two are for the small and large subunits of ribosomal RNA (rRNA).

In most multicellular organisms, mtDNA is inherited from the mother (maternally inherited). In sexual reproduction, mitochondria are normally inherited exclusively from the mother. The fact that mitochondrial DNA is maternally inherited enables researchers to trace maternal lineage. Since mtDNA is not highly conserved and has a rapid mutation rate, it is useful for studying the evolutionary relationships - phylogeny of organisms.

Mitochondrial DNA is involved in many human diseases, *e.g.*, exercise intolerance and Kearns-Sayre syndrome (KSS), Diabetes mellitus and deafness (DAD), Leber's hereditary optic neuropathy (LHON), Leigh syndrome, Neuropathy, ataxia, mitochondrial myopathy, encephalomyopathy etc. are some other mitochondria related disorders.

Procedure

Plant material: This procedure has been tested on different tissues, such as tubers, young leaves, flowers and flower buds.

Preparation of mitochondria

The crucial step is to isolate intact organelles and lyse the nuclei in a medium of high ionic strength (Bookjans *et al.,* 1984). Plastids are eliminated by low speed centrifugations. All steps must be carried out at 4°C either in the cold room or on ice.

1. Portions (1–15 g) of different fresh plant tissues (leaves, flowers, flower buds) are homogenized in a waring blender, for a few seconds at moderate speed, using 5–75 ml of cold Homogenization Buffer A in a cold room.
2. To eliminate plastids, the supernatant was centrifuged twice at 2600 *g* for 15 and 10 min, respectively.

3. To sediment mitochondria, the supernatant is centrifuged at 14,500 *g* for 15 min.
4. Discard supernatant containing lysed nuclei components.
5. Resuspend the mitochondrial pellet in 10 to 30 ml of cold Homogenization Buffer A.
6. Centrifuged at 14,500 *g* for 15 min. This mitochondrial pellet can be stored at -80°C after freezing in liquid nitrogen or directly used for mtDNA isolation.

Mitochondrial DNA isolation

1. To isolate mtDNA, the mitochondria are lysed with a detergent and the resulting mtDNA is dissociated from the proteins by a proteolytic agent.
2. Resuspended the mitochondrial pellet in 0.5–1 ml of Lysis Buffer.
3. Proteinase K is added to a final concentration of 50 mg/ml and the suspension is incubated at 37°C for 1 h with gentle shaking to avoid mechanical breaking of DNA.
4. After the addition of 0.1 volume of 2M ammonium acetate, the nucleic acids are extracted with an equal volume of TE-saturated phenol/chloroform (50:50).
5. Nucleic acids are precipitated by the addition of 2 volumes of ethanol (100%) and overnight incubation at -20°C.
6. After centrifugation at 18,000 *g* for 20 min, the pellet is re-dissolved in 10 to 50 ml of water.
7. At this step, although RNA is still present, the DNA is suitable for restriction enzyme analysis or PCR experiment. If necessary, RNAse A followed by phenol extraction and ethanol precipitation can be performed.

Precautions

1. All steps must be carried out at 4°C.
2. Lysis of nuclei should be done in a medium of high ionic strength.
3. While discarding the supernatant containing lysed nuclei, care should be taken not to disturb the mitochondrial pellet.

3.1.1.8 Agarose gel electrophoresis of DNA

Nucleic acid can be separated electrophoretically according to their size. This separation is done on agarose or polyacrylamide gels. Move-ment of charged particles in an electric field is called electrophoresis. Thus electrophoresis is migration of a charged molecule in an electric field in the direction of an electrode so that cations (positively charged molecules) move in the direction of cathode (negative electrode) and anions (negatively charged molecules) move in the direction of anode (positive electrode). When an electric field is applied to an agarose gel in the presence of a buffer solution which will conduct electricity, DNA which is highly negatively charged at neutral pH, will migrate towards the positive electrode at rates determined by their molecular size and conformation. Electrophoresis with agarose as the support medium is the standard method for the characterization of RNA and DNA of 200–50,000 bp. Agarose is extracted from seaweed and is a linear polymer of D-galactose and 3,6-anhydro-L-galactose. Porosity of the agarose gels is determined by the concentration of the agarose. Higher the agarose concentration, smaller is the pore size and lesser the agarose concentration, larger the pore size. Nucleic acids can be visualized on agarose gels after separation by soaking in a solution of ethidium bromide, a dye that fluoresces under UV light and intercalates between stacked nucleic acids. Most commonly agarose concentrations of 0.3–2 percent are used for nucleic acid separation. The movement of a charged molecule subjected to an electric field is given by,

$$v = \frac{Eq}{f}$$

where,

E = the electric field in volts/cm

q = the net charge on the molecule

f = frictional coefficient, depending on the mass and charge of the molecule

v = the velocity of the molecule.

The movement of charged molecules in an electric field is often expressed in terms of electrophoretic mobility (M) which is defined as velocity per min. of electric field.

$$\mu = v/E = E.q/\ f.E\ (\text{since} = E.q\backslash\ f) = q/f$$

Thus, the electrophoretic mobility (μ) of a molecule is directly propotional to the charge density (charge\mass ratio). Therefore, molecules with different charge\mass ratio migrate at different rates under an electric field and get separated.

The charge\mass ratio of nucleic acids in one, therefore, rate of migration of DNA is inversely proportional to $\log_{10}$ of their molecular weights. Thus smaller DNA molecules will migrate at faster rates than larger ones.

Factors affecting the rate of migration in agarose gel:

1. Molecular size of DNA: Migration rate of linear ds DNA is inversely proportional to $\log_{10}$ of the number of base pairs.
2. Agarose concentration: There is a linear relationship between the logarithm of the electrophoretic mobility of DNA (μ) and the gel concentration (x).

$$\text{Log } \mu = \log \mu_0 - K_r x$$

3. Conformation of DNA: Rate of movement is based on the conformation of DNA as follows,

   Superhelical form > nicked circular form > linear DNA
4. Applied Voltage: At low voltage rate of migration of linear DNA fragment is proportional to applied voltage.
5. Presence of intercalating dye: EtBr reduces the electrophoretic mobility of linear DNA by about 15 percent.
6. Composition of electrophoretic buffer: DNA migrates very slowly in presence of low ionic strength buffer, while in high ionic strength buffer heat is generated which may melt the gel.

Procedure

1. Prepare sufficient amount of electrophoresis buffer to fill the tank (600 ml).
2. Weigh 1 gm agarose and dissolve in 100 ml 1X TAE buffer (1% agarose solution).
3. Heat the contents in boiling water bath till agarose dissolves.
4. Allow the contents to cool to 60°C.
5. Add 3–5 µl of EtBr to the gel.
6. Pour the agarose solutions into the mould.
7. Gel should be 3–5 mm thick, allow the gel to set.
8. Remove the comb from the gel.
9. Pour the buffer into the electrophoresis tank. The level of buffer should be 1 ml above the gel.

Loading of samples

1. Add 6 µl of 6X Sample Loading Buffer to each sample (2:4).
2. Loading buffers are added with the DNA in order to visualize it and sediment it in the gel well. Negatively charged indicators keep track of the position of the DNA. Xylene cyanol and Bromophenol blue are typically used. They run at about 500bp

and 300bp, respectively, but the precise position varies with percentage of the gel. Other less frequently used progress markers are Cresol Red and Orange G which run at about 125bp and 50bp.

3. Record the order of each sample that will be loaded on the gel, the DNA template – what organism the DNA came from, controls and ladder.
4. Carefully pipette 5–15 µl of each sample/sample loading buffer mixture into separate wells in the gel.
5. Pipette 1–2 µl of the DNA ladder standard into at least one well of each row on the gel.

Running of the gel

1. Place the lid on electrophoresis chamber, connecting the electrodes.
2. Connect the electrode wires to the power supply, making sure the positive (red) and negative (black) are correctly connected.
3. Turn on the power supply to about 100 volts. Maximum allowed voltage will vary depending on the size of the electrophoresis chamber – it should not exceed 5 volts/cm between electrodes.
4. Check to make sure the current is running through the buffer by looking for bubbles forming on each electrode.
5. Check to make sure that the current is running in the correct direction by observing the movement of the blue loading dye. This will take a couple of minutes (it will run in the same direction as the DNA).
6. Let the power run until the blue dye approaches the either end of the gel.
7. Turn off the power.
8. Disconnect the wires from the power supply.
9. Remove the lid of the electrophoresis chamber.
10. Using gloves, carefully remove the tray and gel.

Visualizing the gel

Using gloves, remove the gel from the casting tray. As EtBr was added to the gel that will intercalate with the DNA bands and make them fluoresce, DNA bands can be visualized using UV illuminator. Care must be taken to avoid UV exposure to the body. A Gel Documentation system can also be used to visualise the DNA bands and a monitor may be connected to the system that will make easy data recording and storage.

Precautions

1. Gel should be completely immersed in TAE buffer.
2. Proper loading of samples should be done and the wells should not over flow.
3. Gel should be run at proper voltage and current.
4. Gloves should be worn when handling ethidium bromide.
5. Proper protection of eyes should be taken when visualizing the DNA bands under UV light.

3.1.2 RNA Isolation

Obtaining high quality, intact RNA is the first and often the most critical step in performing many fundamental molecular biology experim-ents, including Northern analysis, nuclease protection assays, RT-PCR, RNA mapping, *in vitro* translation and cDNA library construction. To be successful, however, the RNA isolation procedure should include some important steps both before and after the actual RNA purification.

3.1.2.1 Isolation of total RNA

Ribonucleic Acid (RNA) is the transcriptional product of DNA. The total RNA includes three classes–ribosomal, messenger and transfer RNA. Ribonucleic acid occurs as ribonucleoprotein particle in intact cells. There are a number of methods described for the extraction of total RNA. The method selected largely depends upon the source material and the experimental use for which RNA is extracted. Among the two common methods used with plant materials, the phenol-chloroform method is used to recover the total RNA intact. The phenol-SDS method described below is useful for quantitative recovery. The ribonucleoprotein complex is dissociated by SDS into RNA and protein deproteinized by phenol and the free RNA left in aqueous solution is precipitated in the cold after adding alcohol. All glassware should be baked at 120°C overnight. Contamination with any nuclease should be strictly avoided.

Procedure

All operations should be done at 0-4°C.

1. Freeze 0.5–5g of the material in a mortar and pestle with liquid N_2, grind to a fine powder, then to a paste and extract in 10 volumes of extraction buffer.
2. Centrifuge the homogenate at 2,000 x g for 3 min.
3. Transfer the supernatant to a volumetric flask and stir with 0.1 volume of SDS (10%) for 2–3 min.

4. Add an equal volume of buffered phenol (freshly redistilled phenol saturated overnight with 100mM Tris-HCL pH 8.5 as given in Appendix IVC)
5. Partition the content by centrifugation at 2,000 x g for 5 min and collect the upper aqueous phase into a separate flask.
6. Shake the lower and interphase again with an equal volume of extraction buffer for 5 min and centrifuge.
7. Combine the aqueous phase with the first one (step 5) and stir with an equal volume of buffered-phenol for 5 min.
8. Repeat the extraction and centrifugation steps at least five times or till there are no proteins at the interphase.
9. Finally, collect the upper aqueous phase containing RNA, dissolve in 250 mg NaCl, add two volumes of cold ethanol (96%) and leave the flask overnight at -20°C for RNA precipitation.
10. Collect RNA by centrifugation at 2,000 x g for 10 min. Wash the pellet with ethanol (70%), ethanol, ethanol:ether (1:1 v/v) and finally with ether. Dry pellet gently in vacuum for a few minutes.
11. Dissolve the RNA completely in elution buffer if further it is to be used for mRNA isolation.
12. Dilute 20μl aliquot to 2ml with buffer and read the absorbance at 260 nm.

Precautions

1. All the glassware and solutions should be sterile. Any contaminating RNase is inactivated by rinsing the glassware with diethyl pyrocarbonate (1%) solution.
2. Use freshly distilled colourless phenol. It can be stored frozen in small aliquots in brown-coloured bottles. Wear gloves while handling phenol.
3. A variety of extraction medium, chelating agents, deproteinizing agents etc. are used for the extraction of RNA.
4. The denatured proteins gather at the interphase after low speed centrifugation which should be discarded.
5. The volume of aqueous phase will be drastically reduced if the phenol is not fully saturated.
6. Any contaminating DNA will appear just like cotton wool during ethanol addition.
7. Deproteinization using phenol leads to the loss of poly (A)-tail of mRNA considerably.

3.1.2.2 Isolation of messenger RNA

Pure mRNA is required in many molecular biology experiments *e.g.,* RT-PCR, northern hybridization, *in vitro* translation, cDNA library construction, etc. Eukaryotic mRNA, *i.e.,* poly (A) RNA makes up about 1–5 percent of the total cellular RNA and the polyadenylate tail may vary in length from 3-200 residues. It is used for synthesis of probes, construction of random-primed cDNA libraries, detection and quantification of extremely rare mRNAs.

Under high ionic conditions, poly (A) tail of mRNA forms a duplex with the complementary base sequences, such as poly (U) or oligo (dT). The tRNA and rRNA do not form any duplex because they lack poly (A) tail. When the ionic condition is lowered the hydrogen bonding between poly (A) or poly(U) and oligo (dT) is broken and thus the poly (A) containing mRNA is released. This process is based on affinity chromatography in which poly (U) or oligo (dT) is immobilized *i.e.* covalently attached to sepharose or cellulose which is used as the stationary phase.

Now-a-days, oligo (dT)-columns are available which are designed for fast and reliable isolation of poly (A) fraction from total RNA. These columns contain oligo (dT) chains of about 30 nucleotides in length, covalently linked via the terminal 5'phosphate to the cellulose matrix. The matrix has a layer of an anti-clogging resin to maintain a high flow rate. Upto 50 μg of poly A+ RNA can be isolated from a single column from 1–1.5 mg of total RNA. Each column is reusable and can be used upto 10 times. The process is economical and can be completed in 15–20 min.

1. Suspend affinity material oligo (dT) cellulose or sepharose in sterile binding buffer according to instructions given in kits.
2. Prepare column and immediately before use, wash columns with 0.5 ml of binding buffer.
3. Dissolve the total RNA pellet in the binding buffer to a concentration of 1–4 mg RNA/ml.
4. Incubate the solution at 70°C for 5 min. This denatures the secondary structure of RNA and increases retention of poly A+ RNA on the column.
5. Apply the total RNA solution on the column.
6. Collect the elute and again apply it on the column.
7. Wash the column twice with many times with binding buffer or until the absorbance at 260 nm reaches zero. At this step the poly A+ containing mRNA is retained in the column while tRNA and rRNA will pass through.

8. Elute poly A+ RNA with 0.9 ml of the elution buffer into a 2 ml microfuge tube.
9. Wash the column with elution buffer and pre-equlibriate with binding buffer.
10. To the elute add LiCl or NaCl to a final concentration of 0.5M with 2–8 µl of polyacryl carrier.
11. Precipitate RNA by adding one volume of isopropanol or 2.5 volumes of ethanol and incubate for 5 min at room temperature.
12. Centrifuge the RNA-carrier precipitate at 12,000g for 5 min at 4°C.
13. Dissolve the pellet in DEPC-treated water and estimate isolated mRNA spectrophotometrically as given under section 3.1.2.3. or 2.5.4.
14. Finally, after elution of the polyA+ fraction regenerate the column. Wash column with 1.5 ml of SDS (1%), followed by 1 ml of ethanol (75%) and store column wet with ethanol at -20°C. This column can be reused.

Note: *All aqueous reagents and glassware should be made RNase-free as given in Appendix IV C. Ethanol can be prepared in DEPC-treated water.*

3.1.2.3 Agarose gel electrophoresis of RNA

RNA is a single stranded molecule transcribed from DNA. The extracted RNA contains rRNA (75%), tRNA (20%) and mRMA (2–3%). Messenger RNA molecules are heterogeneous in size. RNA molecules undergo aggregation due to single stranded nature and due to occurrence of complimentary stretches. It is therefore, necessary to relive the aggregates by denaturing before using them for further characterization, hence they are treated with denaturing agents, such as formaldehyde, urea etc. before they are separated from each other. Formaldehyde relieves the intra- and intermolecular aggregation of the RNA molecules by interaction.

Procedure

1. These molecules are then subjected to fractionation by electrophoresis under denaturing conditions in a manner similar to electrophoresis of DNA as given under section 3.1.1.8.
2. For gel preparation 5.4 ml of formaldehyde is added to agarose (1.2%) gels.

3. For loading buffer (1.5 ml) - Mix 0.72 ml formamide, 0.16 ml 10X electrophoresis buffer, 0.26 ml of formaldehyde and 0.1 ml glycerol. Mix 2–5 μl of loading buffer to the sample before loading.
4. Prior to loading the samples and after addition of loading buffer incubate the sample at 65°C for 3–5 minutes for denaturing the RNA.

Precautions

1. Formaldehyde vapours are harmful, therefore proper care must be taken while handling.
2. Always wear gloves while handling RNA.
3. All reagents, glasswares and plasticware should be made RNAse free.

3.1.2.4 Quantification of DNA/RNA

Nucleic acids can be quantified spectrophotometrically by reading the absorbance of the isolated DNA or RNA preparation at 260 nm. For this, dissolve a small amount of isolated DNA or RNA in TE buffer and read absorbance at 260 nm against TE buffer as blank. Quantify according to the formula:

1 O.D. corresponds to 50μg of double stranded DNA

1 O.D. corresponds to 40μg of single stranded RNA

The concentration of DNA and RNA can be calculated as follows:

i. For double stranded DNA:

$$\text{Concentration of DNA} = 50 \times A_{260} \times \text{Dilution factor in sample solution (mg\textbackslash ml)}$$

ii. For RNA:

$$\text{Concentration of RNA} = 40 \times A_{260} \times \text{Dilution factor in sample solution (mg\textbackslash ml)}$$

The amount of DNA can also be determined by the diphenylamine method as given under section 2.5.2 and RNA by Orcinol method as given under section 2.5.3.

3.1.2.5 Determination of DNA quality

The quality of DNA can be determined either by spectrophotometric determination or by visualizing the DNA by gel electrophoresis.

i. Spectrophotometric determination

Dissolve a small amount of isolated DNA in TE buffer and read absorbance at 260 and 280 nm against TE buffer as blank. The ratio of 260/ 280 should be 1.8–2.0 for a good quality pure DNA. Proteins are usually the major contaminants in nucleic acid extracts and these have absorpt-ion maxima at 280 nm. The ratio of absorbance at 260 and 280 nm, hence, provides a rough idea about the extent of contaminants in preparation.

However, if the 260/280 ratio is more than or less than 1.8 then it means that the DNA preparation is contaminated either with RNA, traces of proteins or with phenol.

ii. Gel electrophoretic determination

To confirm DNA isolation and quality gel electrophoresis can be performed as given under section 3.1.1.8. or as given below.

Procedure

1. Prepare a solution of agarose (1%) by melting 1 g of agarose in 100 mL of 0.5x TBE buffer in a microwave for approximately 2 min. Allow to cool for a couple of minutes then add 2.5 µl of ethidium bromide, stir to mix.
2. Cast a gel using a supplied tray and comb. Allow the gel to set for a minimum of 20 min at room temperature on a flat surface.
3. Load into separate wells 10 µl 1kb ladder and 5 µl sample + 5 µl water + 2 µl 6x Loading Buffer
4. Run the gel for 30 min at 100 V
5. Expose the gel to UV light and photograph (if necessary)

To confirm DNA quality, presence of a highly resolved high molecular weight band indicates good quality DNA, presence of a smeared band indicates DNA degredation.

3.2 DENATURATION AND RENATURATION OF DNA

3.2.1 Determination of Melting Temperature (Tm value)/Melting Curve of DNA

Principle

When a dilute aqueous solution of double stranded (ds) DNA is heated, the two strands get separated due to disruption of H-bonds between the complementary bases. Such DNA is referred to as denatured DNA and the process is called denaturation. Denaturation is a reversible process. If the above solution is allowed to cool slowly to room temperature, the complementary strands of DNA reassemble to give a duplex DNA. DNA is now said to be reannealed or renatured and process is known as reannealing or renaturation of DNA. Upon denaturation, the absorbance of DNA at 260 nm increases by 30–40% due to exposure of bases. This is called hyperchromic effect. Heat treatment of upto 80°C causes only a very slight increase in absorbance at 260 nm and this is followed by a sharp increase within a narrow temperature range (80–90°C) and then A_{260} become constant when DNA is fully denatured.

The heterocyclic rings of nucleotides absorb light strongly in the UV range (with a maximum at 260 nm). But the absorption of DNA is some 40 percent less than would be displayed by a mixture of free nucleotides of the same composition. This is called the hypochromic effects. Any departure from the duplex state is immediately neglected by adecline in the hypo-chromic effect, *i.e.*, by an increase in optical density towards the value characteristic of free bases. The denaturation can be followed by this hyperchromaticity.

A graph between absorbance at 260 nm and temperature gives a thermal denaturation curve or a melting curve of DNA. The temperature corresponding to the midpoint of the curve is called melting temperature (Tm) of DNA and it denotes the temperature at which 50 percent of the DNA has undergone denaturation.

Chemical Reagents

1. Saline sodium citrate: 0.015 M Sodium citrate + 0.15 M Sodium chloride, pH 7.
2. DNA solution: 50 μg of DNA/ml dissolved in sodium citrate.

Procedure

1. Dissolve 5 ml of DNA solution in 105 ml buffer saline.
2. Take 10 ml of this solution in two test tubes and heated upto 110°C.
3. Read absorbance at 260 nm at temperature 100°C, 90°C, 80°C, 70°C, 60°C, 50°C, 40°C, 30°C and 20°C.
4. Note the absorbance and plot the graph of absorbance v/s temperature.
5. Observe Tm value on the graph.

Precautions

1. Temperature should be maintained at constant value while taking absorbance.

2. While heating the DNA solution care should be taken.
3. The graph should be plotted carefully.

3.2.2. Determination of GC/AT (Base composition) Content of DNA

Principle

Tm value is largely dependent on the (G+C) content of DNA. Higher is the (G+C) content, higher will be the Tm. Under standard conditions of ionic strength and pH (G+C) content is related with Tm in accordance with the following equation.

$$\% \text{ (G+C)} = [\text{Tm} - 69.3] \times 2.44$$

Tm depends on:

i. Proportion of (G+C) content – Each GC pair has 3 H-bonds, it is more stable than an A=T base pair which has two hydrogen bonds. Tm increases 0.4°C with an increase in 1 percent (G+C) content.
ii. Tm is also affected by ionic strength of the solution. Tm increases with 16.6°C with an increase in every tenfold of monovalent cation.
iii. Tm is lowered by performing reaction in presence of reagents such as formamide because this stabilizes the H-bond. For DNAs whose G+C content is in the range of 30–75 percent, the Tm value is depressed by 0.63°C for each percentage of formamide.

Procedure

1. Dissolve 5 ml of DNA solution in 100 ml of buffer saline.
2. Take 10 ml of this solution in two test tubes and heat upto 110°C.
3. Now, at every decrease of 10°C note absorbance at 260 nm.
4. Plot graph of absorbance v/s temperature.

Calculations

If the Tm value is 80°C, then

$$\begin{aligned} \text{(G+C) \%} &= (\text{Tm} - 69.3) \text{ x } 2.44 \\ &= (80 - 69.8) \text{ x } 2.44 \\ &= 33.97\% \\ \text{(A+T) \%} &= 100 - \text{(G+C) \%} \\ &= 100 - 33.97\% \\ &= 66.03\ \% \end{aligned}$$

Precaution

1. Temperature should be noted down carefully.
2. While taking absorbance, temperature should be maintained at a constant value.

3.3 RESTRICTION DIGESTION

3.3.1 Restriction Digestion of DNA

Restriction enzymes recognize specific base sequences in DNA molecule and catalyze the hydrolysis of specific phosphodiester bonds in double stranded DNA. These enzymes have been used for the analysis of DNA and construction of restriction maps. Over 1200 restriction enzymes have been isolated from bacteria. Restriction enzymes are of three types, Type I, II and III. Type II restriction enzyme is most commonly used in genetic engineering since the recognition sites coincide with their cleavage sites. This enzyme recognizes and cuts within a sequence to generate blunt ends or cohesive ends. The Type II restriction enzymes require only Mg^{2+} as cofactors for their activity. The recognition sequences for these enzymes are palindromic sequences or have an axis of rotational symmetry. Since a specific enzyme recognizes a particular sequence of 4–8 bp, so the number of cuts will depend on number of sites present in the DNA molecule. The action of the restriction enzyme EcoRI which is a hexacutter enzyme is given below:

5'....G-G-A-T-C-C....3' EcoRI 5'....G + G-A-T-C-C....3'
3'....C-C-T-A-G-G....5' H_2O 3'....C-C-T-A-G G....5'

Requirements

Restriction enzyme of choice: EcoRI, Hind III

Restriction enzyme buffer (10 X).

DNA samples.

Procedure

1. Take 10 μg plasmid DNA and 10 μg chromosomal DNA in separate microfuge tubes. Add x μl. equivalent to 10 units of restriction enzyme of your choice. Set the digestion mixture as follows :

a.	Sterile water	45- (x + y μl)
b.	10 × R-buffer	5 μl
c.	Enzyme	x μl
d.	DNA	y μl
e.	Total volume	50 μl

2. Mix the tube by gentle tapping and incubate at 37°C for I hour.
3. To stop the reaction heat at 70°C for 3 minute or add 2μl of Quench buffer (0.1 M EDTA, pH 7).
4. Store the reaction mixture on ice until agarose gel electrophoresis.

Precautions

1. Restriction enzymes are very unstable reagents. They can be stored at -20°C in buffer containing glycerol (50%). They should never be stored at room temperature.
2. Incubation should be carried out at the recommended temperatures.
3. Reaction is stopped either by increasing the temperature or adding EDTA which complexes with divalent metal ions essential for nuclease activity.
4. Gloves should be worn when handling restriction enzymes.

3.3.2 Preparation of Agarose Gels and Running of Digested and Undigested DNA (Gel Electrophoresis)

Procedure

1. Weigh 1.5 g agarose powder, *i.e.,* 0.8% (0.16 gm in 20 ml of 1 x TAE buffer. Dilute the 10 x TAE twenty-fold to a final concentration of 1x TAE.
2. Add 150 ml of the 1× TAE to the agarose powder in the bottle and mix.
3. Melt in a microwave until all the powder is dissolved. Regularly take it from the oven, close the lid and give it a good shake. Carefully reopen the bottle and re-microwave.
4. Once everything is dissolved, close the lid tightly and cool the agarose down under a running tap of cold water to 55–60°C.
5. When it has cooled down, release the vacuum by opening the bottle, then add 7.5 μl of ethidium bromide (add ethidium bromide to 1 mg/ml final concentration), mix and carefully pour the gel into the chamber.
6. Tape the edges of a clean and dry template. Place the gel comb in position. Pour the bearably hot agarose solution on it without air bubbles. Allow it to set for 20 – 30 min.
7. After it has set, remove the wedges, or place the gel tray into the running tray. Remove the comb.

8. Add sample buffer to a final concentration of 1/6th of the total volume to the sample, mix and load on the gel or load the DNA samples (cut or uncut) mixed with 2 μl loading dye.
9. Mix 2.5 μl of a suitable Molecular ladder with 7.5 μl of water and 3μl loading buffer. Load all 13 μl of the marker DNA next to the samples.
10. Connect the terminals and switch on the current. Run the gel at high voltage (5–20 V/cm or at 120 volt) for 1 hour or until the bromophenol blue dye reaches the edge/two thirds of the gel.
11. Observe the DNA under UV light and photograph.

Precautions

Same as given under section 3.4.2.

3.4 POLYMERASE CHAIN REACTION (PCR)

PCR is a technique for the rapid amplification of DNA. This techni-que of DNA amplification was discovered by Karry Mullis in 1985. The purpose of a PCR (polymerase chain reaction) is to make a huge number of copies of a gene. This is necessary to have enough starting template for sequencing. PCR technology allows amplification of any sequence of interest from nanogram amounts of DNA and direct visualisation of the amplified product. Although PCR seems promising in the amplification of minisatellites, macrosatellites and sequence tagged site (STS) loci to make genetic mapping easy, the prior sequence information needed to design specific primers is a limiting factor in developing large numbers of genetic markers for any organisms. This amplification of DNA is possible due to a thermostable enzyme, Taq polymerase isolated from the bacterium *Thermus aquaticus*. A given sequence can be amplified upto 4x 10^6 times in 25 cycles.

3.4.1 DNA Amplification by PCR

A PCR reaction involves the following steps.

The cycling reactions: There are three major steps in a PCR, which are repeated for 30 or 40 cycles. This is done on an automated cycler, which can heat and cool the tubes with the reaction mixture in a very short time.

1. **Denaturation** at 94°C: During the denaturation, the double strand melts open to single stranded DNA, all enzymatic reactions stop.
2. **Annealing** at 54°C: The primers (about 20 nucleotides long)

anneal to their complementary bases. Ionic bonds are constantly formed and broken between the single stranded primer and the single stranded template. The more stable bonds last a little bit longer and to this double stranded DNA (template and primer), the polymerase can attach and start copying the template. Once there are a few bases built in, the ionic bond is so strong between the template and the primer that it does not break.

3. **Extension/Synthesis** at 72°C: This is the ideal working temperature for the Taq polymerase. The primers, where there are a few bases built in, already have a stronger ionic attraction to the template than the forces breaking these attractions. The bases (complementary to the template) are coupled to the primer on the 3' side (the polymerase adds dNTP's from 5' to 3', reading the template from 3' to 5' side, adding bases complementary to the template)

Since both strands are copied during PCR, there is an exponential increase in the number of copies of the gene. Suppose there is only one copy of the wanted gene before the cycling starts, after one cycle, there will be 2 copies, after two cycles, there will be 4 copies, three cycles will result in 8 copies and so on.

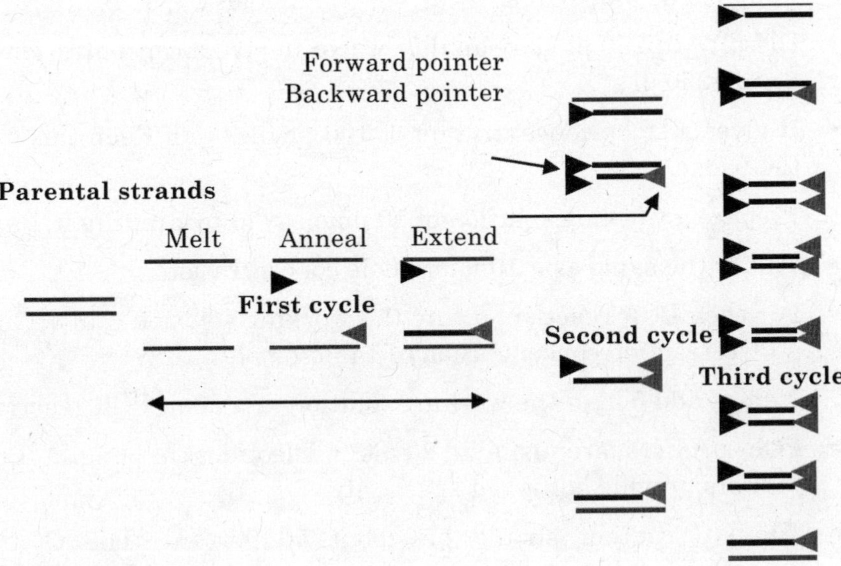

Fig. III. 3.1 : The exponential amplification of the gene in PCR

Requirements

(1) Template DNA

- The DNA must be of good quality and not broken down when visualized on an agarose gel.

- The amount needed for the initial PCR is 10 ng. If possible, determine the concentration of the DNA. If it is impossible, judge the concentration from the DNA loaded on the gel.

Usually amount of template DNA is in the range of 50 pg – 1 ng for plasmid or phage DNA and 0.11 μg for genomic DNA, for a total reaction mixture of 50 μl. Higher amounts of template DNA usually increase the yield of nonspecific PCR products, but if fidelity of synthesis is crucial, maximal allowable template DNA quantities in conjunction with limiting number of PCR cycles should be used to increase the percentage of "correct" PCR product. Template DNA is dissolved in 10 mM Tris-Cl (pH 7.6) containing a low concentration of EDTA (<0.1 mM).

(2) Primers

The primers are available as a powder. This powder is dissolved in sterile good quality water. The stock solution is saved as a 100 micromole concentration, while the PCR reactions are prepared from a 10 fold dilution of the stocks in a separate tube.

Once the working solution is finished, or contaminated, prepare new working solutions from the stocks.

- Supplied as nmole quantities, *e.g.,* 25 nmoles.
- Dissolve in 250 μl good quality water to give a concentration of 25 nmol / 250 μl
- It gives a stock concentration of 100 pmoles / μl. Keep this safe in the fridge.
- Prepare a working solution of 10 pmoles / μl by diluting it 1 in 10.
- This is the same as a 10 micromole concentration.
- For each PCR reaction, dilute the working solution 1 in 10 to give a PCR reaction concentration of 1 micromoles.

  e.g., add 5 μl of the working solution in a 50 μl PCR reaction
- PCR primers are usually 20–30 nucleotides in length. Longer primers provide sufficient specificity.
- The GC content should be about 50 percent. The C and G nucleotides should be distributed uniformly throughout the primer. More than three G or C nucleotides at the 3'-end of the primer should be avoided, as nonspecific priming may occur.
- The primer should not be self-complementary or complementary to any other primer in the reaction mixture, in order to avoid primer-dimer and hairpin formation.

- The melting temperature of flanking primers should not differ by more than 5°C, so the GC content and length must be chosen accordingly.
- All possible sites of complementarity between primers and the template DNA should be noted.
- If primers are degenerate, at least 3 conservative nucleotides must be located at the primer's 3'-end.
- Estimation of the melting and annealing temperatures of primer:

If the primer is shorter than 25 nucleotides, the approx. melting temperature (Tm) is calculated using the following formula:

$$Tm=4(G+C) + 2(A+T)$$

G, C, A, T – number of respective nucleotides in the primer. If the primer is longer than 25 nucleotides, the melting temperature should be calculated using specialized computer programs where the interactions of adjacent bases, the influence of salt concentration, etc. are evaluated. Optimal annealing temperature is generally 5°C lower than the melting temperature of the primer-template DNA duplex.

(3) Buffer/$MgCl_2$ concentration

- It is supplied together with the enzyme as a 10 fold concentration. Dilute it 1 in 10.
- Make sure whether $MgCl_2$ is included in the buffer. If not, the manufacturer will have sent a tube with $MgCl_2$ on its own which can be used for supplement the PCR reaction with $MgCl_2$.
- The final concentration of the $MgCl_2$ is 1.5 mM.

The thermostable DNA Polymerases require free divalent cation usually Mg^{2+}. Since Mg^{2+} ions form complexes with dNTPs, primers and DNA templates, the optimal concentration of $MgCl_2$ is 1.5 mM. Too few Mg^{2+} ions result in a low yield of PCR product, and too many increase the yield of non-specific products. $MgCl_2$ concentration varies from 1–4 mM. If the DNA samples contain EDTA or other chelators, the $MgCl_2$ concentration should be raised proportionally.

(4) dNTPs

The concentration of each dNTPs in the reaction mixture is usually 200 μM. It is very important to have equal concentrations of each dNTPs (dATP, dCTP, dGTP, dTTP), as inaccuracy in the concentration of even a single dNTP dramatically increases the misincorporation level. The dNTPs concentrations of 100–250 μM of each dNTP result in the optimal balance of product yield.

(5) Taq DNA Polymerase

The Taq DNA polymerase with 0.5–2.5 units per standard 25–50 µl reaction is used for routine PCR reactions. The specific activity of most commercial preparations of Taq is around 80,000 units/mg of protein. Standard PCRs therefore, contain 2×10^{12} to 10×10^{12} molecules of enzyme. As the efficiency of primer extension with Taq polymerase is generally about 0.7, the enzyme become limiting when 1.4×10^{12} to 7×10^{12} molecules of amplified product have accumulated in the reaction.

- It is supplied in a 5 units/gI concentration, use 2 units for a start.
- Always add the enzyme last and never remove it from the freezer.

Procedure

1. If possible, determine the concentration of the template DNA to be used in the PCR reactions.
2. Do the planning of the PCR reaction using the Excel program written for the PCR reactions.
3. Label a 0.2 ml eppendorf tube for each reaction. Label by writing on the tag the full name of the reaction, for instance 2003 pcr 01.1 for the first reaction. Place on ice. You will also need a tube for both a positive and a negative control reaction. Label similarly. Use a 1.5 ml eppendorf for the preparation of the stock solution.
4. Prepare the stock solution by mixing the components that are present in all the reactions at the same volumes in the 1.5 ml tube. Add the water first, then the second biggest volume and then finally the smallest. This will always be the Taq-polymerase.
5. Mix well and divide the correct volumes to the individual tubes.
6. Add the missing reagents (*e.g.*, water and DNA that is not the same for all the reactions) that still need to be added.
7. Close the tubes and spin them quickly.
8. Put in the PCR machine and run the appropriate program for 30 cycles.
9. After it is finished, add 8 µl of loading buffer to each 25 µl reaction and 16µl to each 50 µl reaction.
10. Separate 10 µl on an agarose gel along with a suitable molecular weight marker.

Rationale of the controls

In every reaction a positive and negative control must be included. The template DNA for the positive control is 5 μl of a 1 in 100 dilution of pure DNA. The negative control contains everything, but no template DNA is added to the reaction.

Controls, both positive and negative are included in each PCR reaction that is performed. By evaluating the products, or lack thereof, on an agarose gel after the amplification reaction is finished, the validity of each PCR reaction can be determined.

Positive control

- In the first PCR reaction different dilutions of DNA were used to try to amplify the 16S rDNA area from the sample DNA. All four reactions produced a +/- 650 bp band which corresponds well to the expected. Therefore, use the biggest dilution as a positive control in all subsequent PCR reactions. Of this dilution, 5μl is used in the positive control PCR reaction.
- This reaction must always produce a 650 bp band when amplified.

Negative control

- The negative control contains all the components for a PCR reaction, but instead of adding DNA template to it, water is added. There will therefore, be no template DNA available for amplification and when separated on the gel, no band should be visible.
- If there is no band visible, you can believe all the other PCR reactions.
- If there is a band present in the negative control after amplification, it means that one or more of your reagents was contaminated with template DNA. Therefore, any positive reaction found in any of other tubes cannot be believed and must be discarded.
- If such contamination was detected, all the diluted chemicals must be thrown away and new dilutions from the stock must be prepared. The PCR reaction must then be repeated.

Precautions

- Use sterile disposable plasticwares wherever possible.
- Use autoclaved eppendorf tubes and microtips.
- Always wear gloves and lab coat.
- Wipe off surface of laminar air flow, hands, micropipette, etc.
- Don't touch anything with working hands.

3.4.2 RAPD (Random Amplified Polymorphic DNA) Analysis of Different Plant Genotypes

Biochemical or conventional molecular markers include chemo-profiling using biochemical compounds like phenolics, anthocyanins, flavanoids, etc. and enzyme markers (allozymes and isozymes). Chemo-profiling with phytochemical compound are not suitable for high throughput analysis and they suffer from many disadvantages of environment dependence and low reproducibility. The 'DNA markers' detect variability among two individuals based on the polymorphism that is encountered in DNA of the organisms.

Molecular markers are specific locations on a chromosome, which serve as landmarks for genome analysis and genomes are subjected to mutations as a result of normal cellular operations or interactions with the environment, leading to genetic variation (polymorphism). In conjunction with selection and genetic drift, there arises genetic variation within and among individuals, species, and higher order taxonomic groups. DNA marker technology can be applied to reveal these genetic variations and produce characteristics fingerprint or band pattern, and provides information about allelic variation at a given locus.

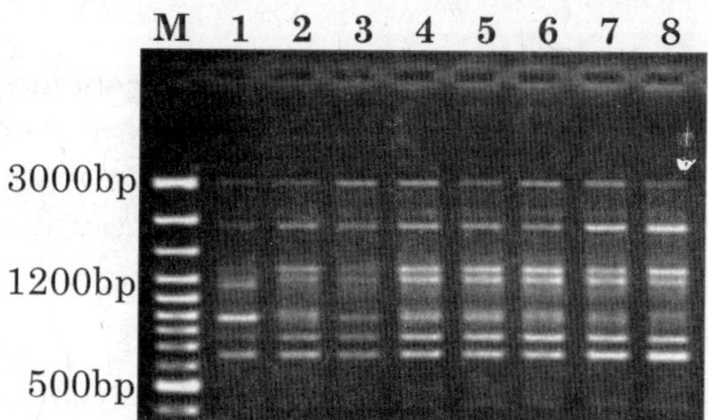

Fig. III. 3.2 : RAPD profile of plant DNA. M contains molecular weight marker.

RAPD

With the advent of RAPD as a molecular marker in early 1990s, the use of RAPDs has become the most potent tool for population studies in plants as the technique is being exploited for more and for both varietal identification and estimation of genetic diversity within the gene pool, and has been applied to many tropical trees. Genetic variation within and among populations is strongly affected by ecological factors and microgeographic differentiation, caused by these factors, is reported for many species. RAPD is also used to analyze genetic differentiation among populations from ecologically different habitat. RAPD procedures were first developed in 1990 using PCR to randomly amplify anonymous segments of nuclear DNA with an identical pair of primers 8–10 bp in length. Since the primers are short and relatively low annealing temperatures (often 36–40°C) are used, the likelihood of amplifying multiple products is great, with each product (presumably) representing a different locus. For the reason that most of the nuclear genome is non-coding, it is presumed that most of the amplified loci will be selectively neutral. Genetic variation and divergence within and between the taxa of interest are assessed by the presence or absence of each product, which is dictated by changes in the DNA sequence at each locus because each band is considered a bi-allelic locus (presence or absence of an amplified product). The potential power is relatively high for detection of polymorphism; typically, 5–20 bands can be produced using a given primer pair, and multiple sets of random primers can be used to scan the entire genome for differential RAPD bands. The high incidence of amplified repetitive DNA in RAPD fingerprints is to be expected given its prevalence in eukaryotic genome. RAPD assay has been widely used in screening for DNA sequence-based polymorphisms because it requires small amounts of DNA (15–25 ng), is a nonradioactive assay and can be performed in a few hours. RAPD fragments linked to a trait of interest can easily be identified by using two pooled DNA samples: one from individuals that express the trait, the other from individuals that do not. Any polymorphism between the two pools is linked to the trait.

The Random Amplified Polymorphic DNA (RAPD) method is based on the Polymerase Chain Reaction (PCR) using short (usually 10 nucleotide) primers of arbitrary sequences. Polymorphism of amplified fragments are caused by: (1) base substitutions or deletions in the priming sites, (2) Insertions that render priming sites too distant to support amplification, or (3) insertions or deletions that change the size of the amplified fragment. Thus RAPD protocol can be used in genetic mapping and finger printing application in plants.

Procedure

The RAPD reaction mixture is prepared as follows:

2.5 µl DNA stock (25ng/µl)

2.5 µl Assay buffer containing 2.5mM $MgCl_2$ (2.5mM)

1 µl $MgCl_2$ stock (1.5mM)

1 µl primer stock (25pmol)

4 µl dNTPS (400µM)

1 µl Taq polymerase (1U)

Add sterile water to make total volume to 25 µl.

For RAPD analysis of different plants different conditions have to be standardized.

The cycling conditions may be modified depending on the thermocycler used

1. For RAPD analysis, plant samples are collected from different locations.
2. Isolate DNA according to the CTAB method given under section 3.1.1.4.
3. PCR is performed using randomly designed primers. About 20–40 primers should be used since lesser number of primers may not give polymorphism. The steps for PCR are performed as given under section 3.4.1.
4. **Gel Interpretation,** Scoring and analysis
   i. Bands are sized and matched directly on gels, or photographic films.
   ii. Note the presence and absence of bands
   iii. Designate a name or a number for each RAPD marker based on the molecular size and primer used.
   iv. Score RAPD bands using a binary system of 0 (in the absence of the band) and 1 (if the band is present).
   v. Data is now ready for analysis using computer software NTSYS / RAPDistance.

Precautions

As given under section 3.4.1.

3.4.3. PCR Amplification of 16s Ribosomal RNA

Application of molecular biology techniques has made possible finer genetic analysis and genetic manipulation of almost all organisms. Analysis of 16S rRNA sequences is now an integral part of classification of microorganisms. It is now possible to analyze member of a microbial community even though many of them may be nonculturable. DNA or RNA can be isolated directly from a sample (soil, water etc.) and rRNA sequence can be amplified by polymerase chain reaction (PCR) or reverse transcriptase-PCR (RT-PCR).

Amplification of 16S rRNA gene

(1) Add reagents to a sterile 0.5 ml microfuge tube in the following order:

	Reagent	Volume (μl)
10x incubation buffer	2.5 μl	5 μl
dNTPs (1mM)	1.0 μl	2 μl
8–27 primer (0.5μg / μl)	1.0 μl	2 μl
1500 primer (0.5μg / μl)	1.0 μl	2 μl
Distilled water	17.5 μl	35 μl
Template DNA	2.0 μl	9 μl
Total volume	25.0 μl	50 μl

Primer sequences: 8-27F:5' - GAGTTTGATCCTGGCTCAG - 3' and

1500 R: 5'- AGAAAGGAGGTGATCCAGCCA-3'

Also include one reaction without template DNA, as a negative control.

(2) Heat tube to 98°C for 3 minutes in a thermal cycler. Add 0.5 μl (5 U/μl) Taq polymerase to the reaction mixture below the oil overlay. Replace tubes in thermal cycler.

(3) Amplify using the following thermal profile:

52°C for 1 minute

72°C for 2 minute

93°C for 1 minute

For 28 cycles

(4) Perform a final extention of 52°C for 1 minute and 72°C for 10 minutes.

(5) Analyse amplification product on agarose gel (1%).

Purification of PCR products

1. Remove the aqueous reaction mixture from below the oil layer and transfer to a new 1.5 ml tube.
2. Add 80 μl of distilled water, 10 μl of 3M sodium acetate (pH 4.6) and 200 μl of absolute ethanol.
3. Keep it on ice for 30 minutes.
4. Centrifuge at 13,000 rpm for 10 minutes, discard the supernatant.
5. Wash pellet with ethanol (70%) twice at room temperature.
6. Air-dry the pellet.
7. Add 15 μl of sterile distilled water to the pellet.

8. Check the presence and amount of purified PCR product by agarose gel electrophoresis. Compare intensity of bands with molecular weight marker of known DNA concentration or as given under section 3.1.1.8.

Precautions

Same as given in section 3.4.1.

3.5 BLOTTING TECHNIQUES AND HYBRIDIZATION

Southern transfer and hybridization (Southern, 1975) is used to study how genes are organized within genomes by mapping restriction sites in and around segments of genomic DNA for which specific probes are available. Genomic DNA is first digested with one or more restriction enzymes, and the resulting fragments are separated according to size by electrophoresis in agarose gel. The DNA is then denatured *in situ* and transferred from the gel to a solid support (usually a nylon or nitrocellulose membrane). The DNA attached to the membrane is hybridized to a labeled DNA, RNA, or oligonucleotide probe, and bands complementary to the probe are located by an appropriate detection system, for example, by autoradiography. By estimating the size and number of the bands generated after digestion of the genomic DNA with different restriction enzymes, singly or in combination, it is possible to place the target DNA with in a context of restriction sites.

Methods of Transferring DNA from Agarose Gels to Solid Supports

The transfer of electrophoretically separated DNA from gels to two-dimensional solid supports is a key step in Southern hybridization. Described below are five methods to transfer fragments of DNA from agarose gels to solid supports (nitrocellulose or nylon membranes).

Upward Capillary Transfer

DNA fragments are carried from the gel in an upward flow of liquid and deposited on the surface of the solid support (Southern, 1975). The liquid is drawn through the gel by capillary action that is established and maintained by a stack of dry absorbent paper towels. The rate of transfer of the DNA depends on the size of the DNA fragments and the concentration of agarose in the gel. Small fragments of DNA (<1 kb) are transferred almost quantitatively from a agarose gel (0.7%) within 1 hour; larger fragments are transferred more slowly and less efficiently. For example, capillary transfer of DNAs >15 kb in length requires at least 18 hours, and even then the transfer is not complete. The efficiency of transfer of large DNA fragments is determined by the fraction of molecules that escape from the gel before it becomes dehydrated. As elution proceeds, fluid is drawn not only from the reservoir, but also from the interstices of the gel itself. This flow reduces the gel to a rubbery substance through which DNA molecules cannot easily pass. The DNA in the gel is exposed to weak acid (which results in partial depurination), followed by strong base (which hydrolyzes the phosphodiester backbone at the sites of depurination). The resulting fragments of DNA (~1 kb in length) can then be transferred rapidly from the gel with high efficiency. The depurination reaction must not proceed too far; otherwise, the DNA will be cleaved into small fragments that are too short to bind efficiently to the solid support. Depurination/hydrolysis can also cause the bands of the final autoradiograph to assume a "fuzzy" appearance, presumably because of increased diffusion of DNA during transfer. Therefore, depurination/hydrolysis is recommended only when it is known ahead of time that the target DNA fragments will be >15 kb in length.

Buffer drawn from a reservoir passes through the gel into a stack of paper towels. DNA eluted from the gel by the moving stream of buffer is deposited onto a nitrocellulose or nylon membrane. A weight applied to the top of the paper towels helps to ensure a tight connection between the layers of material used in the transfer system.

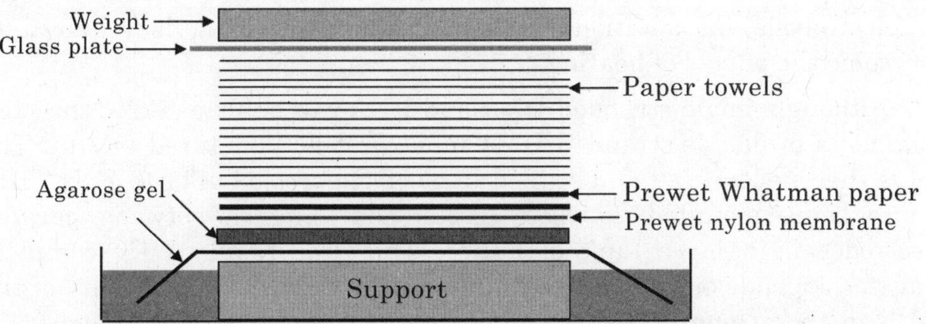

Fig. III. 3.3: Upward Capillary Transfer of DNA from Agarose Gels.

Downward Capillary Transfer

DNA fragments are carried in a downward direction in a flow of alkaline buffer and are deposited onto the surface of a charged nylon membrane. Various arrangements of wicks, reservoirs, and different formulations of transfer buffers have been described to achieve downward transfer. The best results are achieved using 0.4 M NaOH and a setup in which the transfer buffer is drawn from reservoirs to the top of the gel through wicks and pulled through the gel by an underlying stack of paper towels. Transfer of DNA fragments is rapid, and the intensity of signal is ~ 30%

greater than can be achieved by conventional upward transfer. This improvement probably results from a more efficient migration of DNA fragments through the interstices of the gel, which is not under pressure from weights placed on top.

Fig. III. 3.4 : Downward Capillary Transfer of DNA from agarose Gels.

Electrophoretic Transfer

This method is not practical when nitrocellulose is used as the solid support because of the high ionic strengths of the buffers that are required to bind nucleic acids to nitrocellulose. These buffers conduct electric current very efficiently, and it is necessary to use large volumes to ensure that the buffering power of the system does not become depleted by electrolysis. In addition, extensive external cooling is required to overcome the effects of heating.

Although single-stranded DNA and RNA can be transferred directly, fragments of double-stranded DNA must first be denatured *in situ*. The gel is then neutralized and soaked in electrophoresis buffer (*e.g.*, lx TBE) before being mounted between porous pads aligned between parallel electrodes in a large tank of buffer. The time required for complete transfer depends on the size of the fragments of DNA, the porosity of the gel, and the strength of the applied field. However, because even high-molecular-weight nucleic acids migrate relatively rapidly from the gel, depurination/hydrolysis is unnecessary and transfer is generally complete within 2–3 hours. Since electrophoretic transfer requires comparatively large electric currents, it is often difficult to maintain the electrophoresis buffer at a temperature compatible with efficient transfer of DNA. Many commercially available electrophoretic transfer machines are equipped with cooling devices, but others are effective only when used in a cold room.

Membranes used for Southern and Northern Hybridization

For many years, the only available support for immobilization of DNA was nitrocellulose, which was first used in powder form and later as sheets. Northern hybridization was initially carried out exclusively with RNA immobilized on activated cellulose papers. However, it was soon realized that RNA denatured by glyoxal, formaldehdye, or methylmercuric hydroxide binds tightly to nitrocellulose. For several years, nitrocellulose was the support of choice for both Northern and Southern hybridization. Despite its evident success, nitrocellulose is not an ideal matrix for solid-phase hybridization.

Its capacity to bind nucleic acids is low (~50–100 Vg/cm 2) and varies according to the size of the DNA and RNA. In particular, nucleic acids <400 bases in length are retained inefficiently by nitrocellulose. DNA and RNA are attached to nitrocellulose by hydrophobic interactions, rather than covalent interactions, and therefore leach slowly from the matrix during hybridization and washing at high temperatures.

Nitrocellulose membranes become brittle during baking under vacuum at 80°C, which is usually an integral part of the process to immobilize nucleic acids.

Table 3.3 : Types of nitrocellulose membranes and their properties.

Property	Type of Membrane		
	Nitrocellulose	Neutral Nylon	Charged Nylon
Capacity (μg nucleic acid/cm^2)	80-120	~100	400–500
Size of nucleic acid required for maximal binding	>400 bp	>50 bp	>50 bp
Transfer buffer	high ionic strength at neutral pH	low ionic strength over a wide range of pH	
Immobilization	baking at 80°C under vacuum for 2 hrs	baking at 70°C for 1 hour; no vacuum required or mild alkali or UV irradia-tion at 254 nm; damp membranes are generally exposed to 1.6 kj/m2; dried membranes require 160 kf/m^2	
Commercial products		Hybond – N Gene - Screen	Hybond – N$^+$ Zeta-Probe Nytran$^+$ Gene-Screen Plus

3.5.1 Southern Hybridization

Southern hybridization involves first the electrophoresis of the dige-sted DNA samples. The DNA bands are then blotted onto a nitro-cellulose membrane, hybridized to probes and detected accordingly.

Procedure

Digestion and Electrophoresis of the DNA

1. Digest an appropriate amount of genomic DNA with one or more restriction enzymes.
2. If necessary, concentrate the DNA fragments at the end of the digestion by ethanol precipitation. Dissolve the DNAs in 25 μl of TE (pH 8.0).

Make sure that the ethanol is removed from the DNA solution before it is loaded on the gel. If significant quantities of ethanol remain, the DNA "crawls" out of the slot of the gel. Heating the solution of dissolved DNA to 70°C in an open tube for 10 minutes is usually sufficient to drive off most of the ethanol. This treatment also disrupts base pairing between cohesive termini of restriction fragments.

1. Measure the concentrations of the digested DNAs by fluorometry or by the ethidium bromide. Transfer the appropriate amount of each digest to a fresh microfuge tube. Add 0.15 volume of 6x sucrose gel-loading buffer and separate the fragments of DNA by electrophoresis through an agarose gel. Maintain a low voltage through the gel (about <1 V/cm) to allow the DNA to migrate slowly.
2. After electrophoresis is complete, stain the gel with ethidium bromide or SYBR Gold and photograph the gel. The gel may be stored at this stage before the DNA is denatured and transferred to the membrane. Wrap the gel in Saran Wrap and store it on a flat surface at 4°C. Since the bands of DNA diffuse during storage, the gel should not be put aside for more than 1 day before being processed.
3. Denature the DNA and transfer it from the agarose gel to a nitrocellulose or a neutral or charged-nylon membrane using one of the methods described below.

Preparation of the Gel for Transfer

4. After fractionating the DNA by gel electrophoresis, transfer the gel to a glass baking dish. Use a razor blade to trim away unused areas of the gel, including the section of gel above the wells. Be sure to leave enough of the wells attached to the gel so that the

positions of the lanes can be marked on the membrane after transfer of DNA.

5. Denature the DNA by soaking in a denaturing (alkaline) solution as follows:

Preparation of the Membrane for Transfer

6. Use a fresh scalpel or a paper cutter to cut a piece of nylon or nitrocellulose membrane about 1 mm larger than the gel in each dimension. Also cut two sheets of thick blotting paper to the same size as the membrane.
7. Float the membrane on the surface of a dish of deionized H_20 until it wets completely from beneath, and then immerse the membrane in the appropriate transfer buffer for at least 5 minutes. Use a clean scalpel blade to cut a corner from the membrane to match the corner cut from the gel.

Assembly of the Transfer Apparatus and Transfer of the DNA

Neutral transfer buffer (10x SSC or 10x SSPE) is used to transfer DNA to uncharged membranes. Alkaline transfer buffer (0.4 N NaOH with 1 M NaCl) is used to transfer DNA to charged nylon membrane.

8. While the DNA is denaturing, place a piece of thick blotting paper on a sheet of Plexiglas or. a glass plate to form a support that is longer and wider than the gel. The ends of the blotting paper should drape over the edges of the plate. Place the support inside a large baking dish. The support can be placed on top of four neoprene stoppers to elevate it from the bottom of the dish.
9. Fill the dish with the appropriate transfer buffer until the level of the liquid reaches almost to the top of the support. When the blotting paper on the top of the support is thoroughly wet, smooth out all air bubbles with a glass rod or pipette.
10. Remove the gel from the solution in Step 7 and invert it so that its underside is now upper-most. Place the inverted gel on the support so that it is centered on the wet blotting paper.

Make sure that there are no air bubbles between the blotting paper and the gel

11. Surround, but do not cover the gel with Saran Wrap or Parafilm.
12. Wet the top of the gel with the appropriate transfer buffer. Place the wet membrane on top of the gel so that the cut corners are aligned. To avoid bubbles, touch one corner of the membrane to the gel and gently lower the membrane onto the gel. One edge of

the membrane should extend just over the edge of the line of slots at the top of the gel.

13. Wet the two pieces of thick blotting paper in the appropriate transfer buffer and place them on top of the wet membrane. Rolla pipette across the surface of the membrane to smooth away any air bubbles.
14. Cut or fold a stack of paper towels (5–8 cm high) just smaller than the blotting papers. Place the towels on the blotting papers. Put a glass plate on top of the stack and weigh it down with a 400-g weight.
15. Allow the transfer of DNA to proceed for 8–24 hours. Replace the paper towels as they become wet. Try to prevent the entire stack of towels from becoming wet with buffer.
16. Remove the paper towels and the blotting papers above the gel. Turn the gel and the attached membrane over and lay them, gel side up, on a dry sheet of blotting paper. Mark the positions of the gel slots on the membrane softly with a lead pencil or pen.
17. Peel the gel from the membrane and discard the gel.

Instead of discarding the gel, it can be stained (45 minutes) in a 0.5 ug/ml solution of ethidium bromide in H_20 and visualized on a UV transilluminator to view the success of the DNA transfer.

Fixation of the DNA to the Membrane

The sequence of steps from immobilization of DNA to the membrane to subsequent hybridization depends on the type of membrane, the method of transfer and the method of fixation (Table 3.3.2). Since alkaline transfer results in covalent attachment of DNA to positively charged nylon membranes, there is no need to fix the DNA to the membrane before hybridization. DNA transferred to uncharged nylon membranes in neutral transfer buffer should be fixed to the membrane by baking under vacuum or heating in a microwave oven, or cross-linked to the membrane by UV irradiation.

Table 3.4 : Fixing DNA to the Membrane for Hybridization.

Type of Membrane	Type of Transfer	Method of Fixation	Sequence of Steps
Positively charged nylon	alkaline transfer	alkaline transfer	1. Soak membrane in Neutralization buffer II. 2. Proceed to prehybridization.

Type of Membrane	Type of Transfer	Method of Fixation		Sequence of Steps
Uncharged nylon or positively-charged nylon	neutral transfer	UV irradiation (please see Step 21 for details)	3.	Soak membrane in 6x SSC. Fix the DNA by UV irradiation. Proceed to prehybridization
Uncharged nylon or positively charged nylon	neutral transfer	baking in vacuum oven or microwave oven	4.	Soak membrane in 6x SSC. Bake the membrane. Proceed to prehybridization

18. Soak the membrane in one of the following, solutions as appropriate:

    For neutral transfer: 6× SSC for 5 minutes at room temperature.

    For alkaline transfer: Neutralization buffer II (0.5 M Tris-Cl [pH 7.2] with 1 M NaCl) for 15 minutes at room temperature.

19. Immobilize the DNA that has been transferred to uncharged membranes.

Inview of the fact that alkaline transfer results in covalent attachment of DNA to positively charged nylon membranes, there is no need for additional steps to fix the DNA to the membrane.

To fix by baking in a vacuum oven

(a) Remove the membrane from the 6x SSC and allow excess fluid to drain away. Place the membrane flat on a paper towel to dry for at least 30 minutes at room temperature.

(b) Sandwich the membrane between two sheets of dry blotting paper. Bake for 30 minutes to 2 hours at 80°C in a vacuum oven.

Over baking can cause nitrocellulose membranes to become brittle. If the gel was not completely neutralized before the DNA was transferred, nitrocellulose membranes will turn yellow or brown during baking and chip very easily. The background of nonspecific hybridization also increases dramatically.

To cross-link by UV irradiation

(a) Place the damp membrane on a dry piece of blotting paper.

(b) Irradiate at 254 nm to cross-link the DNA to the membrane.

Immobilization of nucleic acids by UV irradiation can greatly enhance the hybridization signal obtained with some brands of positively charged nylon membranes. However, for maximum effect, it is important to make sure that the membrane is not over irradiated. The aim is to form cross-links between a small fraction of the thymine residues in the DNA and positively charged amine groups on the surface of the membrane. Over irradiation results in the covalent attachment of a high proportion of the thyamines, with consequent decrease in hybridization signal. Make sure that the side of the membrane carrying the DNA faces the UV light source. Most manufacturers advise that damp membranes be exposed to a total of 1.5 J/cm^2 and that dry membranes be exposed to 0.15 J/cm^2.

20. Proceed directly to hybridization of immobilized DNA to a probe.

Southern Hybridization of Radiolabeled Probes to Nucleic Acids Immobilized on Membranes

The strength of the hybridization signal obtained in southern hybridization depends on a number of factors, including the proportion of immobilized DNA that is complementary to the probe, the size of the probe and its specific activity, and the amount of genomic DNA transferred to the membrane. Under the best conditions, the method is sufficiently sensitive to detect <0.1 pg of DNA complementary to a probe that has been radiolabeled with ^{32}P to high specific activity (>109 cpm/4). A sequence of 1000 bp that occurs only once in the mammalian genome (*i.e.*, 1 part in 3 million) can be detected in an overnight exposure to conventional X-ray film (or 15–60 minutes on a phosphorimager) if 10 μg of genomic DNA is transferred to the membrane and hybridized to a probe several hundred nucleotides in length. The strength of the signal is proportional to the specific activity of the probe and inversely proportional to its length, thus Southern hybridization reaches the limits of its sen-sitivity when very short probes are used. To obtain a signal from single-copy genomic sequences with, for example, oligonucleotide probes, it is necessary to radiolabel to the highest specific activity possible, to increase the amount of target DNA on the membrane, and to expose the auto-radiograph for several days, or the phosphorimager plate for many hours.

Buffers and Solutions

Please see Appendix IVC for components of stock solutions, buffers, and reagents.

Nucleic Acids and Oligonucleotides

DNA immobilized on membrane

Poly(A) RNA (10 mg/ml) in sterile H_2O

Optional, for hybridization buffers. Prepare solution by dissolving poly (A) RNA in sterile H_20 and store in 100 μl aliquots.

Probe DNA or RNA

For Southern analysis of mammalian genomic DNA, where each lane of the gel contains 10 μg of DNA, use 10–20 ng/ml radiolabeled probe (sp. act. ~ 10^9 cpm/μg). For Southern analysis of cloned DNA fragments, where each band of the restriction digest contains 10 ng of DNA or more, much less probe is required. When analyzing cloned DNA, hybridization is carried out for 6–8 hours using 1–2 ng/ml radiolabeled probe (sp. act. 10^6 to 10^9 cpm/μg). DNA (~10 mg/ml)

Denatured fragmented DNA should be used at a concentration of 100 4g/ml in prehybridization and hybridization solutions involving essentially any type of hybridization experiment.

Method

(1) Float the membrane containing the target DNA on the surface of a tray of 6x SSC (or SSPE) until the membrane becomes thoroughly wetted from beneath. Submerge the membrane for 2 minutes

(2) Prehybridize the membrane by one of the following methods. Hybridization can be done in a heat sealable bag or a plastic container.

Hybridization in a Plastic Container

(1) Place the wetted membrane in a plastic container and add 0.2 ml of prehybridization solution for each square centimeter of membrane.

(2) Seal the box with the lid and place the box on a rocking platform in an air incubator set at the appropriate temperature (68°C for aqueous solvents; 42°C for solvents containing formamide (50%); 65°C for phosphate-SDS solvents).

(3) If the radiolabeled probe is double-stranded DNA, denature it by heating for 5 minutes at 100°C. Chill the probe rapidly in ice water.

Alternatively, denature DNA probes by adding 0.1 volume of 3 N NaOH. After 5 minutes at room temperature, chill the probe to 0°C in an ice-water bath, and add 0.05 volume of 1 M Tris-Cl (pH 7.2) and 0.1 volume of 3 N HC1. Store the probe in ice water until it is needed.

Single-stranded DNA and RNA probes need not be denatured.

(4) To hybridize the probe to a blot containing genomic DNA, carry out one of the following methods.

For hybridization in a plastic container

Transfer the membrane from the container to a sealable bag or a hybridization bottle.

(5) After hybridization, wash the membrane.

(6) After 5 minutes, pour off the first rinse solution into a radioactivity disposal container and add several hundred milliliters of 2× SSC and SDS (0.1%) to the tray. Incubate for 15 minutes at room temperature with occasional gentle agitation.

If hybridization was carried out in a phosphate-SDS buffer, rinse the membrane a total of eight times for 5 minutes each in several hundred milliliters of Phosphate-SDS washing solution 2 at 65°C. Skip to Step 9 after the eighth rinse.

(7) Replace the rinse solution with several hundred milliliters of fresh 0.1× SSC with SDS (0.1%). Incubate the membrane for 30 minutes to 4 hours at 65°C with gentle agitation.

(8) Briefly wash the membrane with 0.1xSSC at room temperature.

(9) Remove most of the liquid from the membrane by placing it on a pad of paper towels. Place the damp membrane on a sheet of Saran Wrap. Apply adhesive dot labels marked with radioactive ink (or phosphorescent dots) to several asymmetric locations on the Saran Wrap. These markers serve to align the autoradiograph with the membrane. Cover the labels with Scotch Tape. This prevents contamination of the film holder or intensifying screen with the radioactive ink.

(10) Cover the membrane with a sheet of Saran Wrap, and expose the membrane to X-ray film for 16–24 hours at –70°C with an intensifying screen to obtain an autoradiographic image.

Precautions

1. Use appropriate gloves and blunt-ended forceps to handle the membrane. A membrane that has been touched by oily hands will not wet.
2. Do not allow the membrane to dry out at any stage during the washing procedure.
3. Do not move the membrane once it has been applied to the surface of the gel. Make sure that there are no air bubbles between the membrane and the gel.
4. Take proper care while handling radioactively labeled probes.

3.5.2 Northern Blotting and Hybridization

Transfer of RNA from electrophoresis medium (gel) to a solid matrix such as nylon membrane is called 'Northern Blotting'. Once the isolated and fractionated RNA fixed to the membrane, can it then be hybridized with a radioactively/non-radioactive labeled DNA as a probe for detecting the relevant transcript on the northern blot. The fractionation of RNAs is carried out in agarose gel in presence of a strong denaturing agent such as formaldehyde and then blotting done similar to that of DNA.

Northern Hybridization is RNA-DNA hybridization. DNA sequence is used as a probe after radioactively labelling it. This hybridization is carried out to study the gene expressions and the level of gene expressions at different conditions/tissues/stages of development. The DNA sequence in a probe molecule binds to the complimentary sequence of the RNA molecule already immobilized on the northern blot under suitable conditions. This RNA-DNA hybrid molecule is detected by tracing the radioactive or non-radioactive molecule of the probe.

Procedure

Blotting of RNA is done similar to that of DNA.

For northern hybridization a DNA probe is used for hybridization with the complementary RNA sequence. This RNA-DNA hybrid is then detected by tracing the radioactively or non-actively labeled probe.

3.5.3 Western Blotting and Hybridization

Western blot analysis is used for detection and characterization of target proteins (proteins of interest) in biological samples. Some of the examples include detection of parasite antigens in human blood, detection of tumor makers, protein in suspected patients (*e.g.*, HIV infection) etc. The technique includes separation of proteins by SDS-PAGE, transfer of resolved protein from gel to nitrocellulose membrane, followed by their detection using antibodies and appropriately labelled (Horse radish peroxidase) second antibody. It can be made more sensitive and quantitative by using radiolabelled antibodies. Polyvinylidene difluoride (PVDF) membranes are not routinely used for Northern or Southern transfers. However, PVDF membranes, by virtue of their higher mechanical strength and greater capacity to bind proteins, are preferred to nitrocellulose for Western blotting. Nylon membranes should not be used for Western blotting because the level of nonspecific absorption of immunological probes is unacceptably high.

Procedure

1. Separation of proteins by sodium dodecyl sulphate polyacrylamide gel electrophoresis (SDS-PAGE)
   (a) Clean the glass plates thoroughly with deionized water.
   (b) Fix the gel apparatus, set up with the spacer and check if the sealing is tight by adding water.
   (c) Pour the separating gel between the plates at an angle. Through one side layer pour 300 μl of water on the gel and cover it with an aluminium foil. Keep the gel at room temperature.
   (d) Pour the stacking gel and put the comb in it.
   (e) Fix the plates to the electrophoretic set up.
   (f) Check for buffer leakage and air bubbles.
   (g) Load the samples and run the gel at a constant voltage of 80–120 V.
   (h) After the gel has run, carefully remove the gel and place it in a staining solution for 3–4 hours.
   (i) Transfer the gel to the destaining solution for removing the extra stain.
   (j) Visualize the bands over transilluminator.
2. Transfer of resolved proteins from the SDS-PAGE to solid support (nitrocellulose membrane).

The resolved proteins in polyacrylamide gels are electroblotted to solid support like nitrocellulose paper. The resolved proteins in the polyacrylamide gels, are electrophoretically transferred to nitrocellulose filter as described below.

(a) One nitrocellulose membrane and four whatman paper (1 mm) sheets are cut equal to the size of the gel. Two sheets of whatman paper, presoaked in transfer buffer are placed on a scouring pad or sponge. Layer gel over it, carefully avoiding entrapment of air bubbles. On this, nitrocellulose membrane (first floated in plain distilled water, then submersed in water followed by rinsing with transfer buffer) is placed. Remove the entrapped air bubbles. Now layer two more sheets of wet whatman paper on it, followed by second scouring pad or sponge. This sandwich is backed by tow perforated plastic sheets from both the sides (to give physical strength) and strung two rubber brands.

(b) The sandwiched assembly is submerged in the tank filled with transfer buffer, with nitrocellulose sheet facing the anode. The blot apparatus is connected with power supply and current (150 mA) is applied for 6 hours.

(c) After transfer, the sandwich is dismantled and a strip of nitrocellulose sheet covering the gel, is stained with amido black for 2–5 minutes and destained with methanol (90%) acetic acid (2%). Rest of the nitrocellulose paper is processed for immuno-detection.

Immunodetection

1. Blocking: Nitrocellulose membrane is incubated in milk (5%) (prepared in TBS containing 0.1% Tween-20) or BSA (3%) for 4 hours at room temperature.
2. Incubation with Primary and secondary antibodies.
   (a) The blocking solution is removed and is followed by incubation with primary antibody diluted in BSA (0.3%, prepared in TBS-Tween-20) for 3 hours with occasional shaking at 50 rpm on a Gyratory shaker at room temperature.
   (b) The blot is washed with TBS-Tween-20, four times, 10 minutes each.
   (c) The washed blot is incubated with secondary antibody (coupled with HRPO), diluted in BSA (0.3%) for 1 hour. Now this blot is washed four times with TBS Tween-20.

Enzyme Linked Visualization

The visualization of bound antibodies is made by incubating the blot with DAB solution (0.6 mg DAB, 1μl H202 in 1 ml 10mM.Tris-Cl, pH 7.6).

Reactive protein band will start appearing from 5 minutes onwards. Stop the reaction by washing off the reagent with distilled water and store the blot at 4°C for photography.

Precautions

Same as given under section 3.5.1.

3.6 LIGATION AND TRANSFORMATION OF BACTERIAL DNA (*E. coli*)

For gene cloning, the gene of interest and vector need to be joined together to form the recombinant DNA. This joining of vector and insert using DNA ligase enzyme is known as ligation. Transformation is the uptake of DNA molecule by a bacterial cell or rather any cell. In order to transform bacteria, they have to undergo some form of physical or chemical treatment which will increase their ability to uptake DNA. Cells which are treated enhance their ability to uptake DNA and are said to be competent. The cells are made competent to take up the plasmid DNA from surroundings.

3.6.1 Ligation

A specific restriction enzyme recognizes and generates cohesive ends or blunt ends. Complementary cohesive ends produced in two different DNA molecules (vector and insert) by the same restriction enzyme can be joined using DNA Ligase. This process is known as ligation. The cloned gene is expressed by transforming the construct in the suitable host.

Requirements

Digested Vector DNA

Digested Passenger DNA

T4 DNA ligase

10× Ligase Mix:

Tris HCI (pH 7.5)	660 mM
$MgCl_2$	66 mM
Dithiothreitol	100 mM

Add 10 ml of 100 mM ATP to 90 ml of 10 x ligase buffer immediately before use.

Procedure

Set the ligation reaction in three sets as follows:

SET 1

- Digested Vector DNA – × μl (equal to 2 μg) Digested DNA to be inserted – y μl (equal to 2 μg)
- 10 × T4 DNA ligase mix – 1.5 ul
- T4 DNA ligase – z μl (equal to 5 units)
- Add sterile water to make final volume 15 μl.

SET 2

- Digested Vector DNA – × μl (equal to 2 μg)
- Digested DNA to be inserted – y μl (equal to 4 μg)
- 10 × T4 DNA ligase mix – 1.5 μl T4 DNA ligase - z μl

- Add sterile water to make final volume 15 μl.

SET 3

- Digested Vector DNA – × μl (equal to 2 μg)
- Digested DNA to be inserted – y μl (equal to 6 μg)
- 10 × T4 DNA ligase mix – 1.5 μl T4 DNA ligase - z μl.
- Add sterile distilled water to make final volume 15 μl.

Procedure

1. Incubate the above reaction mixtures at 16°C for 15–16 hours.
2. Remove 1 μl and check ligation on agarose gel.
3. Ligation can be checked by running digested DNA samples parallely.
4. Transform the cells.

3.6.2 Preparation of Competent Cells

Cells soaked in an ice cold solution of calcium chloride become more efficient at DNA uptake. A solution of 50mM $CaCl_2$ is commonly used, but rubidium chloride, magnesium chloride solutions may also be used. The $CaCl_2$ causes the DNA to precipitate on the outside of the bacterial cells or helps in DNA binding to the cell wall. However, the actual uptake of DNA into competent cells takes place only by raising the temperature briefly or by heat shock treatment at 42°C.

Procedure

1. Pick up a single colony of *E. coli* (DHα or DH10B or JM109) from a plate freshly grown for 16 to 20 hours at 37°C and transfer it into 100 ml of LB both.
2. Incubate the culture for 3 hours at 37°C with vigorous shaking (250 cycles/min in a rotatory shaker).
3. Cell density is monitored by determining OD at 600nm and it should be less than 10^8 cells/ml (OD_{600} = 0.3-0.4).
4. Pellet the cells at 4000 rpm at 4°C for 10 minutes.
5. Discard the media from cell pellets. Stand the tubes in an inverted position for 1 min to allow the last traces of media to drain away.
6. Resuspend each pellet in 10 ml of ice-cold 100 mM $CaCl_2$ and store on ice. Centrifuge at 4000 rpm for 10 min at 4°C.

7. Transfer 200 µl of each suspension of competent cells to a microfuge tube or store frozen (-70°C) until use.

3.6.3 Transformation of *E. coli*

The recombinant vector carrying the foreign DNA or gene of interest is inserted into the bacterial cells *e.g., E. coli*. This process is known as transformation. The competent cells prepared above are transformed with the recombinant DNA.

1. For transformation add DNA (50 µg) to each tube. Mix gently. Store in ice for 30 minutes.
2. Give heat shock of 90 seconds at 42°C.
3. Rapidly transfer the tubes on ice. Allow cells to chill for 1–2 min.
4. Add 1 ml of LB medium to each tube. Incubate at 37°C for 45 min.
5. Transfer the appropriate volume (up to 200 µl per 90-mm plate) to transformed competent cells on to agar LB medium containing appropriate antibiotic. Using a sterile bent glass rod, gently spread the transformed cells over the surface of the agar plate.
6. Invert the plates and incubate at 37°C. Colonies should appear in 12–16 hours.

Precautions

1. Heat shock should be given only for a brief period.
2. Sterile conditions should be maintained in order to avoid contamination.

3.6.4 Transformation of Plant Cells using *Agrobacterium tumifaciens*

Plant transformation mediated by *Agrobacterium tumefaciens*, a soil plant pathogenic bacterium, has become the most used method for the introduction of foreign genes into plant cells and the subsequent regeneration of transgenic plants. *A. tumefaciens* naturally infects the wound sites in dicotyledonous plant causing the formation of the crown gall tumors. *Agrobacterium* is a genus of Gram-negative bacteria that uses horizontal gene transfer to cause tumors in plants. *Agrobacterium* is well known for its ability to transfer DNA between itself and plants, and for this reason it has become an important tool for plant improvement by genetic engineering.

A. tumefaciens causes crown-gall disease in plants. The disease is characterised by a tumour-like growth or gall on the infected plant, often at the junction between the root and the shoot. Tumors are incited by the conjugative transfer of a DNA segment (T-DNA) from the bacterial tumor-inducing (Ti) plasmid. The plasmid T-DNA is integrated randomly into the genome of the host cell and the virulence (vir) genes on the T-DNA are expressed, causing the formation of a gall. The T-DNA carries genes for the biosynthetic enzymes for the production of unusual amino acids called opines, *e.g.,* octopine or nopaline providing a carbon and nitrogen source for the bacteria. The auxin and cytokinin produced by the tumor genes determines the morphology of the tumor.

The process of gene transfer from *Agrobacterium tumefaciens* to plant cells implies several essential steps:

1. bacterial colonization
2. induction of bacterial virulence system
3. generation of T-DNA transfer complex
4. T-DNA transfer and
5. integration of T-DNA into plant genome.

A number of methods are available for transformation using *Agrobacterium tumifaciens,* such as vacuum infilteration, floral dip, spraying etc. Two protocols have been given below. They yield transformants at frequencies ranging up to several percent, most commonly 0.1%–1%.

Reagents

Agrobacterium tumefaciens GV3101::pMP90 (or other suitable strain) containing binary plasmid for expression of gene of interest.

Antibiotics, including 50 μg/ml gentamycin for *Agrobacterium* and the following choices, depending on the binary plasmid:

50 μg/ml kanamycin

100 μg/ml spectinomycin

Co-cultivation medium

LB (Luria-Bertani) liquid medium

MS plates (0.5X) (optional; see Step 14)

500 μg/mL carbenicillin can be added to increase seedling survival.

Seedlings/plants of interest–40, 4-day-old *Arabidopsis* (or tobacco) seedlings or 15, 4-day-old tomato or 5-day-old rice (or switchgrass) seedlings. These numbers are for cocultivation in a 100 x 20-mm Petri dish.

Surface sterilization solution (1% bleach; *i.e.*, 0.05% sodium hypochlorite)

Washing solution for *Agrobacterium*

YEB medium – 1 g/l yeast extract, 1g/l bacto-peptone, 5g/l sucrose, 5g/l Lab-Lemco.

Procedure I

1. One day before cocultivation, inoculate a single colony or frozen glycerol stock of *A. tumefaciens* GV3101::pMP90 into 2 ml of LB medium with appropriate antibiotics (50 μg/mL gentamycin for *Agrobacterium*, plus antibiotic for binary plasmid). Grow at 28°C for 18–24 h.
2. On the day of co-cultivation, dilute the saturated *Agrobacterium* culture into 10 ml of fresh YEB medium without antibiotics to OD_{600} = 0.3. Shake vigorously at 28°C until the OD_{600} reading-reaches >1.5. This step takes ~10–12 hrs.
3. Pellet the *Agrobacterium* cells by centrifugation at 6000*g* for 5 min and discard the YEB medium.
4. Resuspend the *Agrobacterium* cells in 10 ml of washing solution by pipetting.
5. Pellet the *Agrobacterium* cells again by centrifugation at 6000*g* for 5 min and discard the supernatant.
6. Resuspend the *Agrobacterium* cells in 1 ml of washing solution by pipetting.
9. Dilute 100 μl of *Agrobacterium* cell suspension 10 times with washing solution, measure OD_{600} and calculate the OD_{600} of the original cell suspension.
10. To a clean 100 × 20-mm Petri dish, add 20 ml of cocultivation medium and transfer 40, 4-day-old *Arabidopsis* (or tobacco) seedlings or 15, 4-day-old tomato or 5-day-old rice seedlings into the medium. Take care not to mechanically damage the seedlings.
11. Add *Agrobacterium* cell suspension to the cocultivation medium to a final density of OD_{600} = 0.5 and mix well by gentle shaking. Calculate the volume of *Agrobacterium* suspension added to the cocultivation medium based on its starting OD_{600}. For example, if the OD_{600} of *Agrobacterium* suspension is calculated to be 12, add 20 x 0.5/12 = 0.83 ml of *Agrobacterium* suspension to the co-cultivation medium.
12. Wrap the Petri dish with aluminum foil and incubate in a plant growth chamber for 36–40 hr for *Arabidopsis* seedlings, 60 h for tobacco (or tomato) seedlings or 6 day for rice (or switchgrass) seedlings before subsequent analysis. Incubate at the usual

growth temperature for the species used (*e.g.*, 22°C for *Arabidopsis*) in this and the following steps.

13. After the cocultivation period, remove cocultivation solution and replace with surface sterilization solution to remove epiphytic bacteria. Incubate for 10 min, then wash three times with H_2O.
14. (Optional) After surface sterilization, place seedlings on 0.5X MS plates and cultivate for upto 1 week prior to analysis. At this step addition of 500 μg/ml of carbenicillin to remove remaining *Agrobacterium* cells can increase seedling survival.

Procedure II

Floral dip method. This protocol is based upon Clough and Bent (1998).

1. Place the plants in 4-inch pots at a density of 10–15 plants per pot. Grow the plants in short days (9–12 hours of light per day) using ample fertilizer, so that large rosettes are obtained. To help hold the soil in the pot, cover the pots with nylon mesh. If short-day conditions are not available, or if time is limited, grow the plants under long days from the beginning, but increase the density to 15–20 plants per pot.
2. Remove the first inflorescence shoots as soon as they emerge (to encourage the growth of more inflorescences) and move the plants to long days (16–24 hours of light per day). Plants will be ready for transformation after about 1 week, when the secondary inflorescence shoots are ~3 inches tall.
3. Three days prior to plant transformation, inoculate a 5-ml liquid culture of *Agrobacterium* carrying a suitable binary vector and incubate at 28°C with vigorous agitation. Use LB medium containing antibiotics that select for both the Ti and the T-DNA plasmids.
4. After 2 days, inoculate 200 ml of LB medium with 1 ml of the preculture and incubate again with vigorous agitation for an additional 24 hours at 28°C. Use YEP medium for higher *Agrobacterium* density.
5. Stop watering the plants and allow the soil to dry out a little, so that it will be less prone to falling out of the pots during dipping.
6. Pellet *Agrobacterium* by centrifuging at 6000 rpm for 10 minutes at room temperature.
7. Resuspend the cell pellet in 400 ml of infiltration medium.
8. Transfer the *Agrobacterium* suspension to a convenient vessel for dipping plants, *e.g.*, a 400 ml beaker.

9. Invert a pot of plants and dip the inflorescence shoots into the suspension. Rest the pot on the edge of the beaker and allow the plants to soak for ~30 seconds. The same suspension can be used for 10 or more pots. After dipping, lay the pots on their sides in a plastic nursery flat and place a transparent cover over them for the next 24 hours.
10. After 24 hours, remove the cover, rinse the plants with water, and return them to their normal growing conditions.
11. After about 3 weeks, collect seeds.

Precautions

1. All solutions containing *Agrobacterium* and all equipment contacting *Agrobacterium* cultures must be sterilized appropriately.
2. Try to avoid contamination of the soil with *Agrobacterium* in floral dip method, while dipping which produces a rather unpleasant smell.
3. Surface sterilize the plant materials to be transformed.
4. Autoclave and properly dispose of all materials.

3.6.5 Isolation of Plant Protoplasts

Protoplasts are plant cells that had their cell wall removed. Plant protoplasts are required for transformation in a number of gene transformation experiments as well as for isolation of cell organelles. They are prepared from plant tissues, such as leaves, callus and cultures of suspensions. The plant tissue is treated with cell wall lysing enzymes like pectinases, cellulases and hemicellulase which breakdown the cell wall leaving only the cell membrane. Protoplasts loose their original shape and become round after isolation. Sorbitol or mannitol in the digesting medium is used for maintaining the osmotic potential of the cells. The media used for culturing or the nutritional requirements of protoplasts are similar to those of cell suspension cultures. Thus protoplast culture media are usually variants of the media used for cell culture, *i.e.,* MS, B5 media with slight modifications. Ammonium ions are particularly detrimental to protoplast survival and reduced or removed from many media. Proto-plasts from the same or even different species fuse with each other when kept in close contact and in the presence of specific chemicals known as fusogens, *e.g.,* polyethylene glycol (PEG), Ca^{2+} etc. Protoplast fusion is very useful in agriculture biotechnology where new varieties or even hybrid varieties of plants are made.

Procedure

1. Take fresh fully expanded leaves from 4–5 week old aseptically growing plants.
2. Surface sterilize in Ethanol (70%) followed by Mercuric chloride (0.1%) for 1 minute.
3. Wash the leaves thoroughly with sterile distilled water.
4. Peel the lower epidermis of the leaf tissue with fine forceps and cut out the peeled areas with a fine scalpel and place it in 0.5M Mannitol solution for one hour
5. Remove the Mannitol solution and replace it with filter sterilized enzyme solution.
6. Incubate the plates in dark for 16–20 hours at 25µC.
7. Separate the protoplast layer by filtration using 0.45µM nylon membrane. Gently squeeze the leaf pieces with a Pasteur pipette to liberate the protoplasts. Remove the debris by passing through a 60–80 µm mesh sieve.
8. Centrifuge the filtrate at 1000g for 10 minutes at 4°C.
9. Green protoplasts form a band at the top of the suspended solution. Remove the supernatant and resuspend the pellet in cell protoplast washing medium.
10. Centrifuge at 1000g for 10 minutes.
11. Repeat step 9 twice.
12. Finally suspend the protoplast pellet at a density of $0.5–1 \times 10^5$ ml^{-1} in the protoplast culture medium.
13. Count the number of protoplasts in the suspension using a haemocytometer.
14. Plate the protoplasts as fine thin layer in petriplates and incubate at 25°C in dark.

Precautions

1. The leaf pieces for protoplast liberation should be squeezed gently.
2. The protoplasts should be handled very gently.
3. Complete sterilization conditions should be maintained.

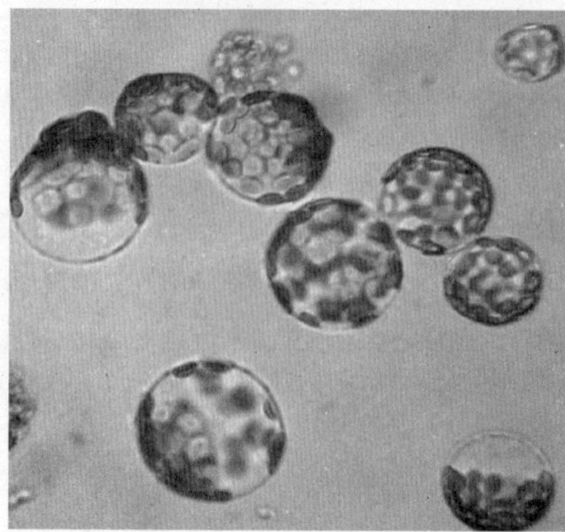

Fig. III. 3.5 : Plant protoplasts isolated from leaves (http://www.biology.Isubfac/nkato/methods/methodsImages/protoplast)

3.7 SELECTION OF RECOMBINANT CELLS

Once recombinant DNA molecules have been formed and appropriate host cells transformed then such transformed cells have to be selected for the presence of the gene of interest. A variety of methods are available for selection of the desired transformants. Specific detection methods such as hybridization with radioactively or non-radioactively labeled probes may also be used for detecting the desired gene in the clones or the product of the cloned gene can be detected. Selection of recombinants is also desired in case of screening of clones in cDNA or genomic libraries. In prokaryotic cells this selection can be done with the help of antibiotic resistant marker genes. However, other selectable marker genes, such as lac Z' gene, GUS gene, etc., may be used for selection depending on the type of cell transformed.

3.7.1 Selection by Insertional Inactivation of Antibiotic Resistance Gene

Plasmid pBR322 carries two antibiotic resistance genes, ampicillin resistance and tetracycline resistance. A recombinant pBR322 molecule, which carries an extra piece of DNA in the Bam HI site located within the tetracycline resistance gene will no longer be able to confer tetracycline resistance on its host, as the gene becomes disrupted by the inserted DNA. Cells containing this recombinant pBR322 will still be resistant to ampicillin, but sensitive to tetracycline ($amp^R tet^S$). For the selection of such recombinant DNA containing *E. coli* cells insertional inactivation of the marker gene, *i.e.,* antibiotic resistance gene is done. Screening of pBR322 recombinants is performed in the following way:

1. After transformation of *E. coli* cells with the recombinant DNA, the cells are plated on to ampicillin containing LB medium and incubated until colonies appear.
2. All colonies will consists of transformants (untransformed cells are amp sensitive so do not produce colonies), but only a few will contain recombinant pBR322 molecules, most will contain the normal plasmid.
3. To distinguish recombinants the colonies are replica-plated on to agar medium that contains tetracycline.
4. After incubation, some of the original colonies will regrow, others will not. Those colonies that grow will consist of cells that carry the normal pBR322 with no inserted DNA and therefore a functional tetracycline resistance gene ($amp^R tet^S$).
5. Colonies on the ampicillin plate that do not grow on the tetracycline plate will consist of recombinants cells ($amp^R tet^S$).

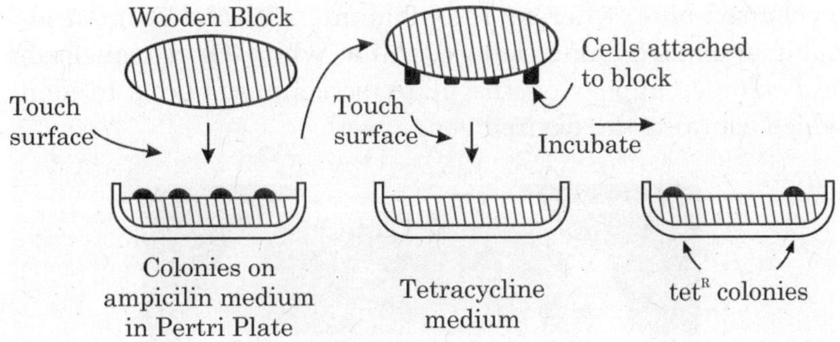

Fig. III. 3.6 : Insertional inactivation of antibiotic resistance gene.

Precautions

1. Maintain sterilized conditions during culture and transfer of *E. coli* culture.
2. Prepare accurate concentrations of antibiotics to be added to the medium.

3.7.2 Lac Z' gene Selection

The plasmid pUC8 carries an ampicillin resistance gene and a gene called *lacZ*', which codes for a part of the enzyme β-galactosidase. Cloning with pUC8 involves insertional inactivation of the *lac*Z' gene, recomb-inants are distinguished by their inability to synthesize β-galactosidase. β-galactosidase is one of a series of enzymes involved in the breakdown of lactose to glucose plus galactose. It is normally coded by the gene *lac*Z, which resides on *the E. coli* chromosome. Some strains of *E. coli* have a modified *lacZ* gene, one that lacks a large segment referred to as *lacZ'* coding for the α-peptide portion of β-galactosidase. Such mutant strains of *E. coli* can synthesize the enzyme only, when they harbour a pUC8 plasmid, that carries the missing *lacZ'* segment of the gene. Cloning with pUC8 plasmid involves selection of transformants on ampicillin agar followed by screening for β-galactosidase activity to distinguish recombinants. Cells that harbour a normal pUC8 plasmid will be amp^R and able to synthesize β-galactosidase however, recombinants will also be amp^R but unable to make β-galactosidase.

Recombinants are screened by plating on agar containing X-gal and IPTG. Screening for β-galactosidase presence or absence involves using a lactose analogue called X-gal (5-bromo-4-chloro-3-indolyl-β-D-galactopyranoside) which is broken down by β-galactosidase to a product that is blue. If X-gal (plus an inducer of the enzyme, such as isopropylthiogalactoside, IPTG) is added to the agar, along with ampicillin, then non-recombinant colonies, the cells of which synthesize β-galactosidase will be coloured blue, whereas recombinants with a disrupted *lacZ'* gene and unable to make β-galactosidase, will be white. Once transformation is done and colonies appear on the plate, screening is done to identify the clone which contains the desired gene insert.

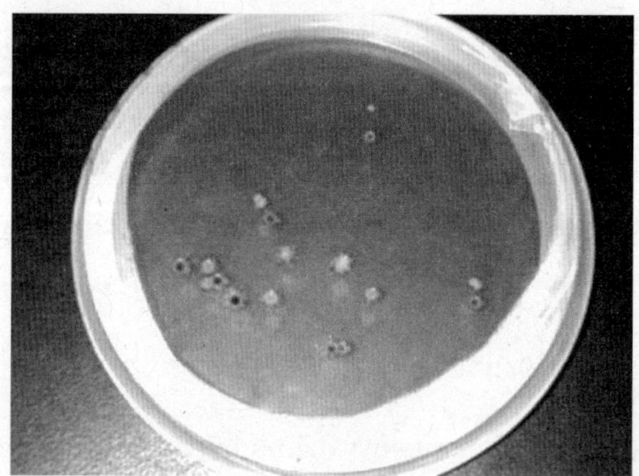

Fig. III. 3.7 : Blue-white colonies

Requirements

X-gal: Prepare 40 mg/ml solution in dimethyl fromamide working concentration is 40 μg/ml. X-gal is light-sensitive. Hence, the stock should be stored in a tube wrapped with aluminium foil and plates in dark.

IPTG: Prepare 24 mg/ml stock solution in sterile water. Filter-sterilize and store in -20°C. Working concentration is 96 μg/ml.

Luria Agar plates containing ampicillin (100 μg/ml).

Procedure

1. Prepare a Luria agar plate. Add X-gal and IPTG to a final concentration of 40 μg/ml and 96 μg/ml, respectively. Spread uniformly with the help of a glass rod and allow it to dry.
2. Seal the plate.
3. Pick up transformants using sterile toothpick and streak on the plate.
4. Incubate at 37°C for 12–14 hours.
5. Look for white colonies. These are putative clones containing the desired insert. Non-recombinants will be blue.

3.7.3 GUS Assay

Screenable marker 'reporter genes' have been developed from bacterial genes coding for easily assayed enzymes, such as chloram-phenicol acetyl transferase (CAT), β-galactosidase, β-glucuronidase (GUS), luciferase (LUX), green fluorescent protein (GFP), nopaline synthase and octopine synthase. Initial steps for genetic transformation involve delivery of a gene cassette into recipient cells followed by analysis of the expression of delivered gene. The results of the above events can be detected by assaying the expression of a reporter gene introduced into plant cell cultures or intact tissues. The reporter genes produce a visible effect, directly or indirectly, due to their activity in the transformed cells. Analysis of reporter gene expression does not require the integration of the transgene into the host genome and is commonly used to test promoter and gene functions.

A protocol for the rapid assay of β-glucuronidase (*gus*) gene from *E. coli* makes selection of transformed plant cells very easy. This gene has emerged as the most widely used scorable marker in plant transformation. GUS enzyme hydrolyzes β-glucuronide compounds and gives reaction products that can be quantified spectrophotometrically or spectrofluorometrically. The *gus* reporter gene system is extremely useful for optimisation of parameters for genetic transformation, due to the availability of a simple histochemical detection procedure. One of the major limitations of *gus* reporter gene system, however, is the destructive nature of its assay. The enzyme β-glucuronidase (*gus*) is not normally found in plants, this means false positive results are rare. During the transformation process only the T-DNA and the GUS gene cloned into it are transferred to the plant cell and integrated into plant genome.

The genetic markers developed for use in plant cells, in general have been derived from either bacterial or plant sources and can be divided into two types: **selectable** and **screenable markers.** Selectable markers are those which allow the selection of transformed cells, or tissue explants, by their ability to grow in the presence of an antibiotic or a herbicide. The most frequently used selectable markers are kanamycin and hygromycin. In addition to selecting for transformants, such markers can be used to follow the inheritance of a foreign gene in a segregating population of plants. Screenable markers encode gene products whose enzyme activity can be easily assayed, allowing not only the detection of transformants but also an estimation of the levels of foreign gene expression in transgenic tissue. Markers such as β-glucuronidase (GUS), luciferase or β-galactosidase allow screening for enzyme activity by histochemical staining of individual cells and can be used to study cell-specific as well as developmentally regulated gene expression. Routinely, the marker genes used have been linked to promoters derived from plant pathogens, such as the T-DNA of Agrobacterium tumefeciens or plant viruses.

The most commonly used screenable markers are, chloramphenicol acetyl transferase (CAT), β-glucuronidase (GUS), β-galactosidase and luciferase. One of the most direct approaches for studying the expression of foreign genes in transformed plant tissue is to measure the abundance, or activity, of the gene products encoded by the transferred genes. Many promoters, and putative gene regulatory sequences have been analysed by making fusions with reporter genes. These are usually the coding sequences of bacterial enzymes for which convenient and sensitive assays are available and whose activities are not normally found in higher plant tissues. Among the most common are octopine and nopaline synthase, CAT, NPT-II and GUS.

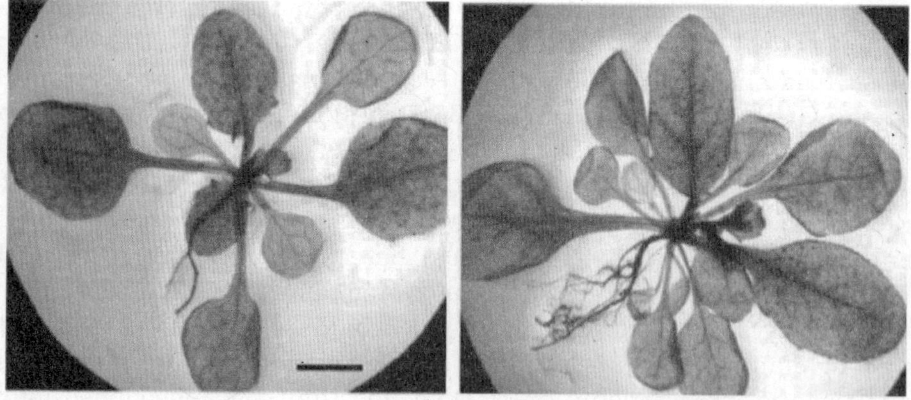

Fig. III. 3.8 : Histochemical detection of GUS

Procedure

The preferred histochemical substrate for tissue localization of GUS is 5-bromo-4 chloro-3-indolyl-ß D-glucuronide (X-gluc). The advantage of these substrates is that the indoxyl group produced upon enzymatic cleavage dimerizes to indigo which is virtually insoluble in an aqueous environment. The histochemical assay for GUS consists of soaking tissue in substrate solution and watching for blue colour to appear. The basic steps of this histochemical staining/fixation method are:

1. Tissue to be stained is fixed for 30 minutes in ice cold fixative, shaking occasionally.
2. Wash for 30–60 minutes in several changes of ice cold 1X buffer.
3. If the tissue floats or will not wet well, Triton-X-100 is added to the staining solution to a final concentration of 0.1 percent. or vacuum infiltrate the X-gluc substrate medium into the objects.
4. The plates wrapped in plastic film to prevent drying and are incubated at 37°C overnight in darkness till blue staining appears (but not more than 24 hrs).
5. Incubate green objects in ethanol (70%) until the chlorophyll is removed, then transfer to distilled water again.
6. After staining, optional step is to place specimens in glycerol (50%) for 1 hr, then transfer to pure glycerol, again leave for 1 hr or more. The tissue is mount in glycerol (100%) and examined under a stereomicroscope for the presence of blue sectors. The tissues can be even mounted in water.

Precautions

1. Solutions should be prepared fresh at the time of use.
2. False positives can be obtained particularly in case of pollen grains.
3. The tissues may become soft and should be handled carefully.

Chapter - 3

D. IMMUNOLOGY

INTRODUCTION

The history of Immunology dates back to 2656 BC on observations of small pox by Chinese and later in 1798 with the experiments of Edward Jenner on cowpox. However, Jenner's observations on cowpox were realised in 1879, when Louis Pasteur in France worked on resistance in Chickens to fowl cholerea caused by *Pasteurella multocida* and coined the term **vaccination** (*vacca* in Latin means "cow"). The term Immunology has been derived from the word *immunis,* which is for "exempt" in Latin. Immunology is an applied science and is closely associated with microbiology. It originated as a branch of medicine to control diseases of humans and animals. Resistance of animals to infection by bacteria, fungi, viruses and parasites is called "immunity". The immune system protects us from infection through various lines of defence. Immune response is the immunological response originating from immune system activation by antigens, including immunity to pathogenic microorganisms and its products, as well as autoimmunity to self-antigens allergies, and graft rejections. There are two types of immunity:

Innate immunity is the first line of defence and is non-specific. The responses are the same for all potential pathogens in innate immunity, no matter how different they may be. Innate immunity includes physical barriers (*e.g.,* skin, mucous membrane, saliva etc.) and cells (*e.g.,* macrophages, neutrophils, basophils, mast cells etc.).

Adaptive (Acquired) immunity is the second line of defence which involves building up memory of encountered infections so that it can mount an enhanced response specific to the pathogen or foreign substance. Adaptive immunity involves antibodies, which generally target foreign pathogens roaming free in the bloodstream. Also involved are T cells, which are directed especially towards pathogens that have colonized cells and can directly kill infected cells or help to control the antibody response. Adaptive immunity is further divided into humoral (or antibody) and cell-mediated components. The humoral (antibody) response is defined as the interaction between antibodies and antigens.

Antibodies are specific proteins released from a certain class of immune cells known as B lymphocytes, while antigens are defined as any foreign substance that elicits the generation of antibodies ("anti" body "gen" erators) which elicits an immune response in the host. Immunology depends on an understanding of the properties of these two biological entities and the cellular response to both.

Antibodies are heterogeneous group of globulins found in the serum. Antibodies being proteins can act as antigens when inoculated into other species. Antibodies of all classes are heavy (~150 kDa) globular plasma proteins and are called Immunoglobulins (Igs). There are 5 types of immunoglobulins: IgG, IgM, IgA, IgD and IgE depending on the type of heavy chain present in the antibody.

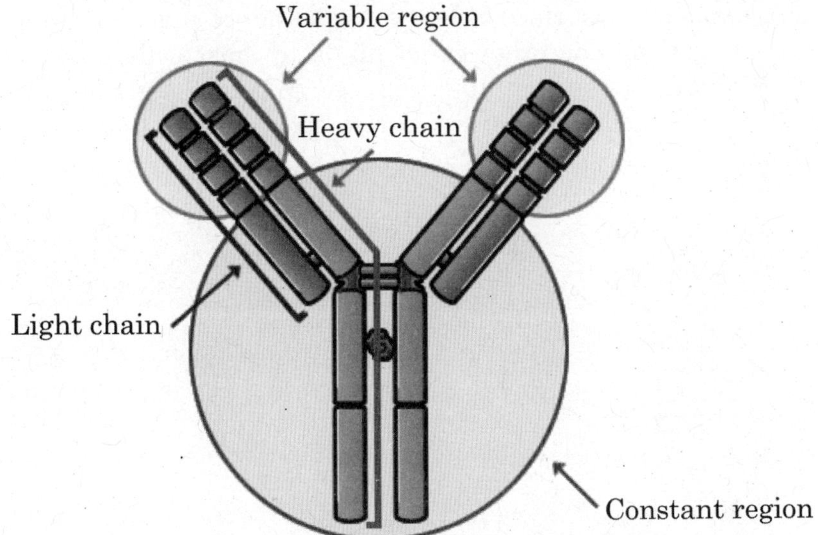

Fig. 4.1 : Basic structure of antibody

Structure of Antibodies: There are four polypeptide chains, two identical heavy chains and two identical light chains connected by disulfide bonds. Light Chain (L) consists of polypeptides of about 22,000 Da, whereas Heavy Chain (H) of larger polypeptides of about 50,000 Da or more (Fig. 4.1). There are five types of Ig heavy chains (in mammal) denoted by the Greek letters: α, δ, ε, γ, and μ. There are two types of Ig light chains (in mammal), which are called lambda (λ) and kappa (κ). An antibody is made up of a **variable region** and a **constant region**, and the region that changes to various structures depending on the differences in antigens is called the variable region, Whereas, the region that has an invariable structure is called the constant region. The classes of immunoglobulins can be divided into subclasses based on small differences in the amino acid sequences in the constant region of the heavy chains. All immunoglobulins within a subclass will have very similar heavy chain in the constant region in the form of amino acid sequences and differences are detected by serological means.

Functions: IgG provides long term protection because it persists for months and years after the presence of the antigen that has triggered their production. It protects against bacteria and viruses, neutralizes bacterial toxins, triggers compliment protein systems and binds antigens to enhance the effectiveness of phagocytosis. Main function of IgA is to bind antigens on microbes before they invade tissues. It aggregates the antigens and keeps them in the secretions so that it is expelled along with the antigen in the secretion. IgA is also first defense for mucosal surfaces such as the intestines, nose, and lungs. IgM is involved in the ABO blood group antigens on the surface of RBCs and enhance ingestions of cells by phagocytosis. IgE on the other hand binds to mast cells and basophils which participate in the immune response. Whereas, IgD is present on the surface of B cells and plays a role in the induction of antibody production.

Table 4.1: Types of Antibodies

The Five Immunoglobulin (Ig) Classes					
	IgM pentamer	**IgG monomer**	**IgA dimer**	**IgE monomer**	**IgD monomer**
			Secretory component		
Heavy chains	μ (mu)	γ (gamma)	α (alpha)	ε (epsilon)	δ (delta)
Number of antigen binding sites	10	2	4	2	2
Molecular weight (Daltons)	900.000	150.000	385.000	200.000	180.000
Percentage of total antibody in serum	6%	80%	13%	0.002%	1%
Crosses placenta	no	yes	no	no	no
Fixes complement	yes	yes	no	no	no
Fc binds to		phagocytes	—	mast cells and basophils	—
Function	Main antibody of primary responses, best at fixing complement; the monomer form of IgM serves as the B cell receptor	Main blood antibody of secondary responses, neutralizes toxins, opsonization	Protect mucus membranes	Antibody of allergy and antiparasitic activity	To signal B cells for activation

4.1 To collect the Serum from blood

(a) Principle

The blood mainly has 2 portions, the **Plasma** and the **Blood Cells.** Blood cells are red blood cells, white blood cells and platelets, whereas plasma is composed of about 92% water, 7% vital proteins such as albumin, gamma globulin, anti-hemophilic factor and other clotting factors, and 1% mineral salts, sugars, fats, hormones and vitamins. Serum on the other hand is a clear, yellowish coloured fluid which does not contain white or red blood cells or a clotting factor. It is the blood plasma without the fibrinogens. Serum contains proteins not used in blood clotting and all the electrolytes, antibodies, antigens, hormones and any exogenous substances.

Blood is centrifuged to remove cellular components soon after the collection of blood, otherwise there may be the risk of cell lysis leading to release of proteolytic enzymes which can degrade antibodies. Coagulated or clotted blood yields serum without fibrinogen.

The preservation of blood is done by addition of sodium azide (0.1%), merthiolate (0.7%) or hydroxyl guanoline sulphate.

(b) Requirements

Human blood, sterilized test tubes, sterilized or disposable syringe, ice cubes, centrifuge.

(c) Procedure

1. Collect the 5 ml blood from a human volunteer with the help of sterilized or disposable syringe into a glass container with ice.
2. Keep the blood at room temperature for 1h and allow clotting.
3. Once the clot has formed, loosen it from the walls of the container to aid retraction.
4. Keep at 4°C and leave it there overnight.
5. Pour the clot free liquid in centrifuge tube and centrifuge at 8000 rpm for 30 minutes at 4°C.
6. Add the liquid thus obtained to above clot free liquid.
7. Centrifuge the clot free liquid again for 15 to 20 minutes at 4000 rpm at 4°C.
8. Collect the straw-coloured supernatant and store at –20°C in containers suitable for long-term storage.

(d) Result

Serum isolated from human blood.

(e) Precautions

1. Always disposable syringe should be preferred.
2. Sterilized glass container should be used to collect the blood as it clots quickly in glass than in plastic.
3. Collected blood should be immediately placed in ice.
4. The serum should be separated from red blood cells within 60 minutes (1 hr) of venipuncture.
5. Repeated freezing and thawing of serum sample during its long term storage should be avoided.

4.2 To prepare the blood film and identify leucocytes

(a) Principle

A blood film is a thin layer of blood smeared on a glass slide and stained to examine the various blood cells microscopically. Blood films are also used to investigate the blood disorder and are routinely employed to identify the pathogens present in blood. White blood cells (WBCs) or leucocytes are the cells of the immune system that are involved in protecting the body against both infectious disease and foreign invaders. White blood cells originate in the bone marrow, but circulate throughout the bloodstream. All white blood cells have nuclei, which distinguishe them from the anucleated red blood cells (RBCs) and platelets. Leucocytes can be classified either by structure (granulocytes or agranulocytes) or by cell lineage (myeloid cells or lymphoid cells). Granulocytes (Neutrophils, Eosinophils and Basophils) are distinguished from agranulocytes (Lymphocytes and Monocytes) by their nucleus shape (lobed versus round) and cytoplasm granules (present or absent, or more precisely, visible on light microscopy or not). On the other hand, the myeloid cells (neutrophils, monocytes, eosinophils and basophils) are distinguished from lymphoid cells (lymphocytes) by hematopoietic lineage (cellular differentiation lineage). Five types of white blood cells can be distinguished by their physical and functional characteristics as discussed below:

(i) Neutrophils

Neutrophils are the most common type of white blood cells in the body constituting 60–70 per cent of the circulating leukocytes and with levels of between 2000 to 7500 cells per mm^3 in the bloodstream. Neutrophils are medium-sized white blood cells and many granules that perform various functions within the cell. They have a multi-lobed nucleus, which consists of three to five lobes connected by slender strands. Neutrophils function by attaching to the walls of the blood vessels, blocking the passageway of germs that try to gain access to the blood through a cut or infectious area. Neutrophils are the first cells to reach an area where a breach in the body has been made. Neutrophils are active in phagocytosing bacteria and are present in large number in the pus of wounds.

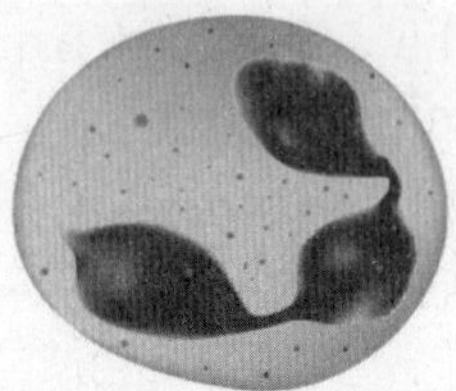

Fig. 4.2 : Neutrophils

(ii) Eosinophil

Eosinophils in the bloodstream are limited in number, only about 40–400 cells per mm^3 of blood and constitute 2-4 per cent of the WBC total. They have large granules that help in cellular functions. Eosinophils are especially important when it comes to allergies and worm infections. In general, their nucleus is bi-lobed. The lobes are connected by a thin strand. The cytoplasm is full of granules that assume a characteristic pink-orange colour with eosin staining.

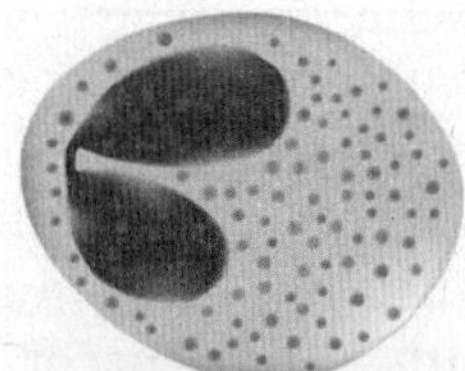

Fig. 4.3 : Eosinophil

(iii) Basophil

Basophils are rarest of the white blood cells (less than 0.5 per cent of the total count) *i.e.,* around 0–100 cells per mm^3 of blood. These are chiefly responsible for allergic and antigen response by releasing the chemical histamine causing the dilation of blood vessels. They are very colourful when stained and looked under the microscope, making them easy to identify. They can be recognized by several coarse, dark violet granules, giving them a blue hue. The nucleus is bi– or tri–lobed, but it is hard to see because of the number of coarse granules that hide it.

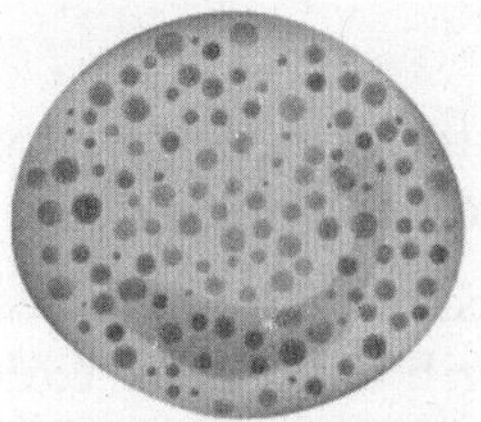

Fig. 4.4 : Basophil

(iv) Monocyte

Monocytes are largest in size (10 to 12 μ), but there are only about 200–800 monocytes per mm^3 of blood. Monocytes eventually leave the bloodstream and become tissue macrophages, which remove dead cell debris as well as attack microorganisms. They have large kidney shaped nucleus and are typically agranulated. They also possess abundant cytoplasm which stains greyish blue.

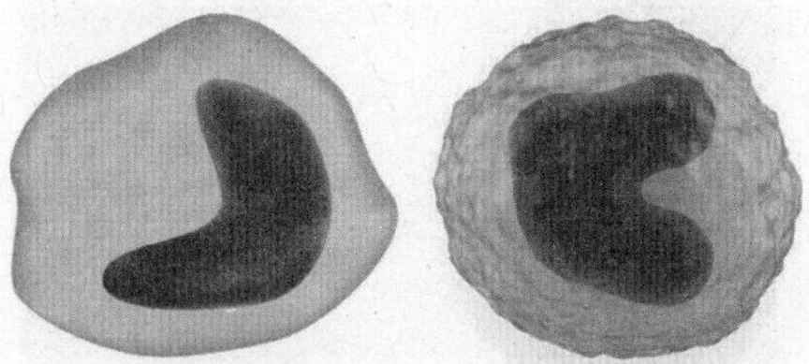

Fig. 4.5 : Monocyte

(v) Lymphocytes

Lymphocytes are much more common in the lymphatic system than in blood. There are about 1300 to 4000 lymphocytes per mm^3 of blood. These include B, T and natural killer cells. These are small, round cells that have a large nucleus within a small amount of cytoplasm. Lymphocytes are distinguished by having a deeply staining nucleus that may be eccentric in location. They play major role in humoral immune system, which is the part of the immune system that relates to antibody production.

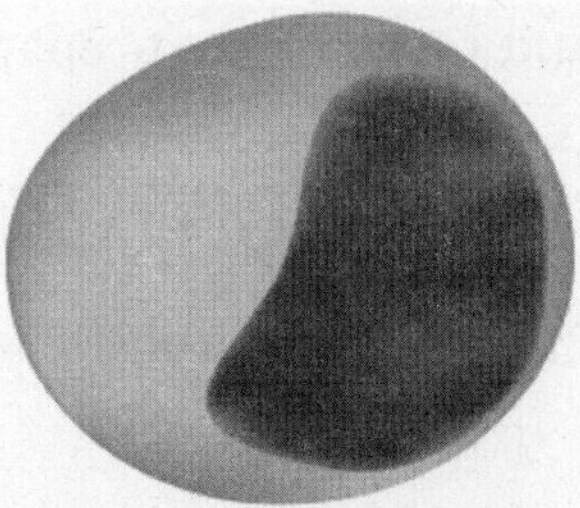

Fig. 4.6 : Lymphocyte

(b) Procedure

1. Prick the finger after sterilizing it with ethanol and collect the blood using disposable needle.
2. Place a drop of blood on one end of a slide, and disperse the blood over the slide's length using a spreader slide.
3. Prepare a thin film of blood which must be uniform and tongue shaped.
4. Leave the slide to air dry, after which fix the blood to the slide by immersing it briefly in methanol.
5. Stain the smear with Leishman stain for about 15 minutes and rinse it with distilled water, then air dry the slide.
6. Examine the slide under a microscope using magnification up to 1000× for individual cells.

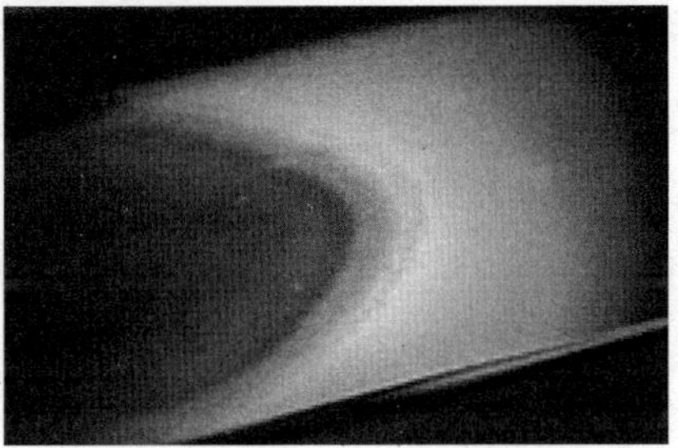

Fig. 4.7 : Blood smear

(c) Result

Different type of leucocytes were observed under the microscope and identified.

(d) Precautions

1. The needle for pricking the finger should either be well sterilized or disposable needle should be used.
2. Finger should be pricked at one go.
3. Blood smear should be properly prepared and the film should be thin.
4. Staining should be done properly.

4.3 To determine ABO blood types by slide agglutination reaction

(a) Principle

The ABO blood group system is the most important blood type system found in humans. It was discovered by the Austrian scientist Karl

Landsteiner. The central principle of the ABO system is that antigens – basically sugars physically exposed on the exterior of red blood cells – differ between individuals, who have immunological tolerance only toward what occurs in their own bodies. The four basic ABO phenotypes are O, A, B, and AB.

The ABO locus, located on chromosome 9, has three main allelic forms: A, B, and O. The A and B alleles encode a glycosyltransferase that catalyzes the final step in the synthesis of the A and B antigen, respectively. The O allele encodes an inactive glycosyltransferase that leaves the ABO antigen precursor (the H antigen) unmodified. The immune system forms antibodies against whichever ABO blood group antigens are found on the individual's RBCs. Thus, a group A individual will have anti-B antibodies and a group B individual will have anti-A antibodies. Blood group O is common, and individuals with this blood type will have both anti-A and anti-B in their serum, whereas blood group AB is the least common, and these individuals will have neither anti-A nor anti-B in their serum.

Slide agglutination or clumping of particles is a qualitative test which is used mostly to determine blood types (ABO types). Direct agglutination is carried out on the slides, by directly mixing the antibody with a antigen particle under the certain concentration of electrolyte. The result is positive when there is visible agglutinate, otherwise negative.

Blood type (phenotype)	Antigen on RBC	Antibody in sera
A	A	Anti–B Ab
B	B	Anti–A Ab
O	H	Anti-A Ab, Anti–B Ab
AB	A, B	–

(b) Requirements

Human peripheral blood, Standard sera: Anti-A and anti-B sera, saline, clean glass slides, cotton ball, sterilized needle, alcohol swabs, tooth pick

(c) Procedure

1. Take a clean and dry slide; plot it into three squares and label as "1", "2" and "3".
2. Put a drop of anti-A and anti-B standard serum on square "1" and square "2", respectively and put a drop of saline on square "3".

1	2	3
anti-A	anti-B	saline

3. Rub the alcohol swab on finger and prick the finger tip with sterilized needle, take the oozing blood using brim of a sterilized slide from the box and mix the blood into the sera and saline on square "1", "2" and "3", respectively.
4. Place a cotton ball at the finger site where it was pricked. Using the thumb, put pressure on the area to stop blood flow.
5. Take a tooth pick and mix the content in each square. Discard the tooth pick after using in one square.
6. After mixing, wait for a while to observe the result. Watch the slide under microscope for agglutination reaction.

(d) Results

Positive result shows particle agglutination, whereas negative result appears uniformly turbid.

(e) Precautions

1. Slides should be sterilized in ethanol (75 per cent).
2. All glassware must be dry and clean.
3. Antisera should be used as per manufacturer's instructions.
4. Antisera should be stored in the refrigerator when not in use.

4.4 To isolate and purify sheep IgG from whole sheep serum and determine its concentration

(a) Principle

Serum contains two types of proteins, **globulins** and **albumins**. It is a good source of immunoglobulins, especially IgG. In order to isolate and partially purify IgG from sheep serum, three techniques are used: salt precipitation, an ion exchange chromatography and dialysis.

(i) **Sodium sulphate precipitation:** In this method when salt concentration of the medium is raised, there is an interference with the interaction of water molecules with charged polar groups on protein molecules. The protein molecules become less hydrophilic allowing greater hydrophobic interaction between the protein molecules. As a result of which the proteins become insoluble. The salt concentration at which each protein precipitates is different and overlaps with similar proteins. For preparing the crude immunoglobulin fraction, the immunoglobulins are precipitated out and the unwanted serum proteins stay in solution. While precipitating the serum, sodium sulphate is added slowly to avoid high local concentrations which would decrease the specificity of the precipitation.

(ii) **Ion exchange chromatography:** The protein molecules bind electrostatically onto an ion exchange matrix that carries an opposite IgG is purified using DEAE cellulose, an ion exchange resin. Cellulose is the support for diethylaminoethyl groups which are positively charged. By providing appropriate equilibrium conditions (pH and ionic strength), DEAE cellulose will bind all the serum proteins (negatively charged) except IgG (positively charged). Therefore IgG is separated in the void volume. Contaminating serum components in the IgG preparation depends upon the ratio of ion exchanger to serum. The greater the amount of DEAE cellulose added per ml of serum, the greater will be the purity of the IgG. However, this leads to loss of IgG.

(iii) **Dialysis:** Dialysis separates proteins based on the size of the dialysis pores. Smaller molecules equilibrate across the dialysis membrane leaving the larger size molecules *i.e.,* proteins/IgG inside the dialysis membrane.

Measurement of protein concentration of crude IgG is estimated by UV light absorption using a spectrophotometer. Most proteins absorb ultraviolet light maximally at 280 nm as they contain tryptophan, tyrosine and phenylalanine residues. Protein solutions obey the Beer-Lambert Law.

(b) Procedure

I. Preaparation of sheep serum:

i. Allowe the fresh blood is allowed to stand at room temperature for 1 hour.

ii. Store the blood overnight at 4°C to allow formation of clot.

iii. Pour the serum is poured off, the clot remains in the bottom of the tube/flask.

iv. Centrifuge at 3000 rpm for 10 min.

v. Transfer supernatant to new tube and repeat the above step till clear homogeneous supernatant free, of RBCs serum is obtained.

II. Sodium sulphate precipitation:

i. Warm lipid free sheep serum sample in a 25°C waterbath for 10 min. (Pretreated serum sample was centrifuged at 10000 rpm for 30 min at 4°C. Sheep serum was incubated at 56°C for 30 min to inactivate the complement system. The serum was centrifuged at 10000 rpm for 30 min at 4°C. The lipids are soluble at low temperature, therefore should be filtered through glass wool to remove the insoluble lipids).

ii. Add 1.8 g sodium sulphate (18 per cent w/v solution). Shake the tube gently to dissolve the contents.

iii. Centrifuge at 10000 rpm for 15 min at 25°C. Transfer the supernatant containing the crude serum to another tube.

iv. Estimate volume of pellet. Make up volume to 10 ml with distilled water. Mix well. Add sodium sulphate powder to make 14 per cent (w/v) solution.

v. Repeat step 3 and resuspend pellet using a spatula in 3 ml distilled water.

III. Dialysis:

i. Remove a piece of dialysis tubing from PBS buffer and tie knots at one end.

ii. Fill the tubing with 3–5 ml of crude IgG and tie a knot at the other end too.

iii. Dialyse overnight in 4 litres PBS, pH 7.2 at 4°C.

iv. Next day change PBS and again dialyse for 2 h.

v. Cut bottom knot of tubing for collecting the sample.

vi. Store samples in small vials at -20°C.

IV. Equilibration of DEAE Cellulose:

i. Add 100 g DE52 DEAE cellulose to 550 ml 0.01 M phosphate buffer, pH 8.0 in a beaker while stirring.

ii. Titrate with 1.0 M HCl back to pH 8.0.

iii. Keep for 30 min to allow the slurry to settle down. Decant and aspirate the supernatant.

iv. Add 0.1M phosphate buffer, pH 6.6, mix and again let the slurry to settle for 30 min and remove supernatant.

v. Repeat wash and filter through two layers of Whatman No. 1 filter paper.

V. Preparation of Sheep IgG:

i. Take 1 ml pre-treated sheep serum in a beaker and place on ice. To this, add 3 ml cold distilled water. Distilled water is added to reduce the ionic concentration of the serum.

ii. Add 5 g DEAE Cellulose to the above solution. Mix using a spatula.

iii. Incubate on ice for 1 hour mixing every 10 min.

iv. Filter the slurry of serum and DEAE cellulose using suction through pre-wet (with PBS) 2 layers of Whatman No. 1 filter paper.

v. Using a micro pipette transfer effluents in vial. This is the purified IgG. Store at -20°C.

VI Measurement of protein concentration:

i. Add 1 ml of PBS, pH 7.2 to an eppendorf tube. To this add 5 μl sheep IgG and mix.

ii. Transfer sample to a cuvette and read absorbance at 280 nm in a spectrophotometer. Determine concentration using the Beer-Lambert Law: A= kcl where A is the Absorbance at 280nm, K is proportionality constant which for IgG= 1.5 ml mg^{-1} cm^{-1}, c is the concentration of the substance in the solution (mg/ml) and l is path of UV light in the solution *i.e.,* 1c m.

(c) Observations

Determine the concentration of crude IgG (mg/ml) for sample at each step viz. Crude, Sodium sulphate precipitation, DEAE cellulose ion exchange and partially purified sheep serum samples.

(d) Results

The partially purified IgG is obtained.

(e) Precautions

1. Proper temperature should be maintained at each step as specified.
2. Care should be taken so that entire sample is aspirated and transferred to the next step.
3. Reading in spectrophotometer should be taken carefully at the Absorbance value specified.

4.5 To perform double Immunodiffusion by Ouchterlony method.

(a) Principle

The method of double immunodiffusion was developed by Ouchterlony in 1966. An antibody is a protein molecule which contains one or more identical sites for combining with a specific antigen. Specific combinations of antigen (Ag) with an antibody (Ab) are the fundamental reaction of immunology. Under specific conditions of antibody and antigen concentration, large complexes or lattices can be formed which are insoluble and precipitate from solution. Precipitaion assays are performed in solution or in gels such as agarose. Immunodiffusion in gels technique is used for the analysis of Ag and Ab. An antigen reacts with a specific antibody leading to formation of an Ag-Ab complex. The Ouchterlony double diffusion is used for comparing different antigen preparations. The antigen containing single antigenic species are allowed to diffuse from separate wells against the antiserum. Depending on the similarity between the antigens, different geometrical patterns are produced between the Ag and antiserum wells.

Applications of double immunodiffusion

1. It is used to analyse the characteristics of antigen or antibody, purity of antigen and antibody or measure the titre of antibody.
2. The homogeneity of antigen-antibody complexes can be determined.
3. It can also be used to follow the purification of an antigenic mixture.
4. The minimum number of antibody-antigen complexes present are determined by this technique.
5. It can be determined whether a given antigen shares structural characteristics (cross-reacts) with other molecules of interest.

(b) Requirements

Saline agar (1.5%) *i.e.,* agar (1.5%) melted in saline solution (0.9%), Antigen: sample serum, Antibody, plastic plates/glass slides, agar puncture, microsyringe/micropipette etc.

(c) Procedure

1. For preparation of agar plate: Clean the slides with ethanol and dry. The clean slides are precoated with agarose (1%) in distilled water to hold the final agarose gel in place.
2. Dissolved the agarose by heating while stirring, then cool to 55°C. Pipette the warm agarose on the slides to cover the entire surface of the slide. Air dry completely. The slides can be stored at room temperature till required.
3. Pour 3.5 ml melted saline agar into a plastic plate on the horizontal table.
4. Punch 7 wells of 3 mm diameter in the solidified agar with 5 mm of interspace to each other. One well is in the centre and the other remaining 6 around the centre well.
5. Clean the microsyringe with clean distilled water. Pour 15 μl sample serum, positive serum and diagnostic serum into three

wells with the microsyringe. The antiserum is placed in the centre well and the antigen in the surrounding wells.

6. Put the plate into a humid box and incubate for 24 hrs at 37°C temperature.

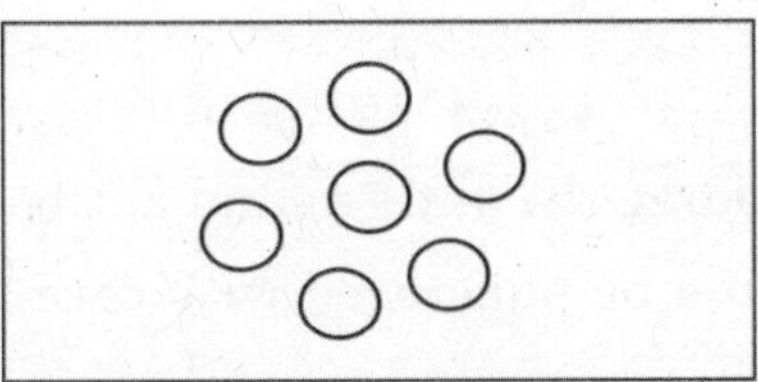

Fig. 4.8 : Wells in solidified agar in equal interspace distance

(d) Observation

Next day the pattern of precipitin lines are observed and pattern drawn on a paper. On the diagram, indicate the contents of each well, including the central well.

A. Reaction of Identity: If an antibody is specific to the corresponding antigen, then there is a complete fusion precipitate band between them. This shows the presence of identical determinants in antigens 1 and 2.

B. Reaction of Non-identity: If two precipitate bands cross each other when they meet without any interaction, it indicates that the two antigens are completely different and so are the antibodies. The proteins do not share any determinants detectable by the antiserum.

C. Reaction of Partial identity: If the precipitate bands accord partially with each other, it indicates that the antibody is a kind of mixture of antibodies and unknown antigen is partially identical with the control antigens. The two precipitation lines have fused, but in addition to that a spur forms towards the well. This implicates that the protein shares some common determinants, but one protein has additional unique determinants detected by the antiserum.

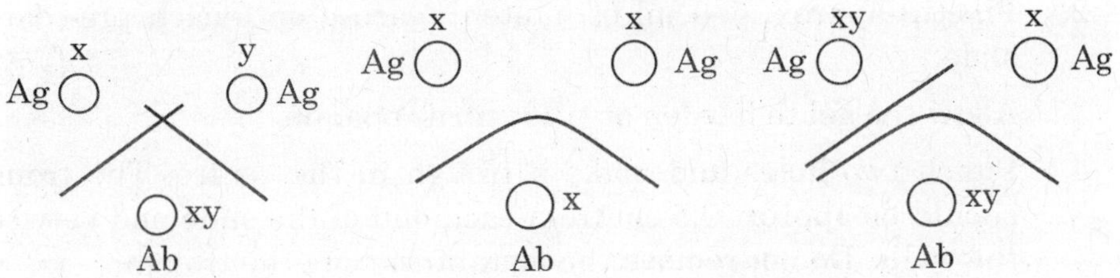

Fig. 4.9 : Pattern of Precipitin lines showing Non-Identity, Identity and Partial Identity.

(e) Results

The result is determined by the existence or nonexistence of characteristic white band of precipitate.

(f) Precautions

1. The micropipette should be washed each time with distilled water when adding the antigen and antibodies.
2. The wells should be punctured with care and at proper distance from each other.
3. While pouring the molten agarose care should be taken that it does not run off the edge of the slide.

4.6 To perform Immunoelectrophoresis

(a) Principle

In order to analyze complex antigen mixtures, the technique of immunoelectrophoresis was developed by Grabar and Williams (1953). This technique involves the placing of the antigen in a gel well and subjecting it to electrophoresis. The antibody to one or more of the antigens in the mixture is loaded into a long well parallel to the electrophoretically separated antigens. The antibody is allowed to diffuse outward and precipitin arcs form when they meet cognate antigen. The serum of suspected disease can be compared to normal humans quickly by using immunoelectrophoresis.

(b) Requirements

Electrophoresis unit, glass slide, agarose, micropipette, Tris-tricine buffer, pH 8.6, distilled water, agar puncture, filter paper etc..

Procedure

1. Prepare agarose (1%) in Tris-tricine and pour on glass slide as in Experiment No. 4.5.
2. Pipette approx. 5–6 ml of molten agarose onto each pre-coated slide.
3. Allow the gel to harden or till it turns opaque.
4. Punch two holes and make a trough in the centre. The trough should be approx. 0.5 cm from each end of the slide and between the wells. Do not remove the agar strip from the trough.
5. Fill one well with 10 μl Sheep IgG and the other well with 10 μl Sheep IgG serum.

6. Attach the electrophoresis unit and place the slide in the unit such that the wells are close to the negative electrode.
7. The running buffer is Tris-tricine, pH 8.6.
8. Electrophorese for 1 h and 45 min at 240 volts.
9. After electrophoresis is completed, remove the agarose strip from the trough.
10. Fill the trough with 60 ml antiserum (anti-sheep whole serum prepared in rabbit).
11. Place the slides in a humid chamber at room temperature overnight.

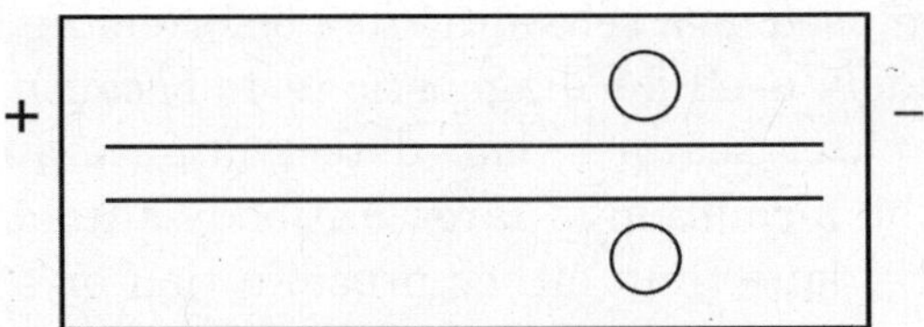

Fig. 4.10 : Pattern of wells and trough for immunoelectrophoresis

(d) Observation

1. Next day observe the pattern of precipitin lines formed.
2. Draw the pattern on a paper and infer the results.

(e) Results

Based on the pattern of the precipitin lines observed draw a diagram and interpret the results can be interpreted.

(f) Precautions

As given above in Experiment No. 4.5

4.7 To perform Enzyme linked immunosorbent assay (ELISA)

Principle

An enzyme-linked immunosorbent assay, also called ELISA or EIA, is an analytical immunoassay that detects and measures antibodies in the blood sample. This test can be used to determine if there are antibodies related to certain infectious conditions. ELISAs were first developed in the early 1970s as a replacement for radioimmunoassays. They remain in wide use in their original format and in expanded formats with modifications that allow for multiple analytes per well, highly sensitive readouts, and direct cell-based output. The basic concept involved in an immunoassay is the interaction of an immunogen with its specific antibody. In ELISA, soluble immunogen is linked to an antibody (which is bound to a solid phase like microtiter plate, plastic bead or strip) or soluble antibodies to immobilized immunogen, in a manner that allows both immunological and enzymatic activity to be retained. The free immunogen (analyte) and the enzyme labeled immunogen compete for the fixed and limited number of specific binding sites of the immobilized antibody. Alternatively, the enzyme labeled antibodies can be made to compete with immobilized immunogen and free immunogen (analyte). An ELISA test may be used to diagnose many diseases that include: HIV, which causes AIDS, pernicious anemia, rotavirus, carcinoma, syphilis, varicella-zoster virus.

ELISA is often used as a screening tool before more in-depth tests are ordered. It is a widely used for diagnostic tests because it is fast and can be read by a computer scanner, can detect minute amount of antigen, does not require the formation of large antibody-antigen complex, and is more sensitive technique than either precipitation or agglutination (Fig. 4.11).

Five steps are involved in ELISA procedure which include

i. Loading of the sample and coating solid surface (of well) with antigen/antibody.
ii. Binding complementary antibody/antigen from test sample.
iii. Binding conjugate.
iv. Addition of enzyme's substrate.
v. Recording of change in absorbance. The net absorbance due to the immunogen or the antibody linked enzyme activity is correlated to the concentration of the analyte on an ELISA reader.

There are two basic methods: The direct ELISA used for detecting antigen against a specific antibody found in a test well and the indirect ELISA used to detect antibodies against an antigen found in the test well. There are four kinds of ELISA.

i. **Direct ELISA to detect antigens:** In a direct ELISA, used for detecting antigens, the small wells of microtiter plate are coated with specific (known) antibody against the antigen to be detected. The specimens such as serum (of unknown bacterium) to be tested are added to the coated wells. If antigen is present in the serum, then it will combine with the coated antibody. Enzyme linked antibody specific for the test antigen is added and bind to the antigen forming sandwich. An enzyme's substrate is added to detect the binding of conjugated antibody to antigen-antibody complex. A positive result is indicated by colour that is intense than the negative results samples. The intensity of the colour also provides a qualitative measure of the amount of antigen in each well.

ii. **Indirect ELISA for antibodies:** In this test, frequently used to detect antibodies in a patient sample (*e.g.*, screening of blood for antibodies to HIV), an antigen (such as inactivated virus) is added to microtiter well followed by addition of on aliquot of patient serum (blood) and is incubated for reaction of antibody to antigen. The well is rinsed to remove the unbound antibodies. To make positive test visible, enzyme linked anti-HISG (*e.g.*, a goat antihuman immunoglobulin antibody) is added to which binds to bound antibody followed by addition of substrate for the enzyme and the reaction produces a sandwich product (of virus/antibody/enzyme linked anti-HISG) that is detected by colour change.

iii. **Competitive ELISA:** The third type of ELISA is the Competition Assay, which involves the simultaneous addition of 'competing' antibodies or proteins. The decrease in signal of samples where the second antibody or protein added gives a highly specific result.

iv. **Sandwich ELISA:** The last type of assay is the sandwich ELISA. Sandwich ELISA involves attachment of a capture antibody to a solid phase support. Samples containing known or unknown antigen are then added in a matrix or buffer that will minimize attachment to the solid phase. An enzyme-labeled antibody is then added for detection.

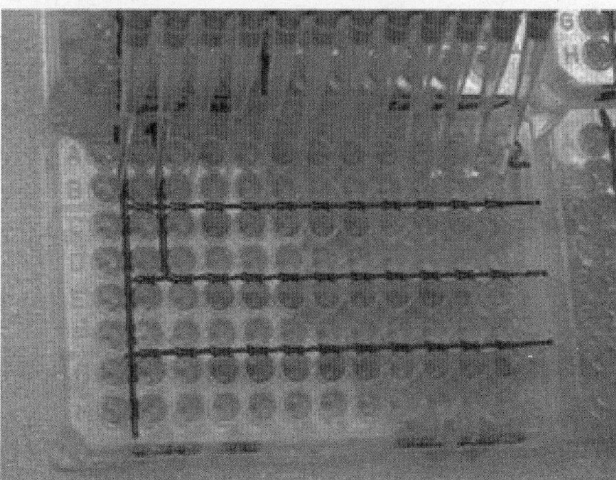

Fig. 4.11 : A typical ELISA output. Darker wells indicate higher levels of analyte in original sample

4.7.1 To detect the antibodies using direct ELISA

(a) Principle

Various concentrations of the immunogen (antigen) are immobilized onto a polystyrene micro-titer plate along with the test immunogen. The bound immunogen is detected by the enzyme labeled specific antibody to the immunogen. The enzyme activity due to the immunogen bound enzyme labeled antibody is directly proportional to the concentration of the immobilized immunogen.

(b) Requirements

Biochemicals: Rabbit IgG, Bovine serum albumin, alkaline phosphatase labelled-goat antibody (IgG) raised against rabbit IgG, Tween -20, diethanolamine, p-nitrophenyl phosphate.

Plasticware: Polystyrene ELISA/microtiter plate (12×8 wells, flat bottomed), microfuge/eppendor tubes (1.5 ml), auto-pipette tips.

Equipment: ELISA Reader, Laboratory incubator; auto-pipettes (variable volume 40-1000μl), spectrophotometer.

(c) Reagents: *Refer to Appendix* IVD: Immunology for preparation.

1. Coating buffer.
2. Blocking buffer .
3. Washing buffer.
4. Substrate buffer (25ml).
5. Sodium hydroxide 2M.

(d) Procedure

1. Coat the wells with 100 μl (100 pg–500 ng of rabbit IgG) reference standard coating buffer.
2. Coat few wells with diluted normal rabbit serum (test sample, 100 μl/well) in coating buffer.
3. Coat few wells with 500 ng of BSA/100 μl in coating buffer, which serves as blank.
4. Dry the plate overnight at 37°C in an incubator.
5. Wash the plate with washing buffer thrice and blot with tissue paper.
6. Block the wells with 150 μl/well with blocking buffer and incubate for 30 min at 37°C.

7. Wash the plate with washing buffer, thrice and blot with tissue paper.
8. Add 100 µl/well an anti-rabbit IgG raised in goat and labeled with alkaline phosphatase (1:2500 diluted), in phosphate buffer (10 mM, pH 7.2) and incubate the plate for 1 hour at 37°C.
9. Wash the plate with washing buffer thrice and blot with tissue paper.
10. Add, 150 µl/ well of the substrate buffer and incubate the plate for 20–30 min 37°C.
11. Stop the enzyme reaction by adding 100 µl of 5M NaOH per well.
12. Record the absorbance at 405 nm in a spectrophotometer, using a micro-cuvette (0.5 ml volume) or in the ELISA reader and compute the concentration of the analyte in the test sample from the calibration plot.

4.7.2 To detect the antibodies using indirect ELISA

(a) Requirements

Biochemicals: Rabbit IgG, Bovine serum albumin, alkaline phosphatase labeled–goat antibody (IgG) raised against rabbit IgG, Tween-20, diethanolamine, p-nitrophenyl phosphate, radish peroxidase.

Plasticware: Polystyrene ELISA/ microtiter plate (12X8 wells, flat bottomed), microfuge/eppendor tubes (1.5ml), auto-pipette tips.

Equipment: Laboratory incubator; auto-pipettes (variable volume 40–1000 µl), spectrophotometer.

(b) Reagents

1. Coating buffer (50 ml): Carbonate-bicarbonate buffer, 100mM, pH 9.6
2. Blocking buffer (25 ml):Phospahte buffer saline (PBS) 10 mM, pH 7.2, containing 0.2% (w/v) bovine serum albumin and 0.5% (w/v) sodium azide.

   Or

   3N H_2SO_4 (to stop the reaction)
3. Washing buffer (1 ml): Phosphate buffered saline–tween (PBS-T), 0.01 M pH 7.2, containing 0.05% (v/v) Tween-20.
4. Substrate buffer (25 ml): Diethanolamine (10% v/v) buffer pH 9.6, containing substrate-p-nitophenyl phosphate (1.25 mg/ml buffer) and 0.05 mM $MgCl_2$

5. Sodium hydroxide 2 M.

Or

Substrate-orthophenylenediamine (OPD) solution: Dissolve 8 mg of OPD in 15 ml of citrate buffer (pH 5.0) followed by addition of 15 ml hydrogen peroxide just before use

6. Citrate phosphate buffer (0.1 M) pH 5: Dissolve 7.3 g of citric acid and 11.86 g of Na_2HPO_4 in 1000 ml of double distill water.
7. Bovine Serum Albumin (BSA) (4%) in PBS-Tween 20.

(c) Procedure

1. Adjust antigen concentration 0.5–5 μg/ml in carbonate bicarbonate buffer.
2. Load 100 μl of diluted antigen to each well of 96 well microtiter plate and incubate at 37°C for 3 hours in humid chamber or at 4°C overnight.
3. Decant the unbound antigen and wash the wells thrice thoroughly with washing buffer and invert the microtiter plate.
4. Load 100μl of BSA (4%) in washing buffer into each well and incubate at 37°C for 60 minutes.
5. Wash the plates with washing buffer thrice and remove all residual fluid.
6. Add 100 μl of diluted serum sample to each well and incubate at 37°C for 60 minutes in a humid chamber.
7. Wash the well thrice with washing buffer and remove all residual fluid.
8. Add 100 μl of enzyme to each well and incubate at 37°C for 60 minutes.
9. Wash the well thrice, add 100 μl of diluted freshly prepared substrate solution into each well and incubate in dark at 37°C for 30 minutes.
10. Add 50 μl of sulphuric acid (3M) to each well to stop the reaction.
11. Observe the microtiter plate for the development of colours in each well and read the results of the reaction at 492 nm on an ELISA reader.

(d) Results

Prepare a standard curve from the data produced from the serial dilutions with concentration on the X-axis (log scale) *vs.* absorbance on the Y-axis (linear). Interpolate the concentration of the sample from this standard curve.

(e) Precautions

1. One should be very careful to avoid spillover from adjacent wells.
2. The plate should not be allowed to dry.
3. Make sure that all incubations are carried out in 100 per cent humid conditions. (wrapping the covered plate in a wet paper towel and then placing in a sealed bag works)
4. The plates must be washed well and carefully with PBS-tween between steps at least 3 times.
5. Fresh buffers at the correct pH shoud be used.

Chapter - 3

E. ENVIRONMENTAL BIOTECHNOLOGY

INTRODUCTION

Environmental protection is an integral component of sustainable development. The environment is threatened every day by the activities of man. With the continued increase in the use of chemicals, energy and non-renewable resources by an expanding global population, associated environmental problems are also increasing. Despite escalating efforts to prevent waste accumulation and to promote recycling, the amount of environmental damage caused by over-consumption, the quantities of waste generated and the degree of unsustainable land use appear likely to continue growing.

Environmental Biotechnology involves specific application of biotechnology to the management of the environment and related socio-economic and developmental issues, keeping in view the concept of sustainable development. Environmental Biotechnology encompasses issues like:

- Environmental monitoring
- Restoration of environmental quality
- Resource/residue/water recovery/utilization/treatment through application of r-DNA technology
- Global changes
- Biological diversity
- Substitution of non-renewable resources by renewable ones
- Microbial strain improvement for degradation of highly toxic pollutants with the production of chemicals
- Risk management

Environmental Detection and Monitoring

A wide range of biological methods is in use to detect pollution and for the continuous monitoring of pollutants. The techniques of biotechnology have novel methods for diagnosing environmental problems and assessing normal environmental conditions so that human beings can be better informed of the surroundings. Applications of these methods are cheaper, faster and also portable. Industrial Pollution Management is one amongst many issues that environmental biotechnology addresses to. It also deals with disposal of solid waste by composting, vermi-culture and methane production. There are now a dozen of companies active in developing and marketing engineered microbes for effluent treatment schemes for waste water of dairy, distillery, tannery, sugar, antibiotic industries including bioremediation of contaminated soil and waste lands.

5.1 To determine the organic carbon content in soil

(a) Principle

Organic matter in soils is widely distributed over the earth surface. Soils and sediments contain a large variety of organic materials ranging from simple sugars and carbohydrates to the more complex proteins, fats, waxes, and organic acids. There are three basic forms of carbon that may be present (i) elemental Carbon, (ii) inorganic Carbon and (iii) organic Carbon. Total organic carbon (TOC) is a measure of one of the chemical components of organic matter that is often used as an indicator of its presence in a soil or sediment. It is an essential part of any site characterization since its presence or absence can markedly influence how chemicals will react in the soil. TOC contents may also be used qualitatively to assess the nature of the sampling location or to normalize portions of the analytical chemistry data set (*e.g.,* equilibrium partitioning).

There are several qualitative and quantitative methods available for determination of organic carbon. Titration method is one of the commonly used methods in laboratories. It is based on the oxidation of soil organic carbon by nascent oxygen produced during the reaction of sulfuric acid with aqueous potassium dichromate. It is estimated by titration with ferrous ammonium sulfate.

(b) Reagents

1. 1N $K_2Cr_2O_7$: Dissolve 49.040 g $K_2Cr_2O_7$ in 1000 ml distilled water.
2. Mohr's Solution: Refer to Appendix IVE for preparation containing 500 ml deionized water. Dissolve and slowly add 50 ml of concentrated sulfuric acid and let it cool. Dilute to the volume to 1 liter with deionized water and mix well. Solution should be prepared fresh.
3. Diphenylamine (DPA) Indicator: Dissolve 0.5 g DPA in 100 ml conc. H_2SO_4 and add 20 ml distilled water

4. Phosphoric acid.

(c) Procedure

1. Collect different soil samples from agriculture land, garden and backyard.
2. Weigh 0.5–1.0 g soil (depending on expected organic carbon content) into a 250 ml Erlenmeyer flask in triplicate.
3. Add 10–15 ml $K_2Cr_2O_7$ and 20 ml conc. H_2SO_4.
4. Allow to stand for half an hour.
5. Dilute the reaction mixture with 200 ml distilled water.
6. Add 10 ml H_3PO_4 and 0.5 ml DPA indicator.
7. Titrate the above solution against Mohr's salt solution taken in a Burette and record 3 concurrent readings.
8. A blank (without soil sample) consisting of 15 ml $K_2Cr_2O_7$ solution is titrated in exactly the same manner.
9. At the end point, the colour will be green.

(d) Observations

Blank titration

S.N.	Initial Volume (ml)	Final volume (ml)	Volume used (ml)
1.	0.0		
2.	0.0		
2.	0.0		
3.	0.0		

Experimental titration (make separate tables for all soil samples)

S.N.	Initial Volume (ml)	Final volume (ml)	Volume used (ml)
1.	0.0		
2.	0.0		
3.	0.0		

(e) Calculations

1 ml of 1 N dichromate solution is equivalent to 3 mg of Carbon

$$\text{Organic Carban content } \% = \frac{V_1}{V_2} \times \frac{N(V_2 - V_3)}{W} \times 0.003 \times 100$$

where, V_1 = Volume of $K_2Cr_2O_7$ used

V_2 = Volume of Mohr's solution used: Blank

V_3 = Volume of Mohr's solution used: Experimental

N = Normality of $K_2Cr_2O_7$ (0.5N)

W = Weight of soil (0.5 or 1.0 g)

(f) Result

Organic Carbon content (%) in garden soil is ...., organic Carbon content (%) in agriculture soil is ...., whereas soil collected from the backyard is .... %.

(g) Precautions

1. Prior to analysis, samples should be collected, handled and weighed carefully.
2. After sample collection, either the estimation should be done immediately or stored at 4°C to avoid microbial degradation, sample drying, oxidation and volatilization.
3. Reagents should be freshly prepared.
4. Titration should be performed carefully.

5.2 Determination of Total Dissolved Solids in Waste Water Sample

(a) Principle

There are 5 forms of solids in waste water *viz.,* total solids, suspended solids, dissolved solids, volatile solids and fixed solids. Total solids (TS) are dissolved solids plus suspended and settleable solids in water. It is the matter that remains as a residue upon evaporation and drying at 103–105°C in an oven. Suspended solids (SS) are the ones which are present on filter paper while filtering through a fine filter such as Whatman's filter paper. SS contain much of the organic matter. Dissolved solids (DS) on the other hand remains in filtrate after filtration and consists mainly of inorganic salts, small amount of organic matter and dissolved gases, whereas Fixed Solids (FS) are represented by residue after the ignition of sample in muffle furnace at 550°C. The last category is of Volatile Solids (VS) which is represented by the difference between the suspended solids and the fixed solids.

Concept Structure

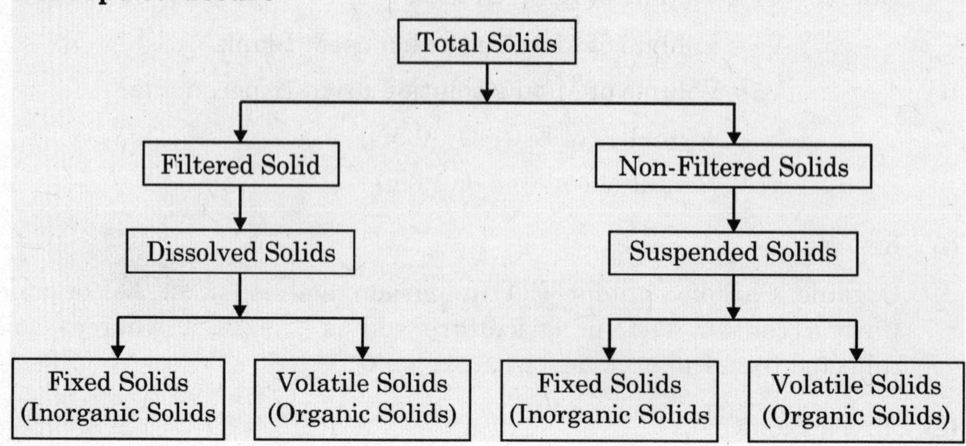

(b) Procedure

Total Solids and Volatile Solids

1. Weigh the clean and dry porcelain dish and record its weight (W_1 g).
2. Take 100 ml of sample in dish.
3. Evaporate to dryness in an oven at 103–105°C for 24 h.
4. Cool the dish and weigh it again (W_2 g).
5. Burn the oven dried sample in furnace at 550–600°C.
6. Record its weight as W_3 in gram.

(c) Suspended Solids

1. Weigh the Whatman filter paper 42 (W_1 g.)
2. Filter 100 ml sample through the above filter.
3. Place the filter paper with residue in an oven to dry.
4. Note the weight of filter paper with residue (W_2 g).

(d) Observations

For Total Solids

S. No.	Water sample	Volume of sample V (ml)	Weight of dish (g)		
			Initial W_1	Final W_2	Difference (W_2-W_1)
1					
2					
3					

For Volatile Solids

S. No.	Water sample Code	Volume of sample V (ml)	Weight of porcelain dish (g)		
			Before Ignition (W_1)	After Ignition (W_3)	Difference (W_3-W_1)
1					
2					
3					

For Suspended Solids

S. No.	Water Sample code	Volume of sample V (ml)	Weight of Filter Paper (g)		
			Initial W_1	Final W_2	Difference (W_2-W_1)
1					
2					
3					

(e) Calculations

Concentration of total Solids (mg/l) =

$(W_2 - W_1) \times 1000 \times 1000$ / Volume of Sample

Concentration of Volatile Solids (mg/l) =

$(W_3 - W_1) \times 1000 \times 1000$ / Volume of Sample

Concentration of Total Suspended Solids (mg/l) =

$(W_2 - W_1) \times 1000 \times 1000$ / Volume of Sample

Concentration of Total Dissolved Solids (TDS) ppm = Concentration of total Solids – Concentration of Suspended Solids

Concentration of Total Fixed Solids = Total Solids – Volatile Solids

(f) Results

The concentration of Total Solids, Volatile Solids, Total Suspended Solids, Total Dissolved Solids and Total Fixed Solids in given sample is ...., respectively.

5.3 To determine the Dissolved Oxygen (DO) in given water sample by Winkler's method

(a) Principle

Oxygen is a necessary element to all forms of life. Oxygen gets into water by diffusion from the surrounding air, by aeration (rapid movement), and as a waste product of photosynthesis. Dissolved oxygen (DO) is a relative measure of the amount of gaseous oxygen (O_2) that is dissolved or carried in a given medium. Adequate dissolved oxygen is necessary for good water quality. A well-mixed body of water will be fully saturated, with approximately 10 mg/l dissolved oxygen at 15°C. If this level in water drops below 5.0 mg/l, aquatic life is put under stress and it becomes difficult for animals to thrive.

In Winkler's method when manganous sulphate is added to the sample containing alkaline potassium iodide, a white precipitate of manganous hydroxide is formed. This precipitate is then oxidized by the dissolved oxygen in the water sample to form a brown precipitate of higher hydroxides such as basic manganic oxide. In the presence of iodine and a strong acid, the basic manganic oxide reverts to the divalent state and liberate iodine equivalent to that of dissolved oxygen originally present in the sample. The liberated iodine is titrated with a standard solution of sodium thiosulphate using starch as an indicator.

$MnSO_4 + 2KOH \rightarrow Mn\,(OH)_2 + K_2SO_4$ (white ppt.)

(Basic manganic oxide, brown coloured)

$$MnO\,(OH)_2 + 2H_2SO_4 \longrightarrow Mn\,(SO_4)_2 + 3\,H_2O_2$$

(Manganic sulphate)

$$Mn\,(SO_4)_2 + 2KI \longrightarrow MnSO_4 + K_2SO_4 + I_2$$

$$2Na_2S_2O_3 + I_2 \longrightarrow Na_2S_4O_6 + 2NaI$$

(b) Requirements

BOD bottles (100–300 ml), conical flask, burette and pipette

(c) Reagents

1. Sodium thiosulphate solution (0.025 N): Dissolve 24.82 g of sodium thiosulphate in previously boiled and cooled distilled water and make up the volume to 1 litre to make 0.1 N stock solution. Add 0.4 g borax as a stabilizer and dilute it 4 times. Keep the solution in brown glass stopper bottle.
2. Manganous sulphate solution: Dissolve 96 g of $MnSO_4.4H_2O$ or 80 g of $MnSO_4.2H_2O$ in 200 ml of pre-boiled distilled water and filter.
3. Alkaline potassium iodide solution: Dissolve separately 75 g of potassium iodide and 350 g of potassium hydroxide in distilled

water. Mix the two solutions and dissolve the chemicals in 200 volume to 500 ml.

4. Starch indicator: *Refer to Appendix* IVE for preparation.
5. Concentrated H_2SO_4.

(d) Procedure

1. Place 50 ml of sample into a glass-stoppered BOD bottle.
2. Add 1 ml each of manganous sulphate and alkali-iodide-azide solution using separate pipettes.
3. Close the bottle and shake well by inverting the bottle repeatedly for about 1 minute.
4. A brown precipitate formed is allowed to settle down. Sample at this stage can be stored for few days if required.
5. Add 2 ml of conc. H_2SO_4 and shake well to dissolve the precipitate.
6. Transfer the whole content or some part into a conical flask for titration.
7. Titrate the contents within 1 hour of dissolution of precipitate against standard sodium thiosulphate solution till the contents turn to straw yellow.
8. Add 2 ml of starch solution. Now the contents get changed to blue colour.
9. Continue the titration till the blue colour disappears. The end point is recorded when blue colour changes to colourless.

(e) Observations

S.N.	Volume of sample (ml)	Sodium thiosulfate rundown (ml)		Titre value (ml) Final–Initial
		Burette Reading Initial	Burette Reading Final	

(f) Calculations

1000 ml of 1N $Na_2S_2O_3$ = 8 gm of O_2

so, 1000 ml of 0.025N $Na_2S_2O_3$ = $\frac{0.025 \times 8}{1}$ g of O_2

Titre value = x ml

so, x ml of 0.025N $Na_2S_2O_3$ = $\frac{x}{1000} \times \frac{0.025 \times 8}{1}$ g of O_2

Volume of sample taken (V) = 50 ml

so, 50 ml of sample = $\frac{x}{1000} \times \frac{0.025 \times 8}{1}$ g of O_2

Then 1000 ml of sample = $\frac{1000}{50} \times \frac{x}{1000} \times \frac{0.025 \times 8}{1}$ g of O_2

= $\frac{x}{50} \times 0.025 \times 8$ g of O_2

DO in (mg/l) = $\frac{x}{50} \times 0.025 \times 8 \times 1000$

(g) Results

The dissolved oxygen in the given sample is ..... mg/l.

(h) Precautions

1. Carefully avoid any sort of bubbling and entrapping of air in the BOD bottles after placing stopper.
2. To get the accurate DO values, the sample is immediately fixed by adding 2 ml of manganous sulphate solution and 3 ml of alkali-iodide-azide to the sample in BOD bottle. Keep the fixed sample at 4°C in dark.

5.4 To determine the Biological Oxygen Demand (BOD_5 at 20°C) of given sample of waste water

(a) Principle

Biological Oxygen Demand (BOD) refers to the amount of oxygen that would be consumed if all the organics in one liter of water were oxidized by microorganisms. Natural sources of organic matter include dead plants, leaf fall, sewage, manure and food waste. Microorganisms such as bacteria and protozoa are responsible for decomposing this organic waste. When organic matter is present in a water supply, the bacteria will begin the process of breaking down this waste. During this process, most of the available dissolved oxygen is consumed by aerobic bacteria creating oxygen deficient conditions for other aquatic organisms.

If there is a large quantity of organic waste in the water supply, the number of decomposing bacteria would also be more generating high oxygen demand, which will lead to high BOD level.

As the waste is consumed or dispersed through the water, BOD levels will begin to decline. Nitrates and phosphates in a body of water can contribute to high BOD levels.

BOD test is widely used to determine the biodegradable organic matter of domestic and industrial waste in terms of oxygen required for its degradation in aerobic conditions.

$$[CH_2O]_{\text{organic matter}} + O_2 + \text{microorganisms} = CO_2 + H_2O$$

Polluted water absorbs oxygen for many months for decomposition of organic matter present in it, but practically it is not feasible to determine the ultimate oxygen demand. Hence, the BOD of water during 5 days at 20°C is generally measured as the standard demand which is denoted as BOD_5 and is about 68–80 per cent of the total demand. The principle of the method involves the determination of the difference in oxygen concentration between the natural sample and after its incubation for 5 days at 20°C.

(b) Requirements

BOD bottles (100–300 ml), conical flasks, burette and pipette, Aerator with compressor, BOD incubator

(c) Reagents

1. Air saturated water: Sufficient quantity of distilled water is aerated with aerator air compressor for 8–10 hours and kept at low temperature until use. This water is used for the preparation of different dilutions of the samples.
2. Seed (Optional): Take fresh sewage and let it settle for 20 hours. Use the supernatant as seed.
3. Phosphate buffer: Dissolve 8.5 g KH_2PO_4, 21.75 g K_2HPO_4, 33.4 g Na_2HPO_4 and 1.7 g NH_4Cl in 500 ml distilled water. Dilute it to 1 litre and adjust the pH at 7.5.
4. Magnesium sulfate: Dissolve 22.5 g $MgSO_4.7H_2O$ in distilled water and make the volume 1 litre.
5. Calcium Chloride solution: Dissolve 27.5 g anhydrous $CaCl_2$ in distilled water and dilute to 1 litre.

6. Ferric Chloride solution: Dissolve 0.25 g $FeCl_3.H_2O$ in distilled water and dilute to 1 liter,
7. Alkali Iodide solution.
8. Concentrated H_2SO_4.
9. Sodium thiosulfate 0.025 N.
10. Starch indicator.

(d) Procedure

A. Preparation of Dilution water.

1. Prepare air saturated dilution water.
2. Add 1 ml each of Phosphate buffer, Magnesium sulfate, Calcium chloride and Ferric chloride solutions to 1 litre dilution water and mix well.
3. Add seed in case sufficient bacterial population is not present in waste water. This step is optional.

B. Dilution of sample

1. Neutralize the pH of sample to 7.
2. Make several dilution of sample as follows: 0.1–1.0% for raw treated waste; 1.0–5.0% for settled waste; 5.0–25.0% for oxidized effluents and 25.0–100% for polluted river water.
3. Prepare desired mixtures by adding dilution water.
4. Fill up BOD bottles with mixture and blank containing dilution water in two sets.
5. Incubate first set of BOD bottles of blank and sample in BOD incubator at 20°C for 5 days.
6. Determine the second set immediately for DO on 1st day.
7. After 5 days of incubation, the DO is determined in the first set.

C. Determination of DO

1. Take BOD bottle containing blank on the first day. Add 2 ml of Manganese sulfate solution by dipping the pipette just below the surface of water.
2. Add 2 ml of azide alkali potassium iodide solution in a similar manner as above.
3. Mix the contents by shaking the bottle.
4. A red precipitate will form in presence of DO in water. Allow the ppt. to settle half way and then mix again.

5. Again allow the ppt. to settle half way and add 2 ml of concentrated H_2SO_4.
6. Let the solution stand for 5 minutes to ensure the formation of I_2.
7. Take 203 ml solution in conical flask.
8. Fill the burette with 0.025 N solution of sodium thiosulfate solution and note the initial reading.
9. Titrate the blank immediately until the yellow colour becomes light.
10. Add 1 ml starch solution. The blue colour will appear.
11. Continue the titration slowly until blue colour disappears. Note down the final burette reading (B_1).
12. Repeat the steps for sample bottle and record the initial and final readings (D_1).
13. Repeat the same procedure for blank (B_2) and sample (D_2) set after 5 days of incubation.

(e) Observations

Dilution Factor (DF) = ..., Volume of sample = 203 ml actual but 200 ml for calculation

S. No.	Source of sample*	Bottle No.	Burette reading		DO (mg/l)	BOD_5 (mg/l)
			Initial reading (ml)	Final reading (ml)	Difference of two readings (ml)	
1	Sewer	.....Blank (1st day)				
2	Sewer	......Sample (1st day)				
3	Sewer	.......Blank (5th day)				
4	Sewer	......Sample (5th day)				
5	Effluent					
6	Effluent					
7	Effluent					
8	Effluent					

* Source may vary

(f) Calculations

$$\text{Dissolved oxygen (DO) mg/l} = \frac{\text{ml of sodium thiosulfate} \times N \times 8 \times 1000}{\text{ml of water sample}}$$

where, N = Normality of sodium thiosulfate *i.e.* 0.025

By using the above formula find out the DO for blank and sample on 1st and 5th day

so, B_1 = ... mg/l; B_2 = ... mg/l; D_1 = ... mg/l; D_2 = ... mg/l

BOD_5 mg/l = $[(D_1\text{-}D_2) - (B_1\text{-}B_2)] \times DF$

where,

B_1 = DO of blank before incubation

B_2 = DO of blank after incubation

D_1 = DO of diluted sample before incubation

D_2 = DO of diluted sample after incubation

D_F = Dilution factor

(g) Results

BOD_5 of the sample from public sewer is ....mg/l, whereas BOD_5 of the sample from industry effluent is .... mg/l

(f) Precautions

1. Do not use organic acid for acidification of sample; use mineral acid only.
2. Samples having DO 9 mg/l or more should be aerated and agitated to reduce the DO level.

5.5 To determine Chemical Oxygen Demand (COD) in waste water sample

(a) Principle

Chemical oxygen demand (COD) is the oxygen required for complete oxidation of biological degradable organic/inorganic matter in reduced state. This is a standard method for indirect measurement of pollution (that cannot be oxidized biologically) in the water sample. The COD test procedure is based on the chemical decomposition of organic and inorganic contaminants, dissolved or suspended in water. The result indicates the amount of water-dissolved oxygen (expressed as ppm or mg/l of water) consumed by the contaminants, during two hours of decomposition from a solution of boiling potassium dichromate. The sample is refluxed with $K_2Cr_2O_7$ and conc. H_2SO_4 in presence of mercuric sulfate to neutralize the effect of chlorides. The excess of potassium dichromate is titrated against ferrous ammonium sulfate (FAS) using Ferroin indicator. The amount of $K_2Cr_2O_7$ used is proportional to the oxidizable organic matter present in the sample.

$$3\{CH_2O\}_{\text{organic matter}} + 16H^+ + 2Cr_2O_7^{-2} \rightarrow 4Cr^{+3} + 3CO_2 + 11H_2O$$

$$6Fe^{+2} + Cr_2O_7^{-2} + 14H^+ \rightarrow 6Fe^{+3} + 2Cr^{+3} + 7H_2O$$

$$6CT + Cr_2O_7^{-2} + 14H^+ \rightarrow 3Cl_2 + 2Cr^{+3} + 7H_2O$$

$$Hg^{+2} + 2Cl^- \rightarrow HgCl_2$$

The higher the chemical oxygen demand, the higher the amount of pollutants in the test sample. For the contaminants that can be oxidized biologically, the biological oxygen demand (BOD) method is used.

The major advantage of the COD test is the short time required to carry out rather than the longer 5 days procedure for BOD. One of the limitations of COD test is its inability to differentiate between biologically inert organic matter and it does not provide any evidence of the rate at which the biologically active material would be stabilized under conditions that exist in nature.

(b) Requirements

Hot plate, COD reflux unit, burette and pipette

(c) Reagents:

1. Standard potassium dichromate ($K_2Cr_2O_7$) 0.25N: Dissolve 12.25 g of $K_2Cr_2O_7$ in distilled water and make up the volume to 1000 ml in a volumetric flask.
2. $HgSO_4$ and $AgSO_4$ crystals.
3. Concentrated H_2SO_4.
4. Ferroin indicator: Dissolve 0.695 g of ferrous sulphate ($FeSO_4,7H_2O$) and 1.485 g of 1:10 phenanthroline in distilled water and make up the volume to 100 ml.
5. Standard ferrous ammonium sulphate [Fe $(NH_4)_2$ $(SO_4)_2$]. 0.1N solution (0.1N=0.1M): Dissolve 39.2 g of ferrous ammonium sulphate (FAS) in distilled water, add 20 ml of conc. H_2SO_4 and make up the volume to 1000 ml with distilled water.

(d) Procedure

1. Place 0.4 g of $HgSO_4$ in the flask of reflux unit and add 20 ml of the sample

2. Add 10 ml of 0.25N $K_2Cr_2O_7$ using a pipette.
3. Add a pinch of $AgSO_4$ followed by slow addition of 30 ml of concentrated H_2SO_4 and a few glass beads.
4. After mixing attach Liebig condenser of reflux apparatus to the mouth of the flask and heat the flask on a hot water bath or heating mantle for at least 2 hours to reflux the contents.
5. Remove the flask from the unit, cool and dilute the contents to about 150 ml by adding distilled water.
6. Add 2–3 drops of ferroin indicator and titrate the contents of the flask with 0.1N ferrous ammonium sulphate solution till bluish green colour changes to reddish brown.
7. Repeat the procedure for blank (without sample) with distilled water in a similar manner.

(e) Observations

S.No.	Ferrous ammonium sulphate rundown (ml)		Titre value (Final–Initial) (ml)
	Initial burette reading	Final burette reading	
Sample			
1			
2			
3			
Blank			
1			
2			
3			

(f) Calculations

1000 ml of 1N FAS = 8 g of O_2

1000 ml of 0.1N FAS = $\frac{0.1}{1} \times 8$ g of O_2

Titre value with sample = A ml O_2

Titre value of blank = B ml

Amount of 0.1N FAS actually utilized for oxidation = (B–A) ml = T ml

T ml of 0.1N FAS = $\frac{T}{1000} \times 8$ g of O_2

Volume of sample taken (V) = 20 ml

so, 20 ml of sample $= \frac{T}{1000} \times 0.1 \times 8$ g of O_2

Then 1000 ml of sample $= \frac{1000}{20} \times \frac{T}{1000} \times 0.1 \times 8$ g of O_2

$= \frac{T}{20} \times 0.1 \times 8$ g of O_2

COD as mg of O_2 L^{-1} $= \frac{T}{20} \times 0.1 \times 8 \times 1000$

(g) Results

The COD value of given sample is .... mg L^{-1}

(h) Precautions

1. The quantity of H_2SO_4 should be equal to that of 'sample plus dichromate' for complete oxidation of organic matters.
2. Contents should be cooled before titration.

5.6 Analysis of total hardness of waste water sample

(a) Principle

Water becomes hard by being in contact with soluble, divalent, metallic cations. The two main cations that cause water hardness are calcium (Ca) and magnesium (Mg). Calcium is dissolved in water as it passes over and through limestone deposits. Magnesium is dissolved as water passes over and through dolomite and other magnesium bearing formations. Total hardness is defined as the sum of calcium and magnesium hardness, in mg/L as $CaCO_3$.

Hardness (mg equivalent of $CaCO_3$ L^{-1}) = 2.497 [(Ca) (mg L^{-1})] + 4.118 [(Mg (mg L^{-1})]

Total hardness in terms of concentration of $CaCO_3$ is determined by estimating calcium and magnesium in the given waste water sample by complexometric titration. Ethylene diamine tetraacetate (EDTA) strongly complexes and forms chelates with a number of polyvalent cations at different pH values. Based on this principle, Calcium can be estimated by titrating the sample with EDTA at pH 12 as at this pH Mg forms insoluble complexes and only Ca be titrated. However, at pH 10.0 Ca+Mg can be estimated as both form complexes with EDTA. Indicators namely murexide in the case of Ca and EBT in the case of Ca+Mg determinations are used to indicate the end point of the titration.

(b) Reagents

1. Ammonium chloride and ammonium hydroxide buffer: Dissolve 67.5 g of ammonium chloride in 570 ml conc. ammonium hydroxide and make the volume to 1 litre with distilled water.
2. Buffer solution (4N NaOH): Dissolve 160 g pure sodium hydroxide in distilled water and make the volume to 1 litre.
3. Eriochrome black T (EBT) indicator: *Refer to* Appendix IVE for preparation
4. EDTA solution 0.01N: Dissolve 2 g of analytical reagent grade versenate (EDTA) in distilled water and make the volume to 1litre. (Standardize the solution with 0.01N calcium solution).
5. Murexide or ammonium perpurate: Weigh 0.2 g of ammonium perpurate and 40 g of potassium sulphate. Mix both the reagents thoroughly in a mortar and pestle and keep it as the powder in a clean dark coloured bottle.

(c) Procedure

(A) For Calcium

1. Take 10 ml of water sample
2. Add 1–2 ml of sodium buffer solution.
3. Add a pinch of murexide indicator till the contents turn to orange red colour.
4. Titrate the contents with standard EDTA (0.01N) till orange red changes to purple or purplish violet.
5. Note down the volume of EDTA run down.

(B) For Calcium+Magnesium

1. Take 10 ml of water sample and add 1–2 ml of ammonia buffer solution
2. Add 2–3 drops of Eriochrome Black T and mix well. The contents turn to wine red colour.
3. Titrate the contents with standard EDTA (0.01N) till wine red colour changes to blue.
4. Note down the volume of EDTA run down

(d) Observations

(A) Observations for Calcium

S.No.	Volume of sample taken	EDTA rundown (ml)		Titre value (ml) Final-Initial
		Initial Burette reading	Final Burette reading	
1	10 ml			
2	10 ml			
3	10 ml			

(e) Calculations

1 lit of 1N EDTA = 20 g of Ca

1 ml of 1N EDTA = 20 mg of Ca

1 ml of 0.01N EDTA = 0.2 mg of Ca

Titre value = X ml

X ml of 0.01N EDTA = X × 0.2 mg of Ca

Volume of sample taken = 10 ml

10 ml of sample contains = X × 0.2 mg of Ca

Then1000 ml of sample contains = $\frac{1000}{10} \times X \times 0.2$ mg of Ca = C

Where, C= calcium in mg L^{-1}

(f) Observations for Ca+Mg

S.No.	Volume of sample taken	EDTA rundown (ml)		Titre value (ml) Final-Initial
		Initial Burette reading	Final Burette reading	
1	10 ml			
2	10 ml			
3	10 ml			

Suppose titre value for calcium + magnesium = A ml

Titre value for calcium = X ml

Then titre value for magnesium = (A-X) ml = Y ml

(g) Calculations

1000 ml of 1N EDTA = 12 g of Mg

1 ml of 1N EDTA = 12 mg of Mg

1 ml of 0.01N EDTA = 0.12 mg of Mg

Titre value = Y ml

Y ml of 0.01N EDTA = $0.12 \times Y$ (mg of Mg)

Volume of sample taken = 10 ml

10 ml of sample contains = $(Y \times 0.12)$ mg of Mg

1000 ml of sample contains = $\frac{1000}{10} \times Y \times 0.12$ mg of Mg = D

D = magnesium in mg L^{-1}

Total hardness of water (mg equivalent to $CaCO_3$ L^{-1})

= 2.497 {Ca (mg L^{-1})} + 4.118 {Mg (mg L^{-1})}.

= $(2.497 \times C) + (4.118 \times D)$

(h) Results

Total hardness of sample is ....

(i) Precautions

1. Samples must be stored at low temperature and analyzed within 24 hours.
2. Soaps, oily water, suspended solids may interfere. Additional time may be given during titration to get the content colour change.
3. EDTA on storage settles down at the bottom, hence should be prepared fresh.

5.7 Detection of coliforms for determination of the purity of potable water

(a) Principle

The microbiological examination of water is a direct measurement of deleterious effects of pollution on human health. Pathogenic bacteria are often removed during water treatment. Water pollution is generally indicated by the presence of coliform bacteria. The coliform bacteria include the genera *Escherichia, Citrobacter, Enterobacter* and *Klebsiella. Escherichia coli* strains are of human origin, but its exclusive estimation is difficult, hence the entire coliforms are used as an indicator. In routine tests for coliforms, the actual number of coliforms is not reported, but they are reported as an approximate count – Most probable Number (MPN).

In this experiment membrane filtration method is explained.

Experimental outline

In the membrane filtration method, water sample is passed through a thin membrane filter (pore size 0.45 μm) which is kept in a special filter apparatus contained in a suction flask. The filter disc that contains the trapped microorganisms is aseptically transferred to a sterile Petri dish having absorbent pads saturated with a selective differential liquid medium and colonies which develop following incubation are counted. This method enables the large volume of water to be tested more economically, the results obtained are more accurate and rapid than by the multiple-tube technique.

(b) Requirements

Water sample, 20 ml tube of M-Endo Broth (Readymade- *Refer to Appendix* IVE for details), sterile Millipore membrane apparatus (base, funnel and clamps), 1 liter suction flask, sterile 0.45 μm filter absorbent pad, sterile Petri dish, alcohol, forceps, sterile rinse water, dissecting microscope.

(c) Procedure

1. Aseptically assemble the filter apparatus and insert membrane in it as follows:
   a. Place the filter holder aseptically in the Buchner flask.
   b. Attach the flask to the vacuum line through a rubber hose.
   c. Using flamed forceps place a sterile membrane filter disc, grid side up on the porous glass support of the filter holder.
   d. Set the funnel on the filter holder and fasten in place.
2. Shake the water sample and pour 100 ml of the water sample into the funnel.
3. Filter under vacuum.
4. When the entire sample has been filtered, wash the inner surface of the funnel with 100 ml of sterile water.
5. Carefully remove the filter from the filter holder using a sterile forceps.
6. Add a sterile absorbent pad to a sterile Petri dish using a sterile forceps.
7. Aseptically transfer the membrane filter onto the medium saturated pad in the Petri dish.

8. Using a sterile pipette, aseptically add 2 ml of M-Endo Broth to the pad to saturate it.
9. Incubate the plate in upright position at 37°C for 24 hrs.
10. Remove the filter disc from the Petri dish and allow drying on absorbent paper for 1 hour.

(d) Result

Examine filter paper disc under a dissecting microscope of coliforms and count their number. On endo-medium, coliforms show pink to red colonies and some may have a green metallic sheen. Express the result per 100 ml of water sample by applying the formula.

$$100\,\text{ml water} = \frac{\text{No. of coliforms as per colony count}}{\text{Volume of sample used}} \times 100$$

Water is considered safe for drinking if it contains fewer than four coliforms per 100 ml of water sample.

(e) Precautions

1. Safety Guidelines: Students must be very careful in these experiments. The water sample should be representative of the bacteriological quality, hence extreme care should be taken to avoid contamination. After water collection the bottle should be immediately closed. Avoid contamination by hand.
2. Pre-lab Preparation: The students are supposed to do the cleaning of the glassware properly and are advised to set the apparatus and check it. Forceps holders should be sterilized.

5.8 Estimation of the pesticide residues in soil sample

(a) Principle

To meet the food requirement for growing population, farmers are using more than 45,000 tonnes of different types of pesticides. But the residues of the applied pesticides stay in the environment (air, soil, ground and surface water) which poses a serious threat to the environment. Microbial degradation process detoxifies pesticide contaminants and can be effectively used to overcome the pollution problem. Soil bacteria with the ability to degrade pesticides have been isolated from soil showing enhanced biodegradation. They include

1. Degrading *Rhodococcus* species (metamitron)
2. Degrading *Flavobacterium* species (chlorpyrofos)

3. Atrazine degrading *Pseudomonas* species
4. Degrading *Arthrobacter* sp.

Pesticide (Organophosphorus pesticide) reacts with 4 (p-nitro benzyl) pyridine in the presence of cyclobenzylamine at 175°–180° to produce coloured complex. This coloured complex is dissolved in ethyl acetate and the absorbance is measured at 540 nm. The concentration of residual pesticide can be calculated with the help of standard curve.

(b) Reagents (for sample preparation)

1. NaCl (5%)
2. Chloroform ($CHCl_3$)
3. Pesticide (parathion, malathion, chlorpyrifos, diazinon, dichlorvos etc.)

(c) Reagents (for estimation of residues of pesticide)

1. Propylene glycol
2. 2% 4(p-nitrobenzyl) pyridine (in acetone)
3. 2% cyclobenzylamine
4. Ethyl acetate
5. Pesticide standard

(d) Procedure

Sample preparation (pesticide degradation)

1. 1 kg soil is mixed with 225 ml of pesticide. The soil is made wet regularly and mixed properly.
2. The mixture is incubated for 48 hrs.
3. 50 gm of above soil sample is collected in a beaker (250 ml).
4. 75 ml acetone is added to above and left overnight.
5. The mixture is filtered with Whatman filter paper.
6. The filtrate is collected and evaporated to dryness in a water bath.

(e) Clean up process

1. The residue is redissolved in a small account of distilled water and transferred to a separating funnel.
2. 25 ml of NaCl (5%) and 40 ml of chloroform is added.
3. It is mixed well and allowed to separate.

4. The residual pesticide is extracted in the chloroform layer which is collected.

(f) Colorimetric estimation of residues of pesticide

1. Prepare different series of standard having different concentrations of pesticide (10, 20, 30, 40, 50, 60 ml of 36% pesticide).
2. 5 ml of sample is taken in a separate test tube. Blank is also prepared that contains 5 ml chloroform.
3. Volume in each test tube except the sample is made upto 5 ml using chloroform.
4. 1–2 drops of propylene glycol is added to each test tube.
5. The contents of all test tubes are evaporated to dryness in a water bath.
6. 0.4 ml of 2% 4(p-nitrobenzyl) pyridine followed by 0.4 ml of 2% cyclohexylamine is added to each test tubes.
7. The tubes are kept in an oil bath at 175°–180°C for 3 min.
8. The tubes are cooled in ice bath.
9. 3 ml of ethyl acetate is added to each test tube.
10. Absorbance is taken at 540 nm.
11. A standard curve is plotted and the amount of residual pesticide is calculated from the standard.

(g) Observation table

S. No.	Vol. of standard (ml)	Vol. of $CHCl_3$ (ml)	Conc. of pesticide (mg/ml)	Propylene glycol	OD at 540 nm

(h) Result

The residual pesticide in 24 hours sample is __ mg/ml.

(i) Precautions

1. Care should be taken while handling pesticides.
2. Reagents should be prepared accurately.
3. Reagents should be added in the exact amount.

5.9 Determination of LD_{50} of a given pesticide

(a) Principle

LD_{50} is a statistically derived expression of a dose that is expected to kill 50 per cent of individuals in a population for a given pesticides. It is expressed in mg/kg of body weight. If LD_{50} value is less then it would be more toxic. Thus the degree of toxicity can be measured by it. LD_{50} value depends upon –

1. Species
2. Mode of exposure
3. Nature of toxicant – organochlorines have higher LD_{50}

There is a range of doses to which individuals in a group respond in a similar way. A frequent response plot that is commonly known as dose response curve can be generated by plotting percentage of pesticides against percentage mortality of individuals.

There are also some measures to estimate the dose required to kill 50 per cent of individuals in the test population, *e.g.,* measurement of mean effective dose and median level dose can give an idea of LD_{50}.

LD_{50} value determination is a mean of estimating error in the value of dose given. It is determined by an experiment.

(b) Requirements

1. Termites
2. Pesticides
3. Petriplates
4. Blotting paper
5. Test tubes
6. Distilled water
7. Pipettes
8. Measuring cylinder

(c) Procedure

1. Different concentrations of pesticides are prepared in distilled water (as shown in following table).
2. 1 ml of different concentrations of pesticide is taken in different Petriplates in duplicates.
3. One pair of the Petriplates are taken as control; plates in which no pesticide was added.
4. 10 termites are placed in each Petriplate and are observed for 30 min.
5. After 30 min. the number of termites (living & dead) is observed in each Petriplate.

(d) Observation Table

S.No.	Pesticide solution	No. of termites (living)	No. of termites dead after 30 min.	Mortality rate (%)

(e) Result

Half of the population of termite was found dead after 30 min. in petriplates with ___ concentration of pesticide.

Hence LD_{50} for given pesticide is _____.

(f) Precautions

1. Care should be taken while transferring termites to Petriplates.
2. Termites in good physical conditions should be taken for the experiment.

5.10 Determination of Fluoride Content in Water Sample

Fluoride (F) occurs in almost all water sources from traces to high concentration. It has been shown to cause significant effects in humans through drinking water (WHO, 2006). Low concentrations of fluoride in drinking water has been considered beneficial to prevent dental carries, but excessive exposure to fluoride in drinking water can give rise to a number of adverse effects. WHO has set a limit value of 1.5 mg L^{-1} for fluoride in drinking water. There is a narrow margin between the desired and harmful doses of fluoride in drinking water, therefore an accurate, simple, rapid and cost-effective analytical method is of high importance. Spectrophotometric methods are widely used in the determination of fluoride because of advantages, such as simplicity, convenience, accuracy and reproducibility. They are based on the reaction of fluoride with coloured metal chelate complexes, producing either a mixed-ligand ternary complex or replacement of the ligand by fluoride to give a colourless metal-fluoride complex and the free ligand with a colour different to the metal-ligand complex. The high solubility of F in soil under acidic condition ($pH<6$) corresponds to the presence of cationic AlF^{+2} complexes. For all the samples where fluoride is found at low level, its determination demands very sensitive method.

(a) Principle

The presence of fluoride can be determined by adding colouring reagents and comparing it with standard colouring solution. Fluoride ion changes the colour of zirconium – Alizarin complex. This colour change is measured spectrophotometrically at 520 nm. The amount of fluoride in the sample is calculated with the help of standard graph. Colour change is inversely proportional to fluoride ion concentration.

The following simplified reaction represents the reaction of the ZrIn (dye) complex and fluoride ions:

$$ZrIn\ (red) + F^- \longrightarrow ZrF + In^- (colourless)$$

When fluoride is added to a solution containing ZrIn, it reacts according to the above equation. This reaction causes the absorbance of the solution to decrease with increasing fluoride concentration.

(b) Requirements and Reagents

1. Standard fluoride solution: Dissolve 221 mg anhydrous sodium fluoride in distilled water and dilute to 1 litre. 1 mL = 100 μg F. This is the stock solution. Pipette 100 mL stock solution and make it up to 1 litre with distilled water to obtain standard solution 1ml = 10 μg F.
2. Zirconyl-Alizarin reagent: Dissolve 300 mg zirconyl chloride octahydrate ($ZrOCl_2.8H_2O$) in 50 mL distilled water in 1 litre glass stoppered volumetric flask. Dissolve 70 mg of 3-alizarin sulphonic acid sodium salt (also called alizarin red S) in 50 mL distilled water and pour slowly into the zirconyl solution while stirring. The resulting solution clears on standing for a few minutes.
3. Mixed acid solution: Dilute 101 mL conc. HCl to approximately 400 mL with distilled water. Add carefully 33.3 mL conc. H_2SO_4 to approximately 400 mL distilled water. After cooling, mix the two acids.
4. Acid-zirconyl-alizarin reagent: To the clear zirconyl-alizarin reagent in 1 litre volumetric flask, add the mixed acid solution and distilled water to the mark and mix. The reagent changes in colour from red to yellow within an hour.
5. Water sample -50 ml
6. Test tubes
7. Pipettes

(c) Procedure

1. If residual chlorine is present, remove the same by adding one drop of arsenite per 0.1 mg Cl and mix.

2. Prepare a series of standard by diluting various volume of standard fluoride solution (1 ml =10 μg F) to 100 ml in tubes. The range should be such that it is between 0 and 1.4 mg/L.
3. To 50 mL of each standard add 10 ml mixed acid-zirconylalizarin reagent.
4. Set the spectrophotometer to a wavelength of 520 nm.
5. Adjust the spectrophotometer to zero absorbance with the reference solution *i.e.*, distilled water with reagent.
6. Plot the concentration along x-axis and absorbance along y-axis and obtain a calibration curve.
7. Take 50 ml of the sample and add 10 ml of mixed acid-zirconyl-alizarin reagent and mix well.
8. Place the solution in the spectrophotometer and read the absorbance.
9. By referring the calibration curve, the concentration for the observed absorbance is read out.
10. Repeat the procedure with dilute samples.

(d) Observation

The observation values are presented in Tables A and B.

Table A : Observation for calibration

Stock F solution (ml)	Fluoride concentration	Absorbance

Table B : Observation for calibration

No.	Absorbance	F concentration in μg from graph	Fluoride concentration in mg

(e) Calculation

$$F \text{ in mg/L} = \frac{A \times B}{V \times C}$$

where,

A = μg F determined

B = sample dilute to this volume

C = portion taken for colour development

V = mL of sample.

(f) Result

The concentration of fluoride in given water sample is _ mg/ml.

Precautions

1. All the reagents should be prepared in deionized water and preferably in plasticware to avoid contamination with F^{-}present in non-ionized water glass.
2. Proper incubation time must be provided to allow the complexometric reaction to take place.
3. Spectrophotometer must be switched on 15 min before taking the reading.

■ ■ ■

REFERENCES

Al-O Mair (2010). Purification and biochemical characterization of acid phosphatase from *Vigna aconitifolia. Am. J. of Plant Physiol.* **5(6):** 361–370.

Bookjans, G., Stummann, B.M. and Henningsen, K.W. (1984). Prep-aration of chloroplast DNA from pea plastids isolated in a medium of high-ionic strength. *Anal. Biochem.* **141:** 244–247.

Capriti J.A., Veda M. and Brown T. (1968). Volumetric method of ethanol estimation by using spectrophotometry. *Am. J. Enol. Vitic.*, **19:** 162–165.

Clough, S.J. and Bent, A.F. (1998). Floral dip: a simplified method for *Agrobacterium* mediated transformation of *Arabidopsis thaliana. Plant J.* **16:** 735–43.

Farrell Jr., R.E. (1993). RNA Methodologies: A Laboratory Guide for Isolation and Characterization. Academic Press, San Diego.

ISTA, (1966). International rule for seed health testing. *Proceedings of International Seed Testing Association* **31:** 1–152.

Li, J.F., Park, E., Von Arnim, A.G. and Nebenfuhr, A. (2009). The Fast Technique: A simplified *Agrobacterium*-based transformation method for transient gene expression analysis in seedlings of *Arabidopsis* and other plant species. *Plant Methods* **5:** 6.

Sambrook, J., Fritch, E.F. and Maniatis, T. (1989). Molecular Cloning, a laboratory manual. Cold Spring Harbor Laboratory Press, New York.

Sambrook, J. and Russell, D.W. (2001) Molecular Cloning: A Laboratory Manual, the third edition, Cold Spring Harbor Laboratory Press, Cold Spring Harbor, New York, 1.31–1.38.

Sharma, R.C. *et al.* (1993) A rapid procedure for isolation of RNA-free genomic DNA from mammalian cells, *BioTechniques,* **14:** 176–178.

Southern, E.M. (1975) Detection of specific sequences DNA fragments separated by gel electrophoresis. *J. Mol. Biol.,* **98:** 503–517.

Towbin, H, Staehelin. T. and Gordan, J. (1979) *Proc. Natl. Acad. Sci.* (USA), **76:** 4350–4354.

Triboush, S.O., Danileuko, N.G. and Davydenko, O.G. (1998). A method for isolation of chloroplast DNA and mitochondrial DNA from sunflower. *Plant Mol. Biol. Reporter,* **16:** 183.

Vitha, S., Benes, K., Philips, J.P. and Gartland, K.M.A. (1995). Histochemical (GUS) Analysis. In *Agrobacterium* Protocols, K.M.A. Gartland and M.R. Davey, eds (Totowa, NJ: Humana Press) pp. 185–193.

Warcup, J.H. (1950). The soil plate method for isolation of fungi from soil. *Nature*, **166:** 117–118.

Yoo, Y.J., Hong, J. and Hatch, R.T. (1987). Comparison of alpha amylase activities from different assay methods. *Biotechnol. Bioeng.* **30:** 147–151.

■ ■ ■

Appendix IV A : MICROBIOLOGY

Culture Media

The appendix includes the media composition to which reference is made in the text and/or commonly used in microbiology, mycology, plant pathology and any specific direction that may be needed.

Dehydrated, powdered, bottled mixtures of a great variety of prepared media or their ingredients are commercially available. These are prepared simply by adding weighed portions of the ingredients to the required amount of water. The instructions are invariably adequately described by the manufacturers. Suppliers of prepared (dehydrated) media include:

Albimia Laboratories, Inc.
16, Clinton Street
Brooklyn, New York 11201, USA

Bailed and Tatlock (London) Ltd.
Freshwater Road,
Chadwell Heath,
Essex, UK

Difco Laboratories Inc.
920, Henry Street
Detroit, Michigan 48200 USA

Oxoid Ltd.
Southwark Bridge Road
London S.E. 1. UK

Baltimore Biological Laboratories
1640 Gorsuch Avenue
Baltimore, Maryland 21203 USA

Hi-Media Laboratories Pvt. Ltd
23, Vadhani Industrial Estate
Bombay 400 086, India

Sigma Chemical Company
P.O. Box 14508, St. Louis
MO 63178, USA

The compositions for specific media are listed below. Unless otherwise noted, dry ingredients are dissolved separately in distilled water in a beaker and the volume is made up to 1 liter. The medium is dispensed into bottles, flasks, test tubes, etc. and sterilized by autoclaving at 121°C and 1.05 kg/cm^2 (15 psi) for 15 minutes.

Agrobacterium isolation medium (pH 7.1)

Mannitol	10.0 g
Sodium nitrate	4.0 g
Magnesium chloride	2.0 g

Calcium propionate	1.2 g
Magnesium phosphate	0.2 g
Magnesium sulphate	0.1 g
Sodium bicarbonate	0.075 g
Magnesium carbonate	0.075 g
Agar	20.0 g
Distilled water	1000.0 ml

Autoclave the constituents, allow to cool to 50–55 °C and aseptically add the following:

Berberine	275 ppm
Sodium selinite	100 ppm
Penicillin G (1,625 units/mg)	60 ppm
Streptomycin sulphate	30 ppm
Cyclohexamide (85-100% active)	250 ppm
Tyrothricin	1 ppm
Bacitracin (65 units/mg)	100 ppm

Highly specific for the isolation of *Agrobacterium tumefaciens* from soil.

Allen and Arnon's medium (modified) (pH 8.5-9.1)

Magnesium sulphate	0.025 g
Calcium chloride	0.05 g
Sodium chloride	0.20 g
Dipotassium hydrogen phosphate	0.35 g
*A5 Trace elements stock solution	1.0 ml
Distilled water	1000.0 ml

*The A5 trace elements stock solution has the following constituents:

Boric acid	2.86 g
Manganese chloride	1.81 g
Zinc sulphate	0.222 g
Molybdenum trioxide (85%)	0.0177 g
Cupric sulphate	0.079 g
Distilled water	1000.0 ml

For nitrogen fixing blue-green algae.

Azotobacter medium (pH 8.3)

Mannitol	15.0 g
K_2HPO_4	0.5 g
MgSO4.7H 20	0.2 g
$CaSO_4$	0.1 g
NaCl	0.2 g
$CaCO_3$	5.0 g
Agar	15 g
Distilled water	1000.0 ml

Beef peptone agar (pH 7.2)

Beef extract	3.0 g
Peptone	5.0 g
Dextrose	10.0 g
Yeast extract	5.0 g
Agar	15.0 g
Distilled water	1000.0 ml

Blood agar (pH 7.3)

Infusion from beef heart	500.0 g
Tryptose	10.0 g
Sodium chloride	5.0 g
Agar	15.0 g
Distilled water	1000.0 ml

Dissolve the above ingredients and autoclave at 121°C for 15 minutes. Cool to 45–50°C and add 50 ml of sterile defibrinated sheep blood, aseptically. Rotate to mix thoroughly, avoiding accumulation of air bubbles and pour immediately into sterile tubes or plates, *i.e.*, before solidification.

Brain heart infusion (pH 7.4)

Infusion from calf brain	200.0 g
Infusion from beef heart	250.0 g
Peptone	10.0 g
Disodium phosphate	2.5 g

Sodium chloride	5.0 g
Dextrose	2.0 g
Agar	1.0 g
Distilled water	1000.0 ml

Bromo cresol purple lactose agar

(a)	Tryptone	10 g
	Yeast extract	5g
	K_2HPO_4	1g
	Lactose	5g
	Distilled water	1000.0 ml
(b)	Bromocresol purple	1.6 g
	Ethyl alcohol (95%)	50 ml
	Distilled water	50 ml

Dissolve bromocresol purple in ethyl alcohol and add water. Add 2 ml of (b) to one litre of (a).

Butter fat agar (pH 7.8)

Yeast extract	5.0 g
Butter fat	50.0 g
Agar	20.0 g
Distilled water	1000.0 ml

Chocolate agar (pH 7.0)

Protease peptone	20.0 g
Sodium chloride	5.0g
Disodium phosphate	5.0 g
Dextrose	0.5 g
Agar	15 g
Distilled water	1000.0 ml

Add 5 percent defibrinated sheep blood to the sterile and molten agar, aseptically. Heat at 80°C until a chocolate colour develops.

Chu's medium no. 10 (modified)

Calcium nitrate	0.04 g

Dipotassium hydrogen phosphate	0.01 g
Sodium carbonate	0.02 g
Magnesium sulphate	0.025 g
Sodium silicate	0.025 g
Ferric citrate	0.003 g
Citric acid	0.003 g
*A5 Trace elements stock solution (optional)	1.0 ml
Distilled water	1000.0 ml

*The A5 Trace elements stock solution has the following constituents:

Boric acid	2.86 g
Manganese chloride	1.81 g
Zinc sulphate	0.222 g
Molybdenum trioxide (85%)	0.0177 g
Cupric sulphate	0.079 g
Distilled water	1000.0 ml

For common freshwater and soil algae.

Citrate agar (pH 6.8)

Magnesium sulphate	0.2 g
Sodium chloride	5.0 g
Ammonium dihydrogen phosphate	1.0 g
Dipotassium hydrogen phosphate	1.0 g
Sodium citrate (or citric acid)	2.0 g
Bromthymol blue	0.08 g
Agar	20.0 g
Distilled water	1000.0 ml

Citrate agar is used to determine the utilization of citrate as the sole carbon source.

Czapek's medium (modified)* (pH 6.4-7.0)

Sodium nitrate	2.0 g
Dipotassium hydrogen phosphate	1.0 g
Potassium chloride	0.5 g

Magnesium sulphate	0.5 g
Ferrous sulphate	0.01 g
** Cellulose	10.0 g
Distilled water	1000.0 ml

*For assaying cellulose

** Substitute cellulose with pectin and autoclave at 10 lb pressure for 15 minutes (to prevent breakdown of pectin) for assaying pectinase.

Czapek's Dox agar (CDA) / broth

Sodium nitrate	3.0 g
Dipotassium hydrogen phosphate	1.0 g
Magnesium sulphate	0.5 g
Potassium chloride	0.5 g
Ferrous sulphate	0.01 g
Sucrose	30.0 g
*Agar	15.0 g
Distilled water	1000.0 ml

*Do not add agar for the preparation of Czapek's Dox broth.

Dissolve all ingredients except phosphate in around half of the water and add the sucrose. Dissolve the phosphate separately and add to the rest. Melt agar in rest of the water and mix all the ingredients. Make volume to 1 litre. Sterilize by autoclaving at 121°C for 15 minutes.

Eosin-methylene blue agar (EMB) (pH 7.2)

Peptone	10.0 g
Lactose	5.0 g
Sucrose	5.0 g
Dipotassium hydrogen phosphate	2.0 g
Eosin Y	0.4 g
Methylene blue	0.065 g
Agar	15.0 g
Distilled water	1000.0 ml

This medium is used for distinguishing lactose-fermenting gram-negative bacteria from non-lactose fermenting types by their appearance. Colonies of *Escherichia coli* have a characteristic metallic sheen.

Gelatin tripticase soy broth (pH 7.3)

Gelatin	40.0 g
Tryptone (trypticase peptone; pancreatic digest of casein)	17.0 g
Soytone (phytone peptone; a papaic digest of soybean meal)	3.0 g
Sodium chloride	5.0 g
Dipotassium phosphate	2.5 g
Distilled water	1000.0 ml

Dissolve all the ingredients except gelatin in 500 ml of distilled water. Add the gelatin to 400 ml distilled water, boil, mix all and make up the volume to 1000 ml. Dispense 5 ml per tube and sterilize by autoclaving at 121°C for 15 minutes. It is used for detecting the microbial hydrolysis of an animal protein (gelatin).

Glucose phosphate broth (pH 7.5)

Peptone	10.0 g
Dipotassium hydrogen phosphate	5.0 g
Glucose	5.0 g
Distilled water	1000.0 ml

Dissolve the ingredients except glucose, in water by heating, filter hot, cool and adjust the pH to 7.5. Add 5 g of glucose, mix well, distribute in 5 ml portions. Sterilize at 115°C for 10 minutes.

Glycerol asparagine agar (pH 7.0)

Glycerol	10.0 ml
Asparagine	1.0 g
Dipotassium hydrogen phosphate	1.0 g
Agar	20.0 g
Tap water	1000.0 ml

Suitable medium for actinomycetes.

Glycerol yeast extract agar (pH 7.0)

Glycerol	5.0 ml
Yeast extract	2.0 g
Dipotassium hydrogen phosphate	1.0 g
Agar	15.0 g

*Aureomycin	0.01 g
Distilled water	1000.0 ml

*Add aureomycin to the sterile, cooled and molten agar, aseptically.

Growth medium for *Bacillus cereus* (pH 6.0)

Glucose	5.00 g
NaCl	3.00 g
Beef extract	5.00 g
Peptone	5.00 g
Yeast extract	0.50 g
$MgCl_2.6H_2O$	0.20 g
$MnCl_2.4H_2O$	0.01 g
$CaCl_2.2H_2O$	0.10 g
Distilled water	1000.0 ml

Jensen's medium (modified) (pH 6.5-7.0)**

$CaHPO_4$	1.0 g
K_2HPO_4	0.2 g
$MgSO_4.7H_2O$	0.2 g
NaCl	0.2 g
$FeCl_3$	0.1 g
*Trace elements stock solution	1.0 ml
Distilled water	1000.0 ml

*A stock solution of trace elements consists of the following:

Bo	0.05%
Mn	0.05%
Zn	0.005%
Mo	0.005%
Cu	0.002%

Add agar at the rate of 15 g/litre when solid medium is needed.

**Nitrogen free medium for growing seedlings to test root nodulation in legumes.

Lactose fermentation broth 1X and 2X* (pH 6.9)

Lactose	5.0 g
Peptone	5.0 g
Beef extract	3.0 g
Distilled water	1000.0 ml

*For 2X broth, double the quantity of all the ingredients, except the distilled water.

Litmus milk medium (pH 6.8)

Skim milk powder	100.0 g
Litmus	0.75 g
Distilled water	1000.0 ml

Autoclave at 12 lb pressure for 15 minutes.

Luria Bertani Broth (pH 7.5)

Tryptone	10 g
Yeast extract	5 g
NaCl	10 g
Distilled water	1000.0 ml

MacConkey's broth single (1X) and double strength (2X)* (pH 7.4)

Peptone	20.0 g
Lactose	10.0 g
Sodium chloride	5.0 g
Bile salt	5.0 g
Neutral red soultion (1% aqueous)	10.0 ml
Distilled water	1000.0 ml

Dissolve the peptone, sodium chloride and bile salt in water by heating. Adjust the pH to 8.0 and boil for 20 minutes. Cool, filter and adjust pH to 7.4. Add the lactose and indicator solution, in 5 ml portions to tubes containing inverted Durham's tubes. Sterilize by autoclaving at 115°C for 15 minutes.

*For double strength broth use twice the quantity of all ingredients, except the distilled water.

MacConkey's agar (pH 7.1)

Peptone	20.0 g
Sodium chloride	5.0 g
Bile salt	1.5 g
Lactose	10.0 g
Neutral red solution (1% aqueous)	10.0 ml
Crystal violet	0.001 g
Agar	13.5 g
Distilled water	1000.0 ml

Malt Extract Agar

Malt extract	15.0 g
Dipotassium hydrogen phosphate	1.0 g
Ammonium chloride	1.0 g
Citric acid N/1	15.0 ml
Agar	20.0 g
Distilled water	1000.0 ml

Recommended for yeasts and moulds. When using as fungal medium take malt extract 30 g; agar 12 g for 1 litre of medium.

Mannitol salt agar (pH 7.4)

Mannitol	10.0 g
Peptone	10.0 g
Sodium chloride	75.0 g
Beef extract	1.0 g
Phenol red	0.025 g
Agar	15.0 g
Distilled water	1000.0 ml

It is used for the selective isolation of pathogenic staphylococci, since most other bacteria are inhibited by the high salt concentration, colonies of pathogenic staphylococci are surrounded by a yellow halo, indicating mannitol fermentation.

Milk Agar (pH 6.9)

Tryptone	5.0 g
Yeast Extract	2.5 g

Dextrose	1.0 g
Skim Milk Powder (antibiotic free)	1.0 g
Agar	12.5 g
Distilled water	1000.0 ml

Malt extract-Yeast extract agar (MYE) medium (pH 7.3)

Yeast extract	4g
Malt extract	10g
Glucose	4g
Agar	20g
Distilled water	1000.0 ml

Mueller-Hinton agar (pH 7.4)

Beef infusion	300.0 g
Casamino acids	17.5 g
Starch	1.5 g
Agar	17.0 g
Distilled water	1000.0 ml

Nutrient broth and agar* (pH 7.0)

Peptone	5.0 g
Beef extract	3.0 g
NaCl	5.0 g
Distilled water	1000.0 ml

*For nutrient agar, add 15.0 agar. Used for isolation of bacteria and actinomycetes.

Nutrient gelatin (pH 6.8)

Peptone	5.0 g
Beef extract	3.0 g
Gelatin	120.0 g
Distilled water	1000.0 ml

Oatmeal agar (pH 6.0)

Oatmeal	20.0 g
Agar	18.0 g

*Trace salt solution	1.0 ml
Distilled water	1000.0 ml

Trace salt solution

$FeSO_4.7H_2O$	0.1 g
$MnCl_2$* $4H_2O$	0.1 g
$ZnSO_4.7H_2O$	0.1 g
Distilled water	100.0 ml

OR glucose agar (pH 7.1)

Peptone	2.0 g
Sodium chloride	5.0 g
Dipotassium hydrogen phosphate	0.3 g
Bromothymol blue (0.2%)	15.0 ml
Agar	3.0 g
D-glucose	10 g
Distilled water	1000.0 ml

PCNB medium (modified)*

Peptone	15.0 g
KH_2PO_4	1.0 g
$MgSO_4.7H_2O$	0.5 g
PCNB (Terraclor, a commercial product is 75% active)	0.5 g
Oxgall	0.5 g
Agar	20.0 g

Pikovskaya medium (pH 7.0)

Glucose	10 g
$Ca_3(PO_4)_2$	5.0 g
KCl	0. 2 g
$(NH_4)_2SO_4$	0.5 g
$MgSO_4.7H_2O$	0.1 g
MnSO4	0.006 g
$FeSO_4$	0.006 g
Yeast extract	0.5 g
Agar	15 g
Distilled water	1000.0 ml

Potato dextrose agar (PDA)

Potato (peeled)	200.0 g
Dextrose	20.0 g
Agar	15.0 g
Distilled water	1000.0 ml

Peel off the skin of potatoes, cut into small pieces and boil (200 g) in 500 ml of water, till they are easily penetrated by a glass rod. Filter through cheese-cloth. Add dextrose to the filtrate. Dissolve agar in water and bring up to the required volume by the addition of water. Autoclave at 15 lb pressure for 15 minutes.

Pringsheim's broth (pH 7.5)

KNO_3	0.02%
$MgSO_4.7H_2O$	0.001%
$(NH_4)_2. HPO_4$	0.002%
$CaCl_2.6H_2O$	0.0005%
$FeCl_3$	0.00005%

This medium is used for isolation of cyanobacteria from soil/water.

Sabouraud's agar (pH 5.6)

Peptone	10.0 g
Dextrose	40.0 g
Agar	15.0 g
Distilled water	1000.0 ml

Sabouraud agar is used to culture fungi.

Sulfur Indole Motility (SIM) agar (pH 7.3)

Peptone	30.0 g
Beef extract	3.0 g
Ferrous ammonium sulphate	0.2 g
Sodium thiosulphate	0.025 g
Agar	3.0 g
Distilled water	1000.0 ml

It is semi solid due to low concentration of agar.

Simmons citrate agar (pH 6.9)

Ammonium dihydrogen phosphate	1.0 g
Dipotassium hydrogen phosphate	1.0 g
Sodium chloride	5.0 g
Sodium citrate	2.0 g
Magnesium sulphate	0.2
Brom thymol blue	0.08 g
Agar	15.0 g
Distilled water	1000.0 ml

Sodium chloride broth (6.5%)

Brain-heart infusion broth*	100.0 ml
Sodium chloride	6.0 g

*Brain-heart infusion broth (obtainable in dehydrated form) contains 0.5 percent sodium chloride. Thus, by adding an additional 6.0 g, the desired concentration (6.5%) is obtained.

Snyder test agar (pH 4.8)

Tryptone	20.0 g
Sodium chloride	5.0 g
Dextrose	20.0 g
Brom cresol green	0.02 g
Agar	20.0 g
Distilled water	1000.0 ml

Starch agar (pH 7.0)

Starch (soluble)	20.0 g
Peptone	5.0 g
Beef extract	3.0 g
Agar	15.0 g
Distilled water	1000.0 ml

Skim milk agar

Skim milk powder	100.0 g
Agar	15.0 g
Distilled water	1000.0 ml

Dissolve agar into 700 ml of distilled water by boiling. Pour into a one-litre flask and sterilize by autoclaving at 121°C for 15 minutes. Dissolve the milk powder into 300 ml of water by heating to 50°C. Sterilize this milk solution at 113–115°C (8 lb pressure) for 20 minutes. Allow the two solutions to cool to 55°C and combine in one flask, aseptically, swirling gently to avoid bubbles.

Starch casein agar (pH 7.1)

Pancreatic digest of casein	17.0 g
Papain digest of soybean meal	3.0 g
NaCl	5.0 g
Dibasic potassium phosphate	2.5 g
Dextrose	2.5 g
Distilled water	1000.0 ml

Thioglycollate fluid (pH 7.1)

Peptone	15.0 g
L-cystine	0.75 g
Dextrose	5.0 g
Sodium chloride	2.5 g
Thioglycollic acid	0.3 ml
Yeast extract	5.0
Resazurin	0.001
Agar	0.75 g
Distilled water	1000.0 ml

Thioglycollate medium (without indicator) (pH 7.2)

Peptone	20.0 g
L-cystine	0.25 g
Glucose	6.0 g
Sodium chloride	2.5 g
Sodium thioglycollate	0.5 g
Sodium sulphite	0.1 g
Agar	0.7 g
Distilled water	1000.0 ml

Note: Store at room temperature.

Triple sugar-iron agar (pH 7.4)

Peptone	15.0 g
Proteose-peptone	5.0 g
Beef extract	3.0
Yeast extract	3.0
Lactose	10.0
Sucrose	10.0
Dextrose	1.0 g
Sodium chloride	5.0
Ferrous sulphate	0.2 g
Sodium thiosulphate	0.3 g
Phenol red	0.024 g
Agar	12.0 g
Distilled water	1000.0 ml

Trypticase nitrate broth (pH 7.2)

Trypticase	20.0 g
Potassium nitrate	1.0 g
Disodium phosphate	2.0 g
Dextrose	1.0 g
Agar	1.0 g
Distilled water	1000.0 ml

Trypticase soy agar (pH. 7.3)

Trypticase (animal peptone)	15.0 g
Phytone (soy peptone)	5.0g
Sodium chloride	5.0 g
Agar	15.0 g
Distilled water	1000.0 ml

Tryptone broth and agar*

Tryptone	10.0 g
Sodium chloride	5.0 g
Calcium chloride (IM)**	1.0 ml
Distilled water	1000.0 ml

** Add agar at the rate of 11.0 g/litre.*

*** Previously sterilized $CaCl_2$ is to be added after autoclaving the medium.*

Tryptone yeast glucose agar

Tryptone	5.0 g
Yeast extract	2.5 g
Glucose	1.0 g
Distilled water	1000.0 ml

Urea agar (pH 6.8)

Peptone	1.0 g
Sodium chloride	5.0 g
Potassium monohydrogen (or dihydrogen) phosphate	2.0 g
Glucose	1.0 g
Phenol red (0.02% solution)*	6.0 ml
Urea (20% aqueous solution)**	100.0 ml
Distilled water	1000.0 ml

Dissolve all the ingredients except glucose, phenol red and urea, autoclave at 15 lb pressure for 15 minutes and allow it to cool to 50°C.

**Add glucose and phenol red to the molten base and steam for 1 hour, cool to 50°C.*

***Add 100 ml urea (filter-sterilized solution) to the basal medium.*

Mix all the constituents well, dispense into sterile culture tubes and/or flasks.

Violet red bile agar with glucose (VRBG) (pH 7.4)

Glucose monohydrate	10.00 g
Pancreatic digest of Gelatin	7.00 g
Sodium chloride	5.00 g
Yeast extract	3.00 g
Bile salts	1.50 g
Neutral red	0.03 g

Crystal violet	0.002 g
Agar	15.00 g

This medium is used for the detection and enumeration of Enterobacteriaceae members.

Water Agar

Agar	20 g
Distilled water	1000.0 ml

Water agar is sometimes used to induce sporulation in fungi and isolating actinomycetes, protozoa and nematophagous fungi.

Yeast extract agar

Mannitol	10.0 g
Dipotassium hydrogen phosphate	0.5 g
Magnesium sulphate	0.2 g
Sodium chloride	0.1 g
Yeast extract	1.0 g
Congo red (1% solution)	2.5 ml
Agar	20.0 g
Distilled water	1000.0 ml

Yeast extract mannitol agar (pH 6.8)

Yeast extract	1.0 g
Mannitol	10.0 g
Dipotassium phosphate	0.5 g
Magnesium sulphate	0.2 g
Sodium chloride	0.1 g
Calcium carbonate	1.0 g
Agar	15.0 g
Distilled water	1000.0 ml

Yeast peptone dextrose medium (YPD)

Yeast extract	10.0 g
Peptone	20.0 g
Dextrose	20.0 g

Agar	20.0 g
Distilled water	1000.0 ml

Store prepared media below 8°C, protected from direct light. Store dehydrated powder, in a dry place, in tightly-sealed containers at 2–25°C.

YPS medium (pH 5-6.5)

Yeast extract	5 g
Soluble starch	5 g
K_2HPO_4	1.2 g
$MgSO_4$	0.5 g
Agar	20 g
Distilled water	1000.0 ml

70 mg chloramphenicol, if needed, is added after sterilization.

Indicators

Bromo cresol green

Bromo cresol green	0.4 g
Ethyl alcohol (95%)	500.0 ml
Distilled water	500.0 ml

Dissolve bromo cresol green in alcohol, add water and filter through filter paper.

Methyl Red Indicator

Methyl red	0.1 g
Ethyl alcohol (95%)	300.0 ml
Distilled water	200.0 ml

Dissolve the methyl red in alcohol. Add water and filter the preparation.

Phenol Red Indicator

Phenol red	0.2 g
Ethyl alcohol (95%)	500.0 ml
Distilled water	500.0 ml

Dissolve phenol red in the alcohol. Add distilled water and filter the preparation.

Phenolphthalein solution

Phenolphthalein	10 g
Methanol	1 ml
Distilled water	10 ml

Ringer's solution

NaCl	8.6 g
KCl	0.3 g
$CaCl_2$	0.33 g
Distilled water	1000.0 ml

It can also be prepared with Ringer Solution Tablets BR0052.

Stains

The appendix contain the list of stains commonly used in microbiology.

Acid-alcohol

Ethyl alcohol (95%)	97.0 ml
Hydrochloric acid (37%)	3.0 ml

Add the acid to the alcohol slowly. Store the solution in a properly labelled, glass-stoppered bottle.

Acid fuschin

Acid Fuchsin	1 g
Glacial acetic acid (HOAc)	1 ml
Distilled water	100 ml

Basic fuchsin

Basic fuchsin (90% dye content)	0.3 g
Ethyl alcohol (95%)	10.0 ml
Water	35 ml

Carbol fuchsin (or Ziehl-Neelsen's Carbol fuchsin)

Solution A

Basic fuchsin (90% dye content)	0.3 g
Ethyl alcohol (95%)	10.0 ml

Solution B

Phenol crystals (C.P.)	5.0 g
Distilled water	95.0 ml

Dissolve the solutes of each solution in the respective solvents indicated. Mix solutions A and B and allow to stand overnight. Filter the final preparation and store in a glass-stoppered bottle.

Cell wall stain

Mordant

5% aqueous solution of tannic acid

Stain

0.2% aqueous solution of crystal violet

Congo Red

Take saturated aqueous solution

Copper sulphate solution (20%)

Copper sulphate ($CuSO_4.5H_2O$)	20.0 g
Distilled water	100.0 ml

Crystal violet (1%)

Crystal violet (85% dye content)	1.0 g
Distilled water	100.0 ml

Cytoplasmic membrane stain

Plasmolysis

10% aqueous solution of potassium nitrate

Bouin's fixative stain

Saturated solution of picric acid	15 parts
Formalin	6 parts
Glacial acetic acid	1 part

Victoria blue stain

1% aqueous solution of Victoria blue 1 part

Water	1.5 part

Ethyl alcohol (95%)

Ethyl alcohol (100%)	95.0 ml
Distilled water	5.0 ml

Gram's stain

Crystal violet (Hucker's)

Solution A	
Crystal violet (90% dye content)	2.0 g
Ethyl alcohol (95%)	20.0 ml
Solution B	
Ammonium oxalate	0.8 g
Distilled water	80.0 ml

Dissolve crystal violet in ethyl alcohol and the ammonium oxalate in distilled water. Mix solutions A and B.

Gram's iodine

Iodine	1.0g
Potassium Iodide	2.0g
Distilled water	300.0 ml

Dissolve iodine and potassium iodide in distilled water.

Giemsa stain

Giemsa stain (powder)	3.8 g
Glycerol	125.0 ml
Methyl alcohol (solvent)	100.0 ml

Dissolve the stain in methyl alcohol. Add glycerol and shake well in bottle with glass beads. Filter if necessary. Keep tightly closed and stoppered at all times. It is a nuclear stain.

Haemotoxylin

Haemotoxylin	2.0g
Glacial acetic acid	10.0 ml
Absolute alcohol	100.0 ml
Glycerol	100.0 ml
Potassium alum	in excess
Distilled water	100.0 ml

Lactophenol cotton blue

Lactic acid	20.0 ml
Phenol crystals	20.0 g
Glycerol	40.0 ml
Distilled water	20.0 ml
Cotton blue (1% aqueous)	2.0 ml

Add lactic acid and glycerol/glycerine to the distilled water and mix thoroughly. Add phenol crystals and heat gently in hot water with frequent agitation until the crystals completely dissolve. Add the dye and mix thoroughly. Store the stain in brown bottle.

Lee's stain

Solution A:

$MgSO_4$	7 g
Tannic acid	8 g
Acetic acid (5%)	40 ml
Distilled water	110 ml

Dissolve the sulphate and tannic acid in distilled water with gentle heat. When solution is formed, add acetic acid.

Solution B:

Basic fuchsin	1.5 g
Ethyl alcohol (95%)	100 ml

Prepare the stain fresh for use by mixing one part of solution B to five parts of solution A. the mixed solution is stable for one day only, however, unmixed solutions can be preserved for indefinite periods.

Leifson's stain (for staining flagella)

Solution A:

NaCl	1.5 g
Distilled water	100 ml

Solution B:

Tannic acid	3.0 g
Distilled water	100 ml

Solution C:

Pararosaniline acetate	0.9 g

Pararosaniline hydrochloride	0.3 g
Ethanol (95%)	100 ml

Mix solution A and B in equal volume and then add its 2 parts in 1 part of solution C.

Light Green

Light green	2.0 g
Acetic acid (2%)	100.0 ml

Litmus solution

Litmus granules	100.0 g
Ethyl alcohol (40%)	500.0 ml
Sodium hydroxide (IN)	100.0 ml

Grind the litmus and transfer to a flask containing 250 ml of ethyl alcohol (40%). Boil the solution for 1 minute. Decant the liquid and add the rest of the ethyl alcohol (*i.e.*, 250 ml) to the granules. Boil for 1 minute. Decant the liquid and add it to the first extract. Make up the volume to 500 ml adding more of 40 percent ethyl alcohol. Add IN sodium hydroxide, drop-wise until the solution becomes purple.

To test the litmus solution, boil a test tube of distilled water and a test tube containing tap water. Add litmus solution, one drop to each tube. The tap water should turn blue and the distilled water, purple.

Loeffler's flagella mordant

20% Aqueous tannic acid	100.0 ml
Ferrous sulphate	20.0 g
10% Basic fuchsin in alcohol	10.0 ml
Distilled water	40.0 ml

Dissolve ferrous sulphate crystals in distilled water by warming and add the remaining ingredients.

Loeffler's flagella stain

1% Basic fuchsin in alcohol	20.0 ml
3% Aniline water	80.0 ml

Loeffler's methylene blue

Methylene blue	0.5 g

Potassium hydroxide solution (1%)	1.0 ml
Ethyl alcohol	30.0 ml
Distilled water	100.0 ml

Warm water to 50°C. Add methylene blue, stir well and then add other ingredients. Filter the contents.

Malachite green (5%)

Malachite green	5.0 g
Distilled water	100.0 ml

Safranin

Safranin 0 (2.5% solution in 95% ethyl alcohol)	10.0 ml
Distilled water	100.0 ml

Methylene blue solution

Methylene blue (90% dye content)	0.3 g
Distilled water	100.0 ml

Methyl red solution

Methyl red	0.04 g
Ethyl alcohol (absolute)	40.0 ml
Distilled water	60.0 ml

Dissolve methyl red in ethyl alcohol and add water.

Nigrosine solution (used for negative staining)

Nigrosine (water soluble)	10.0g
Distilled water	100.0 ml
Formalin	0.5 ml

Dissolve Nigrosine in distilled water and immerse the mixture in boiling water for 30 min. Then add formalin (as preservative). Filter the solution twice through double filter paper. Store in small tubes.

Safranin

Safranin (2.5% solution in 95% ethyl alcohol)	10.0 ml
Distilled water	100.0 ml

Spirochaete stain

Fixative

Glacial acetic acid	1.0 ml
Formalin	2.0 ml
Distilled water	100.0 ml

Mordant

Phenol	1.0 g
Tannic acid	5.0 g
Distilled water	100.0 ml

Ammonical silver nitrate

Ammonia solution (10%)	
Silver nitrate (0.5%. aqueous)	90.0 m1

Remove a few millilitres of silver nitrate. Add ammonia solution to the rest silver nitrate solution until the precipitate formed just dissolves. Add the removed silver nitrate drop by drop, until the precipitate returns and does not redissolve.

Water-iodine solution

Gram's iodine (as in Gram's stain)	10.0 ml
Distilled water	30.0 ml

Immersion oil

Cargille's oil

Mix 1:1 volume of type A (low viscosity) and type B (high viscosity).

■ ■ ■

Appendix IV B : BIOCHEMISTRY

1. Concentrations of Commercial Liquids

Compound	Molecular weight	Molarity	pH of dilute solutions		
			1 M	0.1 M	0.01 M
Acetic acid. glacial	60.05	17.4	2.4	2.9	3.4
Formic acid	46.02	23.4			
Hydrochloric acid, 38%	36.47	11.6	0.1	1.1	2.02
Nitric acid, 70%	63.02	16.0			
Phosphoric acid	98.00	18.1	1.5		
Sulfuric acid	98.08	18.00			
Ammonium hydroxide	35.00	14.8			
Ethanolamine, 99%	61.08	16.5			
Triethylamine	101.19	7.16			
Formaldehyde, 37%	30.30	12.2			
Hydrogen peroxide, 30%	34.02	8.8			
β-Mercaptoethanol	78.13	14.4			

2. Preparation of Acid and Alkali (1 N) Solutions

S.N.	Acid/Alkali	Final volume to be made up to one litre with distilled water
1.	Concentrated sulphuric acid	30 cc
2.	Concentrated hydrochloric acid	100 cc
3.	Concentrated nitric acid	63 cc
4.	Oxalic acid dihydrate	62.96 g
5.	Sodium hydroxide	40.00 g
6.	Potassium hydroxide	56.00 g

3. Molar Absorption Coefficients of Selected Biochemicals

Biomolecule	Absorption maxima	Molar absorption
	(λ_{max}) (nm)	coefficient (ex 10^{-1})
Phenylalanine	257.5	0.19 (in 0. 1 N HCI)
	258.0	0.21 (in 0. 1N NaOH)
Tryptophan	278.0	5.60 (in 0. 1N HCl)

	280.5	5.43 (in 0. 1N NaOH)
Tyrosine	274.5	1.34 (in 0. 1N HCl)
	293.5	2.33 (in 0. 1N NaOH)
NAD^+& $NADP^+$	259	18.0*
NADH & NADPH	339	6.22*
2,4-Dinitrophenol	400	18.3 (at alkaline pH 9–10)
Riboflavin	450	12.2*
Adenine	260	13.3
Guanine	260	7.2*
Cytosine	260	2.6*
Thymine	260	7.4*
Uracil	260	8.2*
ATP	260	15.4*
GTP	260	11.7*
DNP-alanine	360	17.2 (in 1 N NaOH)
DNP-aspartic acid	360	18.2 (in 1 N NaOH)
DNP-tyrosine	360	16.9 (in 1 N NaOH)

* *at pH 7*

4. Prefixes to Modify Units

Prefix	Factor base 10	Symbol	Prefix	Factor base 10	Symbol
Deci	-1	d	Deca	1	da
centi	-2	c	Hecto	2	h
Milli	-3	m	Kilo	3	k
Micro	-6	μ	Mega	6	M
Nano	-9	n	Giga	9	G
Pico	-12	p	Tera	12	T
Femto	-15	f	Peta	15	P
atto	-18	a	exa	18	E

5. Metric calculations

10^{-1} m	1/10th of m	=	1 dm
10^{-2} m	1/100th of m	=	1 cm
10^{-3} m	1/1000th of m	=	1 mm
10^{-6} m	1/1000000 of m	=	1μ
10^{-9} m	1/1000000000 of m	=	1nm

6. Convert between Times Gravity (*x g*) and Centrifuge Rotor Speed (RPM)

Certain procedures necessitate precise centrifugation conditions, which must be specified in terms of relative centrifugal force (RCF) expressed in units of gravity (times gravity or x *g*). Many microcentrifuges only have settings for speed (revolutions per minute, RPM), not relative centrifugal force. Consequently, a formula for conversion is required to ensure that the appropriate setting is used in an experiment. The relationship between RPM and RCF is as follows:

$$g = (1.118 \times 10^{-5})\ RS^2$$

Where, *g* is the relative centrifugal force, R is the radius of the rotor in centimeters, and S is the speed of the centrifuge in revolutions per minute. Values of RCF in units of times gravity (x *g*) for common microcentrifuge rotor radii appear in the following conversion table. As an example, centrifugation of a sample at 5,000 RPM in a microcentrifuge that has a rotor with a radius of 7 cm will deliver a centrifugal force of 1,957 x *g*.

Centrifugation speed and time often are not critical factors in routine sample-handling procedures involving a benchtop microcentrifuge. Usually, as long as speed and time are sufficient to ensure that cells, debris or resin are pelleted effectively, it does not matter if the speed is faster or the time longer than necessary. For this reason, many protocols do not specify a particular centrifugal force to be applied, but instead indicate some general guidelines such as "centrifuge at high speed for 1 minute."

Speed (RPM)	Rotor Radius (from centre of rotor to sample) in centimeters											
	4	5	6	7	8	9	10	11	12	13	14	15
1000	45	56	67	78	89	101	112	123	134	145	157	168
1500	101	126	151	176	201	226	252	277	302	327	352	377
2000	179	224	268	313	358	402	447	492	537	581	626	671
2500	280	349	419	489	559	629	699	769	839	908	978	1048
3000	402	503	604	704	805	906	1006	1107	1207	1308	1409	1509
3500	548	685	822	959	1096	1233	1370	1507	1643	1780	1917	2054
4000	716	894	1073	1252	1431	1610	1789	1968	2147	2325	2504	2683
4500	906	1132	1358	1585	1811	2038	2264	2490	2717	2943	3170	3396
5000	1118	1398	1677	1957	2236	2516	2795	3075	3354	3634	3913	4193
5500	1353	1691	2029	2367	2706	3044	3382	3720	4058	4397	4735	5073
6000	1610	2012	2415	2817	3220	3622	4025	4427	4830	5232	5635	6037
6500	1889	2362	2834	3306	3779	4251	4724	5196	5668	6141	6613	7085

Speed (RPM)	Rotor Radius (from centre of rotor to sample) in centimeters											
	4	5	6	7	8	9	10	11	12	13	14	15
7000	2191	2739	3287	3835	4383	4930	5478	6026	6574	7122	7669	8217
7500	2516	3144	3773	4402	5031	5660	6289	6918	7547	8175	8804	9433
8000	2862	3578	4293	5009	5724	6440	7155	7871	8586	9302	10017	10733
8500	3231	4039	4847	5654	6462	7270	8078	8885	9693	10501	11309	12116
9000	3622	4528	5433	6339	7245	8150	9056	9961	10867	11773	12678	13584
9500	4036	5045	6054	7063	8072	9081	10090	11099	12108	13117	14126	15135
10000	4472	5590	6708	7826	8944	10062	11180	12298	13416	14534	15652	16770
10500	4930	6163	7396	8628	9861	11093	12326	13559	14791	16024	17256	18489
11000	5411	6764	8117	9469	10822	12175	13528	14881	16233	17586	18939	20292
11500	5914	7393	8871	10350	11828	13307	14786	16264	17743	19221	20700	22178
12000	6440	8050	9660	11269	12879	14489	16099	17709	19319	20929	22539	24149
13000	7558	9447	11337	13226	15115	17005	18894	20784	22673	24562	26452	28341
13500	8150	10188	12225	14263	16300	18338	20376	22413	24451	26488	28526	30563
14000	8765	10956	13148	15339	17530	19722	21913	24104	26295	28487	30678	32869

7. Table concerning Ammonium Sulphate Solubility

Initial concentration of Ammonium Sulphate (Percentage saturation)	Final concentration of Amm. Sulphate (Percentage Saturation)																
	10	20	25	30	33	35	40	45	50	55	60	65	70	75	80	90	100
	Grams solid Amm. Sulphate to be added to 1 litre of solution																
0	56	114	144	176	196	209	243	277	351	351	390	430	472	516	561	662	767
10	-	57	86	118	137	150	183	216	251	288	326	365	406	449	494	592	694
20	-	-	29	59	78	91	123	155	189	225	262	300	340	382	424	520	619
25	-	-	-	30	49	61	93	125	158	193	230	267	307	348	390	485	583
30	-	-	-	-	19	30	62	94	127	162	198	235	273	314	356	449	546
33	-	-	-	-	-	12	43	74	107	142	177	214	252	292	333	426	522
35	-	-	-	-	-	-	31	63	94	129	164	200	238	278	319	411	506
40	-	-	-	-	-	-	-	31	63	97	132	168	205	245	285	375	469
45	-	-	-	-	-	-	-	-	31	65	99	134	171	210	250	339	431
50	-	-	-	-	-	-	-	-	-	33	66	101	137	176	214	302	392
55	-	-	-	-	-	-	-	-	-	-	33	67	103	141	179	264	353
60	-	-	-	-	-	-	-	-	-	-	-	34	69	105	143	227	314
65	-	-	-	-	-	-	-	-	-	-	-	-	34	70	107	190	275
70	-	-	-	-	-	-	-	-	-	-	-	-	-	35	72	153	237
75	-	-	-	-	-	-	-	-	-	-	-	-	-	-	36	115	198
80	-	-	-	-	-	-	-	-	-	-	-	-	-	-	-	77	157
90	-	-	-	-	-	-	-	-	-	-	-	-	-	-	-	-	79

8. Amino Acids UAB

Amino Acid	Single letter code	Triple letter code	Molecular weight (pH 7)	Side chain pK	α- NH_2 pK	α - COOH pK
Alanine	A	Ala	89		9.87	2.35
Arginine	R	Arg	174	13.2	9.09	2.18
Asparagine	N	Asn	132		8.8	2.02
Aspartic acid	D	Asp	133	3.65	9.6	1.88
Cysteine	C	Cys	121	8.33	10.78	1.71
Glutamic acid	E	Glu	147	4.25	9.67	2.19
Glutamine	Q	Gln	146		9.13	2.17
Glycine	G	Gly	75		9.6	2.34
Histidine	H	His	155	6.0	8.97	1.78
Isoleucine	I	He	131		9.76	2.32
Leucine	L	Leu	131		9.6	2.36
Lysine	K	Lys	146	10.28	8.9	2.2
Methionine	M	Met	149		9.21	2.28
Phenylalan ine	F	Phe	165		9.24	2.58
Proline	P	Pro	115		10.6	1.99
Serine	S	Ser	105		9.15	2.21
Threonine	T	Thr	119		9.12	2.15
Tryptophan	W	Trp	204		9.39	2.38
Tyrosine	Y	Tyr	181	10.1	9.11	2.2
Valine	V	Val	117		9.72	2.29

9. Some of the Common Laboratory Buffer Solutions

The useful range of these and most other buffers is about 1 unit either side of the pK value.

Buffer	Synonym	pK	Molecular weight
Phosphate (pK_1)		2.12	98.0, free acid
Glycine-HCl		2.34	53
Citrate (pK_1)		3.14	192.1, free acid
Formate		3.75	68.0, Na salt
Carbonate (pK_1)		3.76	106.0

Succinate (pK_1)		4.19	118.1, free acid
Acetate		4.75	82.0 Na salt
Citrate (pK_2)		4.76	214.4, Na salt
Succinate (pK_2)		5.57	162.1, diNa salt
MES[a]		6.15	195.2, hydrate
Carbonate (pK_2)	Bicarbonate	6.36	84.0 bicarbonate
Citrate (pK_2)		6.39	294.1, TriNa salt, dihydrate
PIPES[b]		6.8	302.4
ACES		6.9	182.2
MOPS[c]		7.2	209.7
HEPES[d]		7.55	238.3
Phosphate (pK_2)		7.21	120.0, Na salt, monobasic
TES		7.7	229.2
Barbital	Barbitone veronal	7.78	128.1, barbituric acid
Triethanolamine		7.8	149.2
Tricine		8.15	179.2
Tris[e]		8.3	121.1
Bicine		8.35	163.2
Glycyglycine		8.4	132.1
Borate		9.21	201.2, Na tetraborate
CHES		9.5	207.3
Ethanolamine		9.5	61.1
Glycine-NaOH		9.6	97.1, Na salt hydrate
CAPS		10.4	221.3
Triethylamine		10.7	101.2
Phosphate (pK_3)		12.3	141.9, Na salt, dibasic

[a] 2- (N- morpholino) ethanesulfonic acid

[b] Piperzine –N N' bis- (2- ethanesulfonic acid)

[c] 3- (N-morpholino) propanesulfonic acid

[d] N-2-hydroxyethylpiperazine – N' -2 – ethanesulfonic acid

[e] Tris (hydroxymethyl) aminomethane

Borate forms complexes with sugars and citrate readily combines with calcium.

10. Values of Amino Acids in Paper Chromatography

Amino Acid	Solvents	
	1	2
Histidine	0.07	0.69
Serine	0.10	0.36
Lysine	0.10	0.48
Arginine	0.11	0.59
Aspartic acid	0.13	0.15
Glutamic acid	0.16	0.25
Glycine	0.17	0.40
Alanine	0.22	0.54
Threonine	0.22	0.50
Proline	0.30	0.91
Tyrocine	0.32	0.64
Methionine	0.40	0.80
Valine	0.47	0.77
Tryptophan	0.47	0.83
Isoleucine	0.55	0.86
Phenylalanine	0.58	0.89
Leucine	0.60	0.86

Solvent 1 = n-butyl – acetic acid – water 4:1:5 v/v

Solvent 2 = Phenol – water 4:1 v/v

11. Common Hazardous Chemicals Used in Laboratories

- **Acetic acid (concentrated)** must be handled with great care. It may be harmful by inhalation, ingestion, or skin absorption. Wear appropriate gloves and goggles, and use in a chemical fume hood.
- **Acrylamide** (unpolymerized) is a potent neurotoxin and is absorbed through the skin (the effects are cumulative). Avoid breathing the dust. Wear appropriate gloves and a face mask when weighing powdered acrylamide and methylene-bisacrylamide. Use in a chemical fume hood. Polyacrylamide is considered to be nontoxic, but it should be handled with care because it might contain small quantities of unpolymerized acrylamide.
- **Actinomycin D** is a teratogen and is carcinogenic. It is highly toxic and may be fatal if inhaled, ingested, or absorbed through the skin. It may also cause irritation. Avoid breathing the dust.

Wear appropriate gloves and safely glasses, and always use in a chemical fume hood. Solutions of actinomycin D are light sensitive.

- **Ammonium persulfate,** is extremely destructive to tissue of the mucous membranes and upper respiratory tract, eyes, and skin. Inhalation may be fatal. Wear appropriate gloves, safety glasses, and protective clothing. Use only in a chemical fume hood. Wash thoroughly after handling.
- **Bisacrylamide** is a potent neurotoxin and is absorbed through the skin (the effects are cumulative). Avoid breathing the dust. Wear appropriate gloves and a face mask when weighing powdered acrylamide and methylene-bisacrylamide.
- **Blood** (human) and blood products, and Epstein-Barr virus. Human blood, blood products, and tissues may contain occult infectious materials such as hepatitis B virus and HIV that may result in laboratory-acquired infections. Investigators working with EBV transformed lymphoblast cell lines are also considered a biohazard and should be handled accordingly. Wear disposable appropriate gloves, use mechanical pipetting devices, work in a biological safety-cabinet, protect against the possibility of aerosol generation, and disinfect all waste materials before disposal. Autoclave contaminated plasticware before disposal; autoclave contaminated liquids or treat with bleach [(10%) (v/v) final concentration] for at least 30 minutes before disposal. Consult the local institutional safety officer for specific handling and disposal procedures.
- **Bromophenol blue** may be harmful by inhalation, ingestion, or skin absorption. Wear appropriate gloves and safely glasses. Use in a chemical fume hood.
- **Chloramphenicol** may be harmful by inhalation, ingestion, or skin absorption, and is a carcinogen. Wear appropriate gloves and safety glasses. Use in a chemical fume hood.
- **Chloroform ($CHCl_3$)** is irritating to the skin, eyes, mucous membranes, and respiratory tract. It is a carcinogen and may damage the liver and kidneys. It is also volatile. Avoid breathing the vapours. Wear appropriate gloves and safety glasses and always use in a chemical fume hood.
- **Colchicine** is highly toxic, may be fatal, and may cause cancer and heritable genetic damage; It may be harmful by inhalation, ingestion, or skin absorption. Wear appropriate gloves and safety

glasses and use only in a chemical fume hood. Do not breathe the dust.

- **Coomassie brilliant blue** may be harmful by inhalation, ingestion, or skin absorption. Wear appropriate gloves and safety glasses.
- **Copper sulphate ($CuSO_4$)** may be harmful by inhalation or ingestion. Wear appropriate gloves and safety glasses.
- **3,3'-Diaminobenzidine tetrahydrochloride (DAB)** is a carcinogen. Handle with extreme care. Avoid breathing vapours. Wear appropriate gloves and safety glasses and use in a chemical fume hood.
- **Dimethyl sulfoxide (DMSO)** may be harmful by inhalation or skin absorption. Wear appropriate gloves and safety glasses. Use in a chemical fume hood. DMSO is also combustible. Store in a tightly closed container. Keep away from heat, sparks, and open flame.
- **Dithiothreitol (DTT)** is a strong reducing agent that emits a foul odour. It may be harmful by inhalation, ingestion, or skin absorption. When working with the solid form or highly concentrated stocks, wear appropriate gloves and safety glasses and use in a chemical fume hood.
- **Ethanolamine ($HOCH_2CH_2NH_2$)** is toxic and harmful by inhalation, ingestion, or skin absorption. Handle with care and avoid any contact with the skin. Wear appropriate gloves and goggles and use in a chemical fume hood. Ethanolamine is highly corrosive and reacts violently with acids.
- **Ethidium bromide** is a powerful mutagen and is toxic. Consult the local institutional safety officer for specific handling and disposal procedures. Avoid breathing the dust. Wear appropriate gloves when working with solutions that contain this dye.
- **Formaldehyde (HCHO)** is highly toxic and volatile. It is also a carcinogen. It is readily absorbed through the skin and is irritating or destructive to the skin, eyes, mucous membrane, and upper respiratory tract. Avoid breathing the vapours. Wear appropriate gloves and safety glasses. Always use in a chemical fume hood. Keep away from heat, sparks, and open flame.
- **Glutaraldehyde** is toxic. It is readily absorbed through the skin and is irritating or destructive to the skin, eyes, mucous membranes, and upper respiratory tract. Wear appropriate gloves and safety glasses. Always use in a chemical fume hood.

- **Hydrogen peroxide (H_2O_2)** is corrosive, toxic, and extremely damaging to the skin. It may he harmful by inhalation, ingestion, and skin absorption. Wear appropriate gloves and safety glasses and use only in a chemical fume hood.
- **Iodoacetamide,** can alkylate amino groups in proteins and can therefore, cause problems if the antigen is being, purified for amino acid sequencing. It is toxic and harmful by inhalation, ingestion, or skin absorption. Wear appropriate gloves and safety glasses and rise only in a chemical fume hood. Do not breathe the dust.
- **β-Mercaptoethanol (2-Mercaptoethanol, $HOCH_2CH_2SH_2$)** may be fatal if inhaled or absorbed through the skin and is harmful if ingested. High concentrations are extremely destructive to the mucous membranes, upper respiratory tract, skin, and eyes. β-Mercaptoethanol has a very foul odour. Wear appropriate gloves and safety glasses. Always use in a chemical fume hood.
- **Phenol** is extremely toxic, highly corrosive, and can cause severe burns. It may be harmful by inhalation, ingestion, or skin absorption. Wear appropriate gloves, goggles, and protective clothing. Always use in a chemical fume hood. Rinse any areas of skin that come in contact with phenol with a large volume of water and wash with soap and water; do not use ethanol!
- **Phenylmethylsulfonyl fluoride (PMSF), ($C_7H_7FO_2S$ or C_6H_5 CH_2SO_2F)** is a highly toxic cholinesterase inhibitor. It is extremely destructive to the mucous membranes of the respiratory tract, eyes, and skin. It may be fatal by inhalation, ingestion, or skin absorption. Wear appropriate gloves and safety glasses and always use in a chemical fume hood. In case of contact, immediately flush eyes or skin with copious amounts of water, and discard contaminated clothing.
- **Polyacrylamide** is considered to be nontoxic, but it should be treated with care because it may contain small quantities of unpolymerized material.
- **Polyethyleneglycol (PEG)** may be harmful by inhalation, ingestion, or skin absorption. Avoid inhalation of powder. Wear appropriate gloves and safety glasses.
- **Proteinase K** is an irritant and may be harmful by inhalation, ingestion, or skin absorption. Wear appropriate gloves and safety glasses.

- **Silver nitrate ($AgNO_3$)** is a strong oxidizing agent and should be handled with care. It may be harmful by inhalation, ingestion, or skin absorption. Avoid contact with skin. Wear appropriate gloves and safety glasses. It can cause explosion upon contact with other materials.
- **Sodium azide (NaN_3)** is highly poisonous. It blocks the cytochrome electron transport system. Solutions containing sodium azide should be clearly marked. It may be harmful by inhalation, ingestion, or skin absorption. Wear appropriate gloves and safety goggles and handle with great care. Sodium azide is an oxidizing agent and should not be stored near flammable chemicals.
- **Sodium dodecyl sulfate (SDS)** is toxic, an irritant, and poses a risk of severe damage to the eyes. It may be harmful by inhalation, ingestion, or skin absorption. Wear appropriate gloves and safety goggles. Do not breathe the dust.
- **Streptomycin** is toxic and a suspected carcinogen and mutagen. It may cause allergic reactions. It may be harmful by inhalation, ingestion, or skin absorption. Wear appropriate gloves and safety glasses.
- **Sulfosalicylic acid (dihydrate)** is extremely destructive to the mucous membranes and respiratory system. Do not breathe the dust. Wear appropriate gloves and safety glasses and use only in a chemical fume hood.
- **N,N,N,',N'-Tetramethylethylenediamine (TEMED)** is extremely destructive to tissues of the mucous membranes and upper respiratory tract, eyes, and skin. Inhalation may be fatal. Prolonged contact can cause severe irritation or burns. Wear appropriate gloves, safety glasses, and other protective clothing and use only in a chemical fume hood. Wash thoroughly after handling. Flammable: *Vapour* may travel a considerable distance to source of ignition and flash back. Keep away from heat, sparks, and open flame.
- **Tris** may be harmful by inhalation, ingestion, or skin absorption. Wear appropriate gloves and safety glasses.
- **X-gal** may be toxic to the eyes and skin. Observe general cautions when handling the powder. Note that stock solutions of X-gal are prepared in DMF, an organic solvent.
- **Xylene** is flammable and may be narcotic at high concentrations. It may be harmful by inhalation, ingestion, or skin absorption.

Wear appropriate gloves and safety glasses and use only in a chemical fume hood. Keep away from heat.

- **Zinc chloride, $ZnCl_2$,** is corrosive and poses possible risk to the unborn child. It may be harmful by inhalation, ingestion, or skin absorption. Wear appropriate gloves and safety glasses. Do not breathe the dust.

■ ■ ■

Appendix IV C : MOLECULAR BIOLOGY

1. Reagents and Buffers for Plasmid DNA Isolation

LB (Luria Bertani) media

10g Bacto Tryptone

5g Bacto Yeast extract

10g NaCl

Adjust pH to 7.2–7.5 with 1 N NaOH and make up volume to 1 litre with distilled water.

Distribute the medium among various flasks. Autoclave.

Ampicillin: Dissolve 50 mg ampicillin in 1 ml dH_2O.

Filter sterilize.

TE Buffer

50 mM Tris-HC1, pH 8.0

20 mM ethylenediaminetetraacetic acid (EDTA)

STET Buffer

8% Sucrose

5% Triton X-100

50 mM EDTA

50 mM Tris-HCI, pH 8.0

Lysis Buffer

Glucose	50 mM
Tris-HCI (pH 8.0)	25 mM
EDTA	10 mM

Lytic Mixture

SDS	1%
NaOH	0.2 N
Sodium Acetate Buffer	3M, pH 4.8
r RNase Buffer	10 mg/ml in 0.1 M Na acetate, 0.3 mM EDTA, pH 8.0, preheat to 80°C for 10 min

TE Buffer

Tris-HCI (pH 8.0)	10 mM
EDTA	I mM
Lysozyme	Dissolve 40 mg in I ml sterile water store at –20°C.

2. Reagents for Genomic DNA Isolation

TB (Terrific Broth) media

12g Bactotryptone

24g Bacto Yeast extract

4ml Glycerol

Dissolve the above components in approximately 900ml double distilled water. Autoclave this solution for 20 mins at 121°C at 15 psi. Let the solution cool, add 100ml of TB Salt Solution (described below), and adjust the final volume to 1000ml by adding double distilled water. Add antibiotic, typically Amp at 25 ug/ml or Kan at 50 ug/ml just prior inoculation with cells.

10 X TB Salt Solution (For total volume 100ml)

0.17M KH_2PO_4 = 2.31g KH_2PO_4

0.72M K_2HPO_4 = 12.54g K_2HPO_4

Adjust the volume to 100ml with double distilled water and autoclave.

GET / LYSOZYME solution (for total volume 100ml):

50 mM Glucose=0.9 g d-Glucose

10 mM EDTA, pH 8.0=1 ml of 1M stock

25 mM Tris-HCl, pH 8.0=2.5ml of 1M stock

Make this solution fresh or filter sterilize before storing at 4°C. and add 2 mg/ml Lysozyme (0.2 gms lysozyme/100 ml 1xGET) just before use.

NaOH/SDS (Alkaline lysis solution):

Make this solution fresh. For total volume 100ml:

0.2 N NaOH=20ml of 1N stock

1% SDS=5 ml of 20% SDS <

Make the volume to 100ml with double distilled water.

3.0 M Sodium Acetate (NaOAc), pH 4.8 6) DNA Binding Matrix

Weigh 10g of diatomaceous earth (Sigma D-5384) and resuspend it in 100ml of double distilled water in a 100ml measuring cylinder and let it settle down for 3 hours. Decant the supernatant. If you wish to store the defined matrix, resuspend the pellet in TE:10:1, transfer the matrix to a screw cap storage bottle and store in the cold room. Otherwise, resuspend the pellet in 100ml of 6M guanidine hydrochloride containing 50mM Tris-HCl and 20mM EDTA (final matrix concentration of 100mg/ml assuming no significant losses during the defining step).

Wash Buffer (TE/ethanol) For total volume 1000ml:

10 mM Tris-HCl, pH 8.0 = 10ml of 1M stock

1 mM EDTA, pH 8.0 = 1ml of 1M stock

50% Ethanol = 500ml of absolute alcohol

Make the volume to 1000ml with double distilled water.

Elution Buffer (TE:10:1) – this is Tris-EDTA:10:1 not the usual 10:0.1 molar ratio of Tris-EDTA.

For total volume 1000 ml:

10 mM Tris-HCl, pH 8.0 = 10ml of 1M stock

1 mM EDTA, pH 8.0 = 1ml of 1M stock

Make the volume to 1000ml with double distilled water.

Ethanol Acetate

95% Ethanol = 95ml of absolute ethanol

0.12M NaOAc = 4ml of 3M NaOAc, pH 4.8.

Make the volume to 100ml with double distilled water.

Isopropanol

Acetone

Ethanol (70%)

DNase free RNase A

For 100ml of RNase A buffer:

0.1M Na acetate-0.82g anhydrous

0.3mM EDTA-0.3ml of 100mM stock

Adjust pH to 4.8 with acetic acid. Make the volume to 100 ml with double distilled water. Aliquot in 2 ml amounts and store at -20°C. Add 40mg of pancreatic RNase A to the 2 ml alaquots of RNAse A buffer just before use and inactivate any contaminating DNase by heating at 80°C for 10 mins.

RNase T1 (10 units/ul) made in 50mM Tris-HCl pH 7.6.

5 M Potassium Acetate Solution

1. Stock solution

	volume
5 M Potassium acetate	60 ml
Glacial acetic acid	11.5 ml
ddH_2O	28.5 ml
Total	100 ml

2. Filter sterilize. The resulting solution is 3 M potassium and 5 M acetate and has a pH of about 4.8.

   OR use the following recipe is you don't have KOAc

3. Glacial acetic acid 29.5 ml

Glacial acetic acid	29.5 ml
KOH	several pellets
ddH_2O	to 100 ml
Total	100 ml

Add the KOH pellets until the pH is 4.8.

3. Isolation of Bacteriophage λ DNA

SM Buffer

This buffer is used for storage and dilution of bacteriophage λ stocks. For 1 litre -

NaCl	5.8 g
$MgSO_4.7H_2O$	2g
1M Tris-Cl (pH 7.5)	50 ml
Gelatin solution (2%)	5 ml

DDW to 1 litre

4. Plant DNA Isolation (CTAB method)

CTAB buffer 100ml

2.0 g CTAB (Hexadecyl trimethyl-ammonium bromide)

10.0 ml 1 M Tris pH 8.0

4.0 ml 0.5 M EDTA pH 8.0 (Ethylenediaminetetra Acetic acid Disodium salt)

28.0 ml 5 M NaCl

40.0 ml H_2O

1 g PVP 40 (polyvinyl pyrrolidone (vinylpyrrolidine homopolymer) Mw 40,000)

Adjust all to pH 5.0 with HCl and make up to 100 ml with H_2O.

1 M Tris pH 8.0

Dissolve 121.1 g of Tris base in 800 ml of H_2O. Adjust pH to 8.0 by adding 42 ml of conc. HCl. Allow the solution to cool to room temperature before making the final adjustments to the pH. Adjust the volume to 1 L with H_2O. Sterilize using an autoclave.

5x TBE buffer

54 g Tris base

27.5 g boric acid

20 ml of 0.5M EDTA (pH 8.0)

Make up to 1L with water.

To make a 0.5x working solution, do a 1:10 dilution of the concentrated stock.

Agarose gel (1%)

1 g Agarose dissolved in 100 ml TBE

Preparation of equilibriated phenol

- Distil the phenol carefully at 160°C and store at -20°C.
- Melt the phenol in 65°C water bath.
- Add 8-hydroxyquinoline to a final concentration of 0.1%. This compound is an antioxidant.
- To the melted phenol, add an equal volume of 0.5M Tris-HCl (pH 8.0) buffer at room temperature. Stir the mixture on a magnetic stirrer for 15 min., and then turn off the stirrer. When the two phases get separated, aspirate as much as possible from the upper (aqueous) phase using a glass pipette attached to a vaccum line equipped with traps.
- Add on equal volume of 0.1 M Tris-HCl (pH 8.0) to the phenol. Stir the mixture on a magnetic stirrer for 15 min., and then turn off the stirrer. Remove the upper aqueous phase as described in earlier step.

- Repeat the extraction until the pH of the phenolic phase is greater than 7.8 (as measured with pH paper).
- After the phenol is equilibriated and the final aqueous phase has been removed, add 0.1 volume of 0.1 M Tris-Cl (pH 8.0) containing B-mercaptoethanol (0.2%). The phenol solution may be stored in this form with 100 mM Tris-HCI (pH 8.0) in a brown bottle at 4°C for a period upto two months.

Chloroform-Isoamylalcohol

- Mix chloroform and isoamylalcohol in 24:1 ratio.
- Store at room temperature in brown bottle.

Phenol-chloroform-isoamylalcohol:

- Mix equal volume of phenol and chloroform isoamylalcohol (24:1).
- Sodium Acetate - 3 M, pH 4.8
- Ethanol (70%)- Dilute absolute alcohol to 70%

Note: Very little phenol will be lost during the preparation of the Buffer A saturated phenol. Start with at least 1.2 X the final volume of phenol you will need.

5. Reagents for Genomic DNA Isolation from Blood

EDTA stock (0.5 M, pH 8.0)

Add 186.1g of Disodium EDTA.2H_2O to 800 ml of H_2O. Stir vigorously on a magnetic stirrer. Adjust the pH to 8.0 with NaOH (approx. 20g of NaOH pellets). Dispense into aliquots and sterilize by autoclaving. The Disodium salt of EDTA will not go into solution until the pH of the solution is adjusted to approx. 8.0 by the addition of NaOH.

DNA Gel loading buffer (6X)

0.25% Bromophenol blue (w/v)

0.25% Xylene cyanol FF (w/v)

30% Glycerol in H_2O (v/v)

NaCl stock (5 M)

Dissolve 292g of NaCl in 800 ml of H_2O. Adjust the volume to 1 litre with H_2O. Dispense into aliquots and sterilize by autoclaving. Store the NaCl solution at room temperature.

NaOH stock (10 N)

Slowly add 400g of NaOH pellets to 800 ml of H_2O, stirring continuously. Place beaker on ice. When pellets dissolve completely, adjust the volume to 1 litre with H_2O. Store the solution in a plastic container at room temperature.

SDS Stock (10% w/v)

Dissolve 100g of electrophoresis grade SDS in 900 ml of H_2O. Heat at 68°C and stir with magnetic stirrer to assist dissolution. If necessary, adjust the pH to 7.2 by adding a few drops of concentrated HCl. Adjust the volume to 1 litre with H_2O. Store at room temperature. Sterilization is not necessary.

Sodium Acetate (3M, pH 5.2)

Dissolve 408.3g of Sodium acetate in 800 ml of H_2O. Adjust the pH to 5.2 with Glacial Acetic Acid. Adjust the volume to 1 litre with H_2O. Dispense into aliquots and sterilize by autoclaving.

TAE stock (50X)

Prepare a 50X stock solution with

242g of Tris base

57.1 ml of Glacial Acetic Acid

100 ml of 0.5 M EDTA (pH 8.0) and adjust the volume to 1 litre with H_2O. Sterilize the solution by autoclaving.

TE buffer

10 mM Tris-Cl (pH 8.0)

1mM EDTA (pH 8.0)

Use concentrated stock solution to prepare. If sterile water and sterile stocks are used, there is no need to autoclave. Otherwise, sterilize solution by autoclaving for 20 minutes. Store the buffer at room temperature.

Tris-Cl (1M, pH 8.0)

Dissolve 121.14g Tris base in 800 ml of water. Adjust the pH to 8.0 by adding about 42 ml of concentrated HCl. Make sure that the solution is at room temperature before making final pH adjustments. Bring final volume to 1 litre. Sterilize by autoclaving.

Proteinase K (20 mg/ml)

Take lyophilized powder of Proteinase K, dissolve it at room temperature at a concentration of 20 mg/ml in sterile 50 mM Tris (pH 8.0), 1.5 mM Calcium acetate. Divide the stock solution into small aliquots and store at -20°C. Each aliquot can be thawed and refrozen several times.

RBC lysis buffer (1X)

Ammonium Chloride (NH_4Cl) – 8.3g

Potassium bicarbonate ($KHCO_3$) – 1.0g

0.5 M EDTA (pH 8.0) - 299 μL

Double distilled H_2O upto – 1 litre

Autoclaved and stored at room temperature.

DNA extraction buffer

1 M Tris buffer (pH 8.0) – 5 ml

1 M NaCl – 40 ml

0.5 M EDTA buffer (pH 8.0) – 2 ml

Double distilled water upto – 500 ml

Autoclave in batches of 100 ml and store at room temperature.

Dimethylsulphoxide (DMSO)

Take a high grade DMSO. Divide the contents of a fresh bottle into 1 ml aliquots in sterile tubes. Close the tubes tightly and store at -20°C. Use each aliquot only once and then discard.

Ethidium Bromide (10 mg/ml)

Add 1g Ethidium Bromide to 100 ml of water. Stir on a magnetic stirrer for several hours to ensure proper dissolution of the dye. Wrap the container in aluminum foil of transfer the solution into dark bottle and store at room temperature.

6. Reagents for Agarose Gel Electrophoresis

10 x TAE	500 ml	1000 ml
Tris pH 8.0 (0.4 M)	24.2 g	48.4 g
Glacial acetic acid	5.71 ml	11.42 ml
EDTA (10 mM)	10 ml	20 ml of 0.5 M EDTA pH 8.0 solution

10 x TAE	500 ml	1000 ml
Loading buffer/dye	100 ml	
Bromo phenol blue (0.25%)	0.25 g	
Xylene cyanol (0.25%)	0.25 g	
Ficoll (15%)	15 g	

or

Bromophenol blue (-0.05%)

EDTA pH 8.0, 0.2 M

Glycerol (50%)

Agarose Gel 1% agarose

Ethidium Bromide 1 mg/ml in dH_2O

NB: the reagent is toxic, handle powder and solution with gloves and wear a mask when weighing it.

7. Reagents for Chloroplast and Mitochondrial DNA Isolation

STE buffer: 400 mM sucrose, 50mM Tris, pH 7.8, 20mM EDTA-Na2, BSA (0.2%), β-mercaptoethanol (0.2%).

ST buffer: 400 mM sucrose, 50mM Tris, pH 7.8, 0.1% BSA.

TEN buffer: 100 mM Tris pH 7.2, 50mM EDTA, 100 mM NaCl and β-mercaptoethanol (0.2%).

NETF buffer: 1.25 M NaCl, 50mM EDTA, 50 mM Tris pH 8.0, 50 mM NaF.

Mitochondrial DNA isolation

1. Homogenization Buffer A: 50 mM Tris-HCl (pH 8.0), 1.3 M NaCl, 25 mM EDTA (pH 8.0), BSA (0.2%). Add cysteine (0.05%) and 56 mM β-mercaptoethanol immediately before use.
2. Homogenization Buffer B: 100 mM Tris-HCl (pH 8.0), 2.6 M NaCl, 50 mM EDTA (pH 8.0), BSA (0.4%). Add cysteine (0.1%) and 56 mM β-mercaptoethanol immediately before use.
3. Lysis Buffer: 25 mM Tris-HCl (pH 8.0), 20 mM EDTA (pH 8.0), SDS (0.5%)
4. Proteinase K: 10 mg/ml
5. Ammonium acetate: 2 M
6. TE: 10 mM Tris-HCl (pH 8.0), 1 mM EDTA

7. TE-saturated phenol/chloroform (50:50)
8. Water-saturated phenol/chloroform/isoamyl alcohol (50:49:1)
9. RNAse A: 10 mg/mL
10. RNA Extraction Buffer: 4 M guanidine thiocyanate, 25 mM sodium citrate (pH 7.0), sarcosyl (0.5%). Add β-mercaptoethanol (0.1%) immediately before use.
11. LiCl: 4 M

8. RNA Isolation

RNase Free Apparatus

For using RNA, it is necessary to keep reagents and apparatus RNase free. Treatment for RNase free depends on the material of apparatus.

Glassware

I. Diethyl pyrocarbonate (DEPC) treatment

DEPC is highly carcinogenic. Make sure to put on a pair of gloves and ventilate the room when it is used. Be careful not to breath vapour in.

(1) Immerse glassware in distilled water containing DEPC (1.0%) (DEPC solution). Otherwise, fill such glassware as medium bottle with this solution and leave them at 37°C for 2 hours or all night.
   - DEPC cannot be dissolved in distilled water soon and it stays like oil drops in the bottom. Stir fully so that DEPC will be uniformly dispersed.
   - Shake the medium bottle with the cap well.

(2) Throw the DEPC solution away and rinse with new DEPC solution.
   - A little amount of DEPC solution is enough as it is valuable.

(3) Dry the glassware at 121°C for 20 min. by autoclave.
   - Excessive DBPC is removed by autoclave. If the removal is not enough, sterilize again by autoclave.

II. Dry sterilization

Wash and dry apparatus as usual, then conduct dry sterilization at 180°C for more than 8 hours or at 250°C for more than 30 min.

III. Plastic apparatus

Basically, unused sterilized apparatus is RNase free. Plastic apparatus such as electrophoretic chamber is treated by following method. It is convenient to keep the treated apparatus separate from others as treatment is not required every time as long as they are not contaminated.

Procedure

(1) Wash plastic apparatus with water thoroughly.
- Make sure to use a pair of gloves so as not to attach RNase to the apparatus by hand. Do not scrub the apparatus strongly with a brush.

(2) Immerse them in H_2O_2 (5%) at room temperature for 10 min.

(3) Rinse well with DEPC solution.

(4) Dry them unless required immediately.

Extraction buffer (pH 9.0)

Tris-HCL	1.21g
NaCl (0.075M)	0.44g
EDTA Na_2 (0.005m)	0.19g
Water to	100ml

Preparation of Buffer Saturated Phenol for RNA isolation

Materials

- Redistilled Phenol, molecular biology grade: Stored in aliquots at -20°C.
- 8-hydroxyquinoline
- 0.5 M Tris·Cl (pH 8.0)
- 0.1 M Tris·Cl (pH 8.0)
- TE buffer - 10 mM Tris·Cl (pH 8.0), 1 mM EDTA
- Buffer A
- 50 mM NaOAc, 10 mM EDTA

Procedure

Aliquoting the phenol for storage:

1. Heat a water bath to 65°C in the fume hood. Place the bottle of phenol in the fume hood to warm to room temperature.
2. Place the bottle of phenol in the 65°C water bath to melt the crystals.

3. Once the phenol has melted, aliquot it into 15 or 50 ml polypropylene tubes or 100 ml glass bottles using a RNase free glass pipette. Seal the individual aliquots and wrap them in aluminum foil. Store at -20°C.

Precautions

1. EDTA, a chelating agent, removes from solutions metal ions that are essential for the activity of certain nucleases. It thus protects DNA from digestion with nucleases.
2. Ethidium bromide is a potent mutagen. Disposable vinyl gloves should be worn when handling this chemical.
3. Use polyethylene pipettes, since DNA may stick to the walls of glass pipettes.
4. Since visible light may cause single-stranded nicks in the DNA in the presence of ethidium bromide, darken the room or wrap each container in foil before addition of the dye. Ethidium bromide intercalates with DNA, thereby decreasing its density in CsCl. Covalently closed circular (CCC) DNA binds much less with ethidium bromide than open circular (OC) and linear DNA. Therefore, CCC-DNA can readily be separated from other types of DNA on the gradient.
5. At a pH of 12.0 to 12.5, linear and chromosomal DNA are selectively denatured, while covalently closed circular (CCC) DNA is not affected. However, at greater pH values CCC-DNA may become irreversibly denatured. Proteins are also denatured at pH 12.5 thereby reducing the possibility of enzymatic degradation of plasmid DNA. The SDS lyses the cells and denatures proteins.
6. Phenol extraction: Phenol denatures proteins and selectively removes denatured DNA from aqueous solutions. It is therefore frequently used to "clean-up" cell lysates. The phenol used is saturated with a solution of the major components of the aqueous phase (*e.g.*, salts, buffer, H_2O), so as to prevent their excessive loss from solution during the extraction.
7. Phenol is volatile and caustic, prepare the phenol in the fume hood.
8. Wear gloves and eye protection.
9. All solution should be RNase free. Use RNase free glassware and plasticware for RNA isolation procedures.

Dissolve the above components in double distilled water. Adjust the volume to a litre and autoclave at 121°C/15 lbs for 20 mins. Cool the solution and add the appropriate amount of antibiotic just before use. Typically Amp at 25 ug/ml or Kan at 50 ug/ml.

9. Reagents for Southern Blotting and Hybridization

Phosphate-SDS washing solution 1

40 mM sodium phosphate buffer (pH 7.2)

1 mm EDTA (pH 8.0)

SDS 5% (w/v)

Fraction-V-grade bovine serum albumin 0.5% (w/v)

Phosphate-SDS washing solution 2

40 mM sodium phosphate buffer (pH 7.2)

1 mM EDTA (pH 8.0)

SDS 1% (w/v)

Prehybridization/hybridization solutions

Prepare the prehybridization solution appropriate for the task at hand. Approximately 0.2 ml of prehybridization solution is required for each square centimeter of membrane. Smaller volumes (~0.1 ml/cm^2) can be used when hybridizing in roller bottles. For advice on which solution to use, please see the introduction to Protocol 32 in Chapter 1.

Prehybridization/hybridization solution may be prepared with or without poly(A) RNA. When ^{32}P-labeled cDNA or RNA is used as a probe, poly(A) RNA may be included in prehybridization or hybridization solutions to prevent the probe from binding to thymidine-rich sequences commonly found in eukaryotic DNA. Poly(A) RNA should be added to aqueous and formamide hybridization buffers at a final concentration of 1μg/ml.

Solution for hybridization in aqueous buffer

6x SSC (or 6x SSPE)

5x Denhardt's reagent

SDS 0.5% (w/v)

1 μg/ml poly(A)

100 μg/ml salmon sperm DNA

6x SSPE contains EDTA, which is a better chelator of divalent metal ions (*e.g.*, Mg^{2+}) than citrate, and in turn will more efficiently inhibit DNase activity that can decrease probe and target DNA concentrations. After thorough mixing, filter the solution through a 0.45–4m disposable cellulose acetate membrane.

Solution for hybridization in formamide buffers

6x SSC (or 6x SSPE)

5x Denhardt's reagent

SDS 0.5% (w/v)

1 μg/ml poly(A)

100 μg/ml salmon sperm DNA

Formamide 50% (v/v)

After thorough mixing, filter the solution through a 0.45-gm disposable cellulose acetate membrane (Schleicher & Schuell Uniflo syringe membrane or equivalent). To decrease background when hybridizing under conditions of reduced stringency (*e.g.*, 20–30% formamide), it is important to use formamide that is as pure as possible.

Solution for hybridization in phosphate-SDS buffer

0.5 M sodium phosphate (pH 7.2)

1 mM EDTA (pH 8.0)

SDS 7% (w/v)

Bovine serum albumin 1% (w/v)

Use an electrophoresis grade of bovine serum albumin (BSA). No blocking agents or hybridization rate enhancers are required with this particular prehybridization/hybridization solution.

Sodium phosphate (1 m, pH 7.2)

0.1 x SSC

0. 1 x SSC with 0.1% (w10 SDS)

2x SSC with 0.1% WO SDS

2x SSC with 0.5% (w/v) SDS

6x SSC or 6x SSPE

10. Western blotting and Hybridization

Reagents for SDS-PAGE

5X SDS / Electrophoresis Buffer

15.1 gm Tris base

72.0 gm Glycine

5.0 gm SDS.

Add water to 1000 ml

Don't adjust the pH of the stock solution as pH will be 8.3 when diluted to 1 litre.

2X SDS/Sample Buffer

To 4 ml distilled water add 0.152 gm. of Tris base

2 ml glycerol

0.2 gm SDS

0.2 ml 2-mercaptoethanol

0.1 mg Bromophenol blue

Adjust pH to 6.8 with 1N HCI. Add glycerol and β-mercaptoethanol after adjusting the pH.

Make up the volume to 0.10 ml with distilled water

4X Tris Cl/SDS pH 6.8

To 40 ml distilled water add 6.05 gm of Tris base

0.4 gm of SDS

Adjust pH 6.8 with 1 N HCl. Add distilled water to 100 ml.

4X Tris. Cl/ SDS pH 8.8

To 300 ml distilled water add 91 gm of Tris base

2 gm of SDS

Adjust pH to 8.8 with 1N HCI. Add distilled water to 500 ml.

Acrylamidel (30%) & Bisacrylamide (0.8%) Acrylamide stock solution

30 gm Acrylamide

0.8 gm bisacrylamide

Add distilled water to 100 ml

Filter through 0.45 μm filter and store at 4°C in the dark

(Discard the solution after 30 days as acrylamide hydrolyse to acrylic acid and ammonia).

Staining solution

Coomassie Brilliant Blue - 0.1% (W/V), Methanol - 25% (V/V), Acetic Acid-10% (V/V)

Destaining solution

Methanol - 5% (V/V), Acetic Acid-7% (V/V)

Transfer buffer

Add 8.7 gm glycine, -17.4 gm. of Tris-base, 1.0 gm SDS and 600 ml of methanol made upto 3000 ml with deionised water.

Nitrocellulose membrane

Amido black stain (0.01 amido black, 45% methanol, 10% acetic acid)

TBS (Tris-buffered saline): 25 mM Tris-cl. pH 8.0, 0.87% NaCl) Tween-20

BSA (Bovine serum albumin)

Skimmed Milk

10 mM Tris-Cl pH 7.0

10mM Tris-Cl pH 8.0

H_2O_2 (hydrogen peroxide) (30%)

Diamino benzidine tetrahydrochloride (DAB)-

Transblot Apparatus

Power Pack

11. Protoplast Isolation

- Ethanol (70%)
- Mercuric Chloride (0.1%)
- 0.5M Mannitol
- Enzyme solution

  Cellulase (1%), Macerozyme R-10 (0.25%)

  27.2 mg/l KH_2PO_4

  101 mg/l KNO_3

  1480 mg/l $CaCl_2.2H_2O$

246 mg/l $MgSO_4.7H_2O$

0.5M Mannitol (pH 5.6)

- Cell Protoplast Washing Medium

  0.5M Mannitol (pH 5.6)

  27.2 mg/l KH_2PO_4

  101 mg/l KNO_3

  1480 mg/l $CaCl_2.2H_2O$

  246 mg/l $MgSO_4.7H_2O$

Protoplast Culture Medium

S.No.	Constituents	Amount
	A Mineral Salts	
1	Ammonium Nitrate	82.5 gm/l
2	Potassium Nitrate	95.0 gm/l
3	Boric acid	1190 mg/l
4	Potassium Dihydrogen Phosphate	34 mg/l
5	Potassium Iodide	166 mg/l
6	Sodium Molybdate	250 mg/l
7	Cobalt Chloride Dihydrate	25.0 mg/l
8	Calcium Chloride Dihydrate	88.0 gm/l
9	Magnesium Sulphate Heptahydrate	74.0 gm/l
10	Manganese Sulphate tetrahydrate	4460 mg/l
11	Zinc Sulphate Heptahydrate	1720 mg/l
12	Copper Sulphate Pentahydrate	25.0 mg/l
13	Ethylene Diamino Tetra Acetic acid Disodium salt	7.45 gm/l
14	Ferrous Sulphate heptahydrate	5.57 gm/l
	B Sugars	
1	Glucose	30.0 gm/l
2	Ribose	0.5 gm/l
3	Mannose	0.5 gm/l
4	Sorbitol	0.5 gm/l
	C Organic Acids	
1	Sodium Pyruvate	5.0 mg/l
2	Citric acid	10.0 mg/l
3	Malic acid	10.0 mg/l

S.No.	Constituents	Amount
	D Vitamins	
1	Thiamine Hydrochloride	100 mg/l
2	Nicotinamide	500 mg/l
3	Pyridoxine Hydrochloride	500 mg/l
4	Calcium pantothenate	250 mg/l
5	Biotin	100 mg/l
6	Ascorbic acid	75.0 mg/l
7	Riboflavin	50.0 mg/l
8	Inositol	100 gm/l
	E Hormones	
1	2,4-Dichloro Phenoxy Acetic Acid (2,4-D)	0.5 mg/l
2	Benzyl Amino Purine (BAP)	1.0 mg/l

12. Histochemical Detection of GUS

1. 2X Buffer: phosphate buffer, pH 7.0 (made of 0.1 M NaH_2PO_4 and 0.1 M Na_2HPO_4) or citrate-HCl buffer, pH 7.0 (made of 0.1 M sodium citrate and 0.2 N Hydrochloric acid).
2. Fixative: formaldehyde (4%), prepared fresh from paraformaldehyde, in 1X buffer.
3. X-gluc substrate solution: Dissolve 1 mg 5-bromo-4-chloro-3-indolyl β-D-Glucuronide (X-Gluc) in 0.1 ml methanol, add 1 ml 2X buffer, 20 μl 0.1M potassium ferrocyanide, 20μl 0.1 M potassium ferricyanide, 10μl 10% (w/v) solution of Triton X-100, 0.85 ml water.
4. Ethanol 70% (v/v).
5. 50% (v/v) and 100% glycerol.

Preparation of some important Solutions used in Molecular Biology

1 M Tris pH 7.4	50 ml	100 ml	500 ml	1000 ml
Tris powder	6.055 g	12.11 g	60.55 g	121.11 g

Dissolve the powder in two-thirds water of the final volume, adjust the pH with HCI, make up to the final volume with water in a volumetric flask and transfer to a Schott bottle, label with the concentration, pH and date of preparation.

0.5 M EDTA pH 8.0	50 ml	100 ml	500 ml	1000 ml

The reagent is toxic, handle powder and solution with gloves and wear a mask when weighing it.

EDTA powder	9.3 g	18.61 g	93.1 g	186.1 g

Add two-thirds of the water and put on a stirrer, place the pH electrode in and adjust the pH to 8.0 by adding solid NaOH pellets. The powder will only go into solution once the pH approach 8.0. Transfer to a Schott bottle, label with the concentration, pH and date of preparation.

SDS (10%)	**10 ml**	**25 ml**	**50 ml**	**100 ml**

The reagent is toxic, handle powder and solution with gloves and wear a mask when weighing it.

SDS powder	1 g	2.5. g	5 g	10g

Dissolve in two-thirds of the final volume of water, make up to the final volume in a volumetric flask and transfer to a Schott bottle, label with the concentration, pH and date of preparation.

Sarcosine (10%)	**10 ml**	**25 ml**	**50 ml**	**100 ml**

The reagent is toxic, handle powder and solution with gloves and wear a mask when weighing it.

Lauroyl sarcosine powder	1 g	2.5g	5g	10g

Dissolve in two-thirds of the final volume of water, make up to the final volume in a volumetric flask and transfer to a Schott bottle, label with the concentration, pH and date of preparation.

3 M Na-acetate pH 5.2	**10 ml**	**25 ml**	**50 ml**	**100 ml**
Na-acetate powder	4.08 g	10.2 g	20.4 g	40.8 g

Dissolve in two-thirds of the final volume of water, adjust the pH to 5.2 with glacial acetic acid, make up to the final volume in a volumetric flask and transfer to a Schott bottle, label with the concentration, pH and date of preparation.

Proteinase K

The reagent is toxic, handle powder and solution with gloves and wear a mask when weighing it.

The enzyme is bought from Sigma and is presented as a 500 unit powder quantity with a concentration of 34.6 units/mg. Dissolve the proteinase K in 572 ml of 50 mM Tris pH 7.4, 2 mM Ca-acetate to a final concentration of 500 units/572 ml (0.875 units/ml). We need 0.1 mg proteinase K per 200 ml blood sample, therefore using the original concentration, that amounts to 3.46 units. Since we have extracted DNA from 400 ml blood, we therefore need 7 units proteinase K, and using the diluted solution, it amounts to 8 ml of the 0.875 units/ml proteinase K solution.

Extraction buffer	**400 µl**	
SDS solution (10%)	40 µl (final concentration is 1%)	
Sarcosine (10%)	40 µl (final concentration is 1%)	
Proteinase K	8 µl (final concentration is 0.2 mg)	
Water	312 µl	
Ethanol (70%)	**50 ml**	**100 ml**
Ethanol absolute	35 ml	70 ml
PCR water	15 ml	30 ml

Functions of chemicals

Tris

- Balances the pH of solutions to ensure that no pH fluctuations could occur that might damage the proteins or DNA at the extreme pH values.

EDTA

- Acts as a chelating agent in that it can remove heavy metals, such as cofactors, from a solution. That will then inactivate any protease or nuclease in a solution, thereby protecting your DNA or proteins.
- Carries the current within the gel to ensure that the DNA will separate within the gel.

SDS

- It denatures all protein inside a solution, making it lose its function.
- By denaturing the proteins that are part of the cellular membrane, the integral structure of the membranes are disrupted and the cells will break open.

Sarcosine

- The function is basically the same as SDS.

Proteinase K

- It is an enzyme responsible for the complete breakdown of all proteins. Especially important is its breakdown of proteinases, DNases and Rnases. The removal of the enzymes will protect the nucleic acids during extraction.

Ethanol (100%)

- It dehydrates the DNA molecules by removing any water molecules that are bound to the DNA structure, thereby making the DNA insoluble and making it precipitate from the solution.

Sodium acetate

- It helps in the precipitation of DNA, but must be removed later on.

Ethanol (70%)

- When you precipitate DNA, the ethanol also causes the salt to co-precipitate. This salt can inhibit further reactions, therefore, we must remove it. By adding the solution that contains the 30% water, the salt dissolve in the water, while the DNA remains precipitated, thereby effectively desalting the DNA.

Acetic acid

- Carries the current within the gel to ensure that the DNA will separate within the gel.

Ficoll / Sucrose / Glycerol

- Makes the DNA heavy so that it would sink down in the well.

Bromophenolblue / Xylene Cyanol

- Act as a dye to follow the separation of the DNA inside the gel.

Ethidium bromide

- Binds to the DNA inside the gel and when exposed to ultraviolet light, if fluoresces and makes the DNA visible.

Table : Some common *E. coli* strains used in experiments.

S.N.	Strain	Properties	Reference
1.	DH5α	A recombination-deficient suppressing strain used for plating and growth of plasmids and cosmids. The ξ80 lacZ𝔑M15 permits α-complementation with the amino terminus of β-galactosidase encoded in pUC vectors.	Hanahan, 1983 Bethesda Research Laboratories, 1986.

2.	HB101	A suppressing strain commonly used for large scale production of plasmids. It is an *E. coli* K12x *E.coli* B hybrid that is highly transformable.	Boyer and Roulland-Dussoix, 1969 Bolivar and Backman, 1979.
3.	JM105	A strain that will support growth of vectors carrying amber mutations and will modify, but not restrict transfected DNA.	Yanisch-Perron *et al.*, 1985
4.	JM 109	A recombination-deficient strain that will support growth of vectors carrying amber mutations and will modify, but not restrict transfected DNA.	Yanisch-Perron *et al.,* 1985
5.	NM5 38	A strain used for assay and propagation of bacteriophage.	Frischauf *et al.*, 1983
6.	LE392	A suppressing strain commonly used to propagate lambda phage vectors and their recombinants.	Borck *et al.*, 1976 Murray *et al.*, 1977

Genetic Code UAB

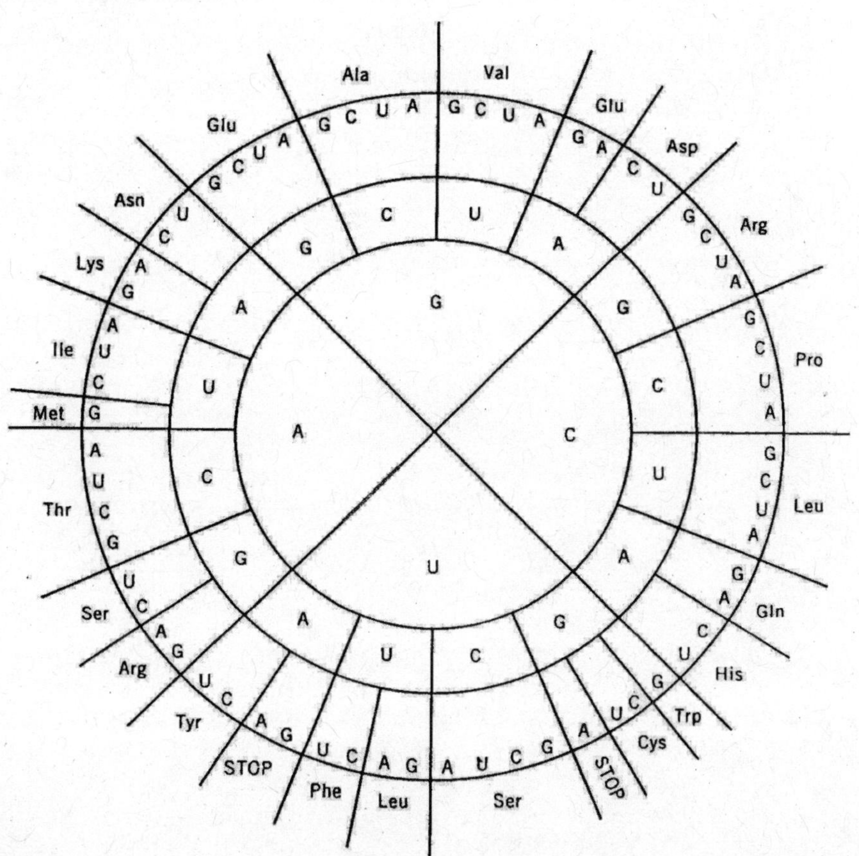

The first base in each mRNA codon is in the centre of the circle, the second base is in the second circle from the center, and the third base is in the third circle from the centre. The standard three-letter abbreviation for the amino acid encoded by each codon is on the outside (UAG, UAA, and UGA stop codons are designated by the word STOP).

Codon acts GA

Codon Redundancy

Amino acids	Number of synonymous codons
W, M	1
N,D, C, Q, E, H, F, Y	2
I	3
A, G, P, T, V	4
R, L, S	6

Termination codons

UAA	ochre
UAG	amber
UGA	opal

■ ■ ■

Appendix IV D : IMMUNOLOGY

1. To isolate and purify sheep IgG from whole sheep serum

PBS (Phosphate buffer saline, pH 7.4)

Prepare by 8 g NaCl, 0.20 g KCl, 1.45 g disodium hydrogen phosphate and 0.20 g potassium dihydrogen phosphate in distilled water and make final volume to 1 litre with distilled water.

PBS-Tween 20 (PBST)

Add Tween 20 to PBS so that its final concentration is 0.05% (v/v)

2. To perform double Immunodiffusion by Ouchterlony method

Saline Solution, 0.9% Sodium Chloride (NaCl)

1. Dissolve 9 g NaCl (mw 58.44) in 700 ml deionized or distilled water in clean container.
2. Add water to bring total solution volume to 1000 ml.
3. Make 10 ml aliquots in sterile 15 ml culture tubes.

Tris-Tricine buffer pH 8.6 (10X)

121.1 g Tris base and 179.2 g Tricine with dH20 to 800 mL pH to 8.6 (with HCL). Add dH20 to 1L. Store at Room Temperature.

3. Enzyme Linked Immunosorbent Assay (ELISA)

1. Coating buffer (50 ml): Carbonate-bicarbonate buffer, 100 mM, pH 9.6
2. Blocking buffer (25 ml): Phosphate buffer saline (PBS) 10 mM, pH 7.2, containing 0.2% (w/v) bovine serum albumin and 0.5% (w/v) sodium azide.
3. Washing buffer (1 l): Phosphate buffered saline–tween (PBS-T), 0.01M pH 7.2, containing 0.05% (v/v) Tween -20.
4. Substrate buffer (25ml): Diethanolamine (10% v/v) buffer pH 9.6, containing substrate-p-nitophenyl phosphate (1.25mg/ml buffer) and 0.05 mM $MgCl_2$
5. Sodium hydroxide 2M.

6. Substrate-orthophenylenediamine (OPD) solution: Dissolve 8 mg of OPD in15ml of citrate buffer (pH-5.0) followed by addition of 15ml hydrogen peroxide just before use
7. Citrate phosphate buffer (0.1M) pH5: Dissolve 7.3g of citric acid and 11.86g of Na_2HPO4 in 1000ml of double distill water.
8. 4% bovine serum albumin (BSA) in PBS-Tween 20.

■ ■ ■

Appendix IV E : ENVIRONMENTAL BIOTECHNOLOGY

Determination of Organic Cabon Content in soil

1. M-Endo broth

M-Endo broth is used for the estimation of coliform bacteria in water samples Endo Medium was first developed by Endo to differentiate between lactose-fermenters and non-fermenters. The medium employs sodium sulphite and basic fuchsin instead of bile salts to achieve inhibition of gram-positive bacteria.

Casein enzymic hydrolysate, tryptose, peptone special and yeast extract provide essential nutrients especially nitrogenous for the coliforms. Lactose is the fermentable carbohydrate. Sodium sulphite, sodium deoxycholate and basic fuchsin inhibit the growth of gram-positive organisms. Phosphates buffer the medium. Coliforms ferment lactose and the resulting acetaldehyde reacts with sodium sulphite and basic fuchsin to form red colonies and similar colouration of the medium. Lactose nonfermenters form colourless colonies.

Composition (Ingredients gms/ litre)

Tryptose or polypeptone	10.0 g
Thiopeptone or thiotone	5.0 g
Casitone or trypticase	5.0 g
Yeast Extract	1.5 g
Lactose	12.5 g
Sodium chloride, NaCl	5.0 g
Dipotassium hydrogen phosphate, K_2HPO_4	4.375 g
Potassium dihydrogen phosphate, KH_2PO_4	1.375 g
Sodium lauryl sulfate	0.05 g
Sodium desoxycholate	0.10 g
Sodium sulfite, Na_2SO_3	2.10 g
Basic fuchsin	1.05 g
Agar (optional)	15.0 g
Reagent grade water	1 L

1. Agar preparation – Rehydrate all product in 1L water containing 20 mL 95% ethanol. Heat to near boiling to dissolve agar, then

promptly remove from heat and cool to between 45 and 50°C. Dispense 5- to 7- ml quantities into 60-mm sterile petri dishes. If any other size petri dish is used, adjust quantity to give an equivalent depth. Do not sterilize by autoclaving. Final pH should be 7.2 ± 0.2. A precipitate is normal in Endo-type media.

Refrigerate finished medium in the dark and discard unused agar after 2 weeks.

2. Broth preparation – Prepare as above, omitting agar. Dispense liquid medium (at least 2.0 mL per plate) onto absorbent pads and carefully remove excess medium by decanting the plate. Refrigerated broth may be stored up to 4 days.

   Caution: Basic fuchsin is a potential carcinogen and care should be taken to avoid inhalation of the powdered dye and contamination of the skin.

Determination of Organic Carbon Content in soil

1. Mohr's Solution

0.4 N FeSO4 $[(NH_4)_2\ (Fe(SO_4)_2 \cdot 6H_2O]$: Weigh 11.2 g of ferrous ammonium sulfate inti 1 litre volumetric flask containing 500 ml deionized water. Dissolve and slowly add 50 ml of concentrated sulfuric acid and allow it to cool. Dilute to the volume of 1 litre with deionized water and mix well. Solution should be prepared fresh.

Determination of Dissolved Oxygen by Winkler's Method

1. Starch Indicator

Dissolve 1 g of starch in 100 ml of warm distilled water and addfew drops of toluene or formaldehyde as the preservative.

Analysis of total hardness of Waste water sample

1. Eriochrome Black T (EBT)

Weigh 0.5 g of EBT indicator and 1.5 g hydroxylamine hydrochloride and dissolve on 100 ml of ethanol or methanol.

Water Quality Standards

(Indian Standard for Drinking Water - Specification IS 10500: 1991)

Colour, Hazen Units

Standard	Desirable: 5 Hz. Permissible: 25 Hz.
Risks or effects	Visible tint, acceptance decreases
Sources	Tannins, Iron, Copper, Manganese Natural deposits
Treatment	Filtration, Distillation, Reverse osmosis, Ozonisation

Odour

Standard	Unobjectionable
Risks or effects	Rotton egg, Musty, Chemical
Sources	Chlorine, Hydrogen sulfide, Organic matter, Septic contamination, Methane gas
Treatment	Activated carbon, Air stripping, oxidation, Filtration

pH

Standard	Desirable: 6.5 – 8.5, Permissible: No relaxation
Risks or effects	Low pH – corrosion, metallic taste High pH – bitter/soda taste, deposits
Sources	Natural
Treatment	Increase pH by soda ash Decrease pH with white vinegar / citric acid

Total Dissolved Solids (TDS)

Standard	Desirable: 500 mg/l, Permissible: 2000 mg/l
Risks or effects	Hardness, scaly deposits, sediment, cloudy colored water, staining, salty or bitter taste, corrosion of pipes and fittings
Sources	Livestock waste, septic system Landfills, nature of soil Hazardous waste landfills Dissolved minerals, iron and manganese
Treatment	Reverse Osmosis, Distillation, deionization by ion exchange

Hardness

Standard	Desirable: 300mg/l, Permissible: 600 mg/l
Risks or effects	Scale in utensils and hot water system, soap scums

Sources	Dissolved calcium and magnesium from soil and aquifer minerals containing limestone or dolomite
Treatment	Water Softener Ion Exchanger, Reverse Osmosis

Alkalinity

Standard	Desirable: 200mg/l, Permissible: 600 mg/l
Risks or effects	Low Alkalinity (i.e. high acidity) causes deterioration of plumbing and increases the chance for many heavy metals in water are present in pipes, solder or plumbing fixtures.
Sources	Pipes, landfills Hazardous waste landfills
Treatment	Neutralizing agent

Metal Ions

Iron, Fe	
Standard	Desirable: 0.3mg/l, Permissible: 1.0 mg/l
Risks or effects	Brackish color, rusty sediment, bitter or metallic taste, brown-green stains, iron bacteria, discolored beverages
Sources	Leaching of cast iron pipes in water distribution systems Natural
Treatment	Oxidizing Filter, Green – sand Mechanical Filter
Manganese, Mn	
Standard	Desirable: 0.1mg/l, Permissible: 0.3 mg/l
Risks or effects	Brownish color, black stains on laundry and fixtures at 0.2 mg/l, bitter taste, altered taste of water – mixed beverages
Sources	Landfills, Deposits in rock and soil
Treatment	Ion Exchange, Chlorination, Oxidizing Filter, Green-sand Mechanical Filter
Sulphate, SO_4	
Standard	Desirable: 200mg/l, Permissible: 400 mg/l
Risks or effects	Bitter, medicinal taste, scaly deposits, corrosion, laxative effects, "rotten-egg" odor from hydrogen sulfide gas formation
Sources	Animal sewage, septic system, sewage By-product of coal mining, industrial waste Natural deposits or salt
Sulphate Treatment	Ion Exchange, Distillation, Reverse Osmosis

Nitrate, NO_3^-	
Standard	Desirable: 45mg/l, Permissible: 100 mg/lit
Risks or effects	Methemoglobinemia or blue baby disease in infants
Sources	Livestock facilities, septic systems, manure lagoons, fertilizers Household waste water, fertilizers, Natural Deposits
Treatment	Ion Exchange, Distillation, Reverse Osmosis
Chloride, Cl.	
Standard	Desirable: 250mg/l, Permissible: 1000 mg/l
Risks or effects	High blood pressure, salty taste, corroded pipes, fixtures and appliances, blackening and pitting of stainless steel
Sources	Fertilizers, Industrial wastes, Minerals, seawater
Treatment	Reverse Osmosis, Distillation, Activated Carbon
Fluoride, F	
Standard	Desirable: 1.0mg/l, Permissible: 1.5 mg/l
Risks or effects	Brownish discoloration of teeth, bone damage
Sources	Industrial wastes, Geological
Treatment	Activated Alumina, Distillation, Reverse Osmosis, Ion Exchange
Arsenic, As	
Standard	Desirable: 0.05mg/l, Permissible: No relaxation
Risks or effects	Weight loss; Depression; Lack of energy; Skin and nervous System toxicity
Sources	Previously used in pesticides (orchards) Improper waste disposal or product storage of glass or electronics, Mining, Rocks
Treatment	Activated Alumina Filtration, Reverse Osmosis, Distillation, Chemical Precipitation, Ion exchange, lime softening
Chromium, Cr	
Standard	Desirable: 0.05mg/l, Permissible: No relaxation
Risks or effects	Skin irritation, skin and nasal ulcers, lung tumors, gastrointestinal effects, damage to the nervous system and circulatory system, accumulates in the spleen, bones, kidney and liver
Sources	Septic systems Industrial discharge, mining sites, Geological
Treatment	Ion Exchange, Reverse Osmosis, Distillation

Copper, Cu	
Standard	Desirable: 0.05mg/l, Permissible: 1.5 mg/l
Risks or effects	Anemia, digestive disturbances, liver and kidney damage, gastrointestinal irritations, bitter or metallic taste; Blue-green stains on plumbing fixtures
Sources	Leaching from copper water pipes and tubing, algae treatment Industrial and mining waste, wood preservatives Natural deposits
Treatment	Ion Exchange, Reverse Osmosis, Distillation
Cyanide	
Standard	Desirable: 0.05mg/l, Permissible: No relaxation
Risks or effects	Thyroid, nervous system damage
Sources	Fertilizer, Electronics, steel, plastics mining
Treatment	Ion Exchange, Reverse Osmosis, Chlorination
Lead, Pb	
Standard	Desirable: 0.05mg/l, Permissible: No relaxation
Risks or effects	Reduces mental capacity (mental retardation), interference with kidney and neurological functions, hearing loss, blood disorders, hypertension, death at high levels
Sources	Paint, diesel fuel combustion Pipes and solder, discarded batteries, paint, leaded gasoline Natural deposits
Treatment	Ion Exchange, Activated Carbon, Reverse Osmosis, Distillation
Mercury, Hg	
Standard	Desirable: 0.001mg/l, Permissible: No relaxation
Risks or effects	Loss of vision and hearing, intellectual deterioration, kidney and nervous system disorders, death at high levels
Sources	Fungicides, Batteries Mining, electrical equipment, plant, paper and vinyl chloride Natural deposits
Treatment	Reverse Osmosis, Distillation
Zinc, Zn	
Standard	Desirable: 5mg/l, Permissible: 15 mg/l
Risks or effects	Metallic taste
Sources	Leaching of galvanized pipes and fittings, paints, dyes Natural deposits
Treatment	Ion Exchange Water Softeners, Reverse Osmosis, Distillation

Total Coliform Bacteria	
Standard	95% of samples should not contain coliform in 100 ml 10 coliform / 100ml
Risks or effects	Gastrointestinal illness
Sources	Livestock facilities, septic systems, manure lagoons Household waste water Naturally occurring
Treatment	Chlorination, Ultraviolet, Distillation, Iodination
***E. coli* Bacteria**	
Standard	Nil / 100ml
Risks or effects	Gastrointestinal illness
Sources	Livestock facilities, septic systems, manure lagoons Household waste water Naturally occurring
Treatment	Chlorination, Ultraviolet, Distillation, Iodination

INDEX

Index for Microbiology

Acid-fast Staining 77
Aeromycoflora 89
agar plates 53, 63, 86, 89, 92, 94, 97, 98, 99, 100, 101, 104, 107, 112, 114
agar slants 53, 61, 67, 68
Agaricus bisporus 125
Alcohol Production 120
Alpha-amylase Biosynthesis 122
Anaerobic Bacteria 97
Antibiotic Sensitivity 114
Antibiotics 67, 114
Aquatic Fungi 95
Arsenomolybdate colour reagent, 139
Aspergillus 83, 86, 107, 108, 111, 113, 116, 117, 118
Assay method 172
Autoclave 29, 52, 53, 58, 106, 228
autoclaving 4, 18, 21, 58, 228
Bacillus cereus 78, 79, 80, 123
Bacillus subtilis 74
Bacterial Growth Curve 106
Bacterial Smear 73
basic dye 74
bioconversion 119
Blotter method 89
broth 2, 50, 51, 71, 73, 79, 94, 95, 100, 105, 106, 108, 109, 110, 111, 113, 123
Calibration 14, 15, 47, 48
Candle jar method 97, 99
Capsule Staining 80
carbolfuchsin 78
Centrifuge 30, 31, 91, 109, 124, 163, 175, 180, 181, 182, 192, 207, 214, 216, 221, 223
Citric Acid 116
colony forming units (cfu) 66, 87
congo red 80
contamination 1, 8, 20, 27, 53, 54, 57, 61, 67, 89, 126, 171
Continuous streaking 60
cotton plug 53, 54, 55, 71, 121, 218
Counter stain 77, 78, 79
counting chamber 102, 103
crystal violet 72, 73, 74, 75, 76, 77
Cultivation 97, 127
culture medium 50, 110, 111
culture tubes 53, 54, 67, 68
Cyanobacteria 94, 95
Czapek's Dox Agar 51, 89, 92, 107
decolourization 76
Differential Media 51, 55
dilution factor 66, 87, 124, 159, 161, 162, 176
Dry cell weight estimation 109
E. coli 55, 56, 76, 77, 106, 107, 109, 111, 113, 114
Endospore Staining 78
Enterobacter aerogenes 55, 80
Enumeration of Microorganisms 91
Eosine Methylene Blue (EMB) medium 55
Erlenmeyer flasks 65, 86, 106, 113
Fermentation Technology 114

filtration 57, 109, 163, 167, 169
170, 171, 179, 232, 238
Freeze drying 70
gallic acid 119, 120
Gaseous Sterilization 59
Gaspak anaerobic jar method 98
generation time 106
Gram Staining 75
Gram-negative 76, 77
Gram-positive 76, 77
hemocytomer 103
Hot Air Oven 40
Hot air sterilization 57
Incineration 57
Incubator 38, 39
Inoculating loop 59
Isolation of Microorganisms 93
lactic acid 83, 85, 115, 116, 120
laminar Air Flow 27
Lee's method 81, 82
Leifson's method 82
liquid medium 52
Liquid Nitrogen 69
Luria Bertani Broth 51
Lyophilization 70
Maintenance of Culture 67
Malachite green 78
Mannitol Salt Agar 56
Measurement of Fungal Growth 107
108
methylene blue 55, 72, 73, 74, 78, 83
98, 100, 101
Microbial Growth 106
Microscope 41, 42, 47
Mineral Oil 68
Moist heat 57
Mushroom Cultivation 125
Mycobacterium 77, 78, 106
MYE broth 110
ocular micrometer 47, 48, 49, 50
Orcinol Method, 210
Osazone Test, 135
Packed cell volume (PCV)
determination 110
Paraffin Method 68
Percent frequency 87, 90
Percent Occurrence 87
Petri plates 53, 55, 59, 60, 61, 62, 63
64, 66, 67, 69, 70, 71, 81, 87, 89, 90
92, 93, 94, 95, 96, 97, 98, 99, 100
104, 114
pH Meter 36, 37
phosphate solubilizers 56
Pikovskaya Agar 56
Potato Dextrose Agar 50, 52
Pour Plate Technique 61
Presumptive Test for Coliforms 100
protozoa 70, 93
Quadrant streaking 60
Radiant Streaking 60
radiation 25, 53, 57, 58
Rhizosphere 91
Sabouraud agar medium 87
Saccharomyces cerevisae 120, 121
Seed Microflora 88
Selective Media 51
Seliwanoff's reagent, 133
Serial Dilution Technique 64, 104
Simple Staining 74
Skim milk 100
solid medium 53, 59, 107
spawn 125, 126, 128
spectrophotometer 15, 26, 27, 44, 105
154, 155, 170, 172, 175, 187, 190
192, 213, 216, 218, 219, 220
Spirillum sp. 74
Spread Plate Technique 63
stage micrometer 47, 48, 49
Staining of Arbuscular Mycorrhiza 84
Staining of Fungi 83

Staining Techniques 72
Staphylococcus aureus 55, 56, 74
Steam under pressure 58
sterilization 6, 7, 41, 55, 57, 58, 59, 97, 113
Streak Plate Method 59
Streaming steam 58
sub culturing 72
Tannic Acid 118
Tyndallization 58
viable plate count method 64
Warcup method 86, 87
Weighing Balance 22
Yeast 51, 96, 121

Index for Biochemistry

2,6-Dichlorophenol indophenol 224
acid phosphatase 173, 174, 177, 178, 179, 180, 183, 184
Acrolein Test 200
ammonium sulphate 144, 166, 168
Anthrone reagent 136, 140
Arsenomolybdate colour reagent 139
Barfoed's Test 132
Benedict's test 134
Bial's Test 135
Bicinchoninic Acid 159
Biuret test 149
BSA 153, 155, 156, 212
Casein 92, 162
Dialysis 165, 169, 182, 183
Diphenylamine 209, 210
EDTA 20, 21, 158, 167, 180, 182, 183, 186, 187, 194, 212, 215, 249
Ehrlich's test 145
enzyme activity 124, 167, 169, 171, 172, 173, 174, 176, 177, 178, 179, 188, 190, 191
enzyme unit 173, 174
Fehling's Test 133
Folin's reagent 156
Grease Spot Test 200
Guaiacol Solution 189
Hopkins Cole test 146
iodine solution 76, 77, 132, 193, 204
isatin reagent 148
Isatin test 148
Isoelectric point 151
Liebermann-Burchard Test 201
Lowry method 155, 157, 159, 160, 176, 222
Lycopene 219
Millon's test 148
molar absorbance coefficient 154, 155
Molisch's Test 131
Nelson-Somogyi method 138
Ninhydrin test 144
Orcinol Method 210
Osazone Test 135
papain 194, 228
p-nitrophenyl phosphate 174, 175
Polyethyleneimine 169
Polyphenol Oxidase 190
Sackett's Method 207
Sakaguchi Test 147
Seliwanoff's reagent 133
Sodium acetate buffer 162
Sodium nitroprusside 147
Stock acrylamide solution 244
sulphosalicylic acid 144, 149, 150
Wij's solution 203, 204
Xanthoproteic test 145

Index for Molecular Biology

Agarose gel electrophoresis 278, 285
  DNA 278
  RNA 282
Agrobacterium tumifaciens 318
Antibiotic resistance 324, 325
Base composition 289
Competent cells 317
DNA
  Chloroplast 275
  genomic 259, 260, 265, 270, 272, 273, 295, 303, 307, 311, 312, 313, 326
  mammalian 271, 272, 311, 312
  Mitochondrial 277, 278
  Plant 278, 299, 320, 324, 325
  plasmid 259, 260, 261, 263, 264, 266, 291, 295, 317, 320, 321, 326, 327
GC/AT 289
GUS Assay 328
Hybridization 302
  Southern 306
  Northern 313
  Western 313
lacZ' 325, 326
Ligation 315
Nucleic acids 257, 278, 279, 286
PCR 278, 282, 284, 292, 293, 294, 295, 296, 297, 299, 300, 301, 302
Protoplasts 322
Quantification 286
  DNA 286
  RNA 286
RNA 282
  mRNA 256, 283, 284, 285
  rRNA 256, 277, 284, 285, 300
  tRNA 256, 277, 284, 285
RAPD 298, 299, 300
Renaturation 287
Restriction Digestion 290
Taq polymerase 292, 293, 296, 299, 301
Tm value 287, 288, 289
Transformation 318
  E.coli 318
  Plant 318

Index for Immunology

ABO blood group system 338
ABO locus 339
Adaptive immunity 330
antibodies 330, 334, 339, 345, 346, 347, 348, 349, 350, 351
antigen 332, 336, 339, 343, 344, 345, 346, 348, 349, 350, 352
Basophils 335, 336
blood film 335
Competitive ELISA 349
DEAE cellulose 341, 342, 343
Dialysis 341, 342
direct ELISA 348, 350
EIA 347
ELISA Reader 350
Enzyme linked immunosorbant assay (ELISA) 347
Eosinophils 335, 336
humoral 330, 337
Immunodiffusion 343
immunogen 347, 348, 350
immunoglobulins 331, 340
Indirect ELISA 349
Innate immunity 330
Leishman stain 338
leukocytes 335

Lymphocytes 335, 337
Monocytes 335, 337
neutrophils 330, 335
Ouchterlony method 343
platelets 334, 335
protein concentration 341, 343
Reaction of Identity 345
Reaction of Non-identity 345
Reaction of Partial identity 345
red blood cells 334, 335, 339
Sandwich ELISA 349
Serum 334, 335, 340, 352
Sheep IgG 342, 346
sodium azide 334, 350, 351
Sodium sulphate precipitation 340, 341, 343
Tris-tricine buffer 346
white blood cells 334, 335, 336

Index for Environmental Biology

4(p-nitrobenzyl) pyridine 375, 376
Biological Oxygen Demand 362
Chemical Oxygen Demand 366
coliforms 372, 374
Dilution Factor 365
EDTA 369, 370, 371, 372
Eriochrome black T (EBT) indicator 370
Ferroin indicator 367
fixed solids 357
Fluoride 378, 380
LD_{50} 377, 378
membrane filtration method 372, 373
Mohr's Solution 355
Most probable Number 372
Organic matter 355
organochlorines 377
Organophosphorus pesticide 375
pesticides 374, 376, 377
Starch indicator 361, 364
suspended solids 357, 372
total hardness 369
Total organic carbon 355
volatile solids 357
Winkler's method 359, 360
zirconium – Alizarin complex 379

■ ■ ■